STEVEN   WASSERSTROM

responded   to   Just Ruler
at   AAR   1988 (CHICAGO)
with   interchange   with A.S.

The Just Ruler
(*al-sultān al-'ādil*)
in Shī'ite Islam

The Just Ruler

(*al-sultān al-ʿādil*) in Shīʿite Islam

The Comprehensive Authority
of the Jurist
in Imamite Jurisprudence

Abdulaziz Abdulhussein Sachedina

New York   Oxford
OXFORD UNIVERSITY PRESS
1988

Oxford University Press

Oxford   New York   Toronto
Delhi   Bombay   Calcutta   Madras   Karachi
Petaling Jaya   Singapore   Hong Kong   Tokyo
Nairobi   Dar es Salaam   Cape Town
Melbourne   Auckland
and associated companies in
Berlin   Ibadan

Copyright © 1988 by Abdulaziz A. Sachedina

Published by Oxford University Press, Inc.,
200 Madison Avenue, New York, New York 10016

Oxford is a registered trademark of Oxford University Press

Library of Congress
Library of Congress Cataloging-in-Publication Data

Sachedina, Abdulaziz Abdulhussein, 1942–
The just ruler (al-sultān al-ʿādil) in Shīʿite Islam:
the comprehensive authority of the jurist in Imamite
jurisprudence / Abdulaziz Abdulhussein Sachedina.
p.   cm.
Bibliography: p.
Includes index.
ISBN   0-19-505363-X
1. Islam and State.   2. Temporal power of religious rulers.
3. Shīʿah—Doctrines.   I. Title.   II. Title: Sultān al-ʿādil in
Shīʿite Islam.
JC49.S162 1988
297'.1977—dc19   87-28591
CIP

2 4 6 8 10 9 7 5 3 1

Printed in the United States of America
on acid-free paper

Dedicated to

The Awaited Just Ruler
(*al-sultān al-ʿādil al-muntazar*)

PREFACE

I began research on this book in 1978–79 when I was writing sections of my earlier work, *Islamic Messianism,* in Mashhad, Iran. This was when Iran was going through the Islamic revolution. Under those circumstances I had the rare opportunity to participate in much of the academic as well as the popular discussion that was centered around the subject of Islamic government and the authority of the Imamite jurist in the absence of the twelfth Imam. I had drawn up a rough plan of research on the Just Ruler in Twelver Shī'ism. For this research I returned to Iran in January 1983.

I had realized in 1980 that library research on the subject would be incomplete without engaging in discussion with Imamite jurists in Iran. Accordingly, in 1983, for almost six months, I spent my mornings in the library of the late Āyatullāh Mīlānī, housed at his son Sayyid Muhammad 'Alī's house in Mashhad, going through the Imamite jurisprudence in its entirety with the Āyatullāh's grandson, Sayyid Fādil Mīlānī. During these morning hours I joined the scholars who visited the late Āyatullāh's residence regularly to raise some of the issues that were relevant to the concept of *wilāyat al-faqīh* (the discretionary authority of the jurist). In the afternoons I used the recently established Turāth Ahl al-Bayt Library and Research Center, where I found a number of young students of Islamic theology and jurisprudence engaged in research, with whom I had the opportunity to sharpen my comprehension of the Shī'ī worldview. With them I traveled in the realm of an intellectual and idealized Islamic vision of justice and of the necessity of the Just Ruler to establish that just social order. Those months in Iran prepared me to undertake the writing of the present work.

The research in Iran would not have materialized without the affiliation granted by the University of Mashhad, which enabled me to procure my temporary residence permit and facilitated the use of other libraries in Mashhad, especially the Āstāne Quds-i Radāwī Library. In this connection Professors Khwājawiyān and Jahāngīrī of the Dr. 'Alī Sharī'atī Faculty of Letters and Humanities were extremely helpful. Professor Nathan Scott, chairman of the Department of Religious Studies, University of Virginia, using his discretionary authority, assigned me to research leave during the spring of 1983. In addition, Mr. Sajjad Ibrahim, president of Muhammadi Islamic Trust in Toronto, provided me with a travel grant. I am indebted to all of them.

In the fall of 1985, I was awarded the Sesquicentennial Fellowship by the Center for Advanced Studies, University of Virginia, to do research in comparative Islamic law at the University of Jordan in Amman, Jordan. In Amman I was affiliated with the Faculty of Sharīʿa where I discussed parts of the book to check some substantial references made to Sunnī schools of law by Shīʿī jurists in their sections on political jurisprudence.

My initial reaction in 1980 proved to be immature, because in Sunnī Islam, after the termination of the ideal caliphate of the "rightly guided" early associates of the Prophet, it was the jurist—like Shāfiʿī or Mālik—who became the Imam of the *umma,* securing the arm of the state in the religious interests of the community. In a sense, their role was similar to that of the Shīʿī jurists who assumed the *wilāyat al-faqīh* and perpetuated the sense of autonomous religious community under the sacred law of Islam. However, when I asked several professors of the Faculty of Sharīʿa to explain the Sunnī viewpoint on the authority of the jurists during the absence of the ideal caliph of God, they understandably could not place the concept of *wilāyat al-faqīh* in the Sunnī juridical corpus, where such an authority of the jurist can be detected by implication only.

On further probing I discovered that there was general ignorance or misinformation (mostly of a polemical nature) about Shīʿism and the basic notion of the *wilāya* of the Imam and the jurist among faculty members. For instance, they believed that Shīʿites maintained that Āyatullāh Khumaynī was *al-imām al-maʿsūm* ("the infallible Imam") whose obedience was incumbent upon all the Shīʿa. Nevertheless, my aim was to investigate whether the jurist was assigned any role in Sunnī jurisprudence similar to that of the Shīʿī jurist. What emerged out of these discussions is now part of the present work, which I completed writing in the peaceful environment of the Faculty of Arts at the University of Jordan. Professor Yāsin al-Darādka, a prominent scholar of comparative Sunnī law, shared his time generously with me. Other members of the Faculty of Sharīʿa also from time to time allowed me to discuss with them some of the relevant issues of this book.

In addition to the Sesquicentennial Fellowship, the research in Jordan was made possible by a travel grant from the dean of the Faculty of Arts and Sciences, Dr. Kelly Hugh, and the Summer Research Grant from the provost's office, University of Virginia.

My research in connection with this book at the University of Jordan would have been impossible without the assistance I received from several professors and administrators, especially, Professors Mahmoud al-Samra, Muhammad al-Hammouri, Awad Khleifat, and Kamil al-Saʿid. They all helped me in different ways to settle down with my family in Amman. The staff of the library of the university was extremely accommodating in facilitating my research in Sunnī jurisprudence. To facilitate my writing my brother Salim Sachedina made a generous gift of an IBM PC/XT computer with a word processing program. This made the writing and numerous subsequent revisions easy. I am grateful to all of them.

As for the contents of the book, my mentors and friends, Professors G. M. Wickens and R. M. Savory, at the University of Toronto, read the work completely, offering incisive criticisms to improve it in many places. Without Professor Wickens's painstaking reading of the earlier drafts of the work, many parts consisting of dense and complex discussions would have remained in need of further clarification. I cannot find appropriate words to express my gratitude to him. My colleague at the

Department of Religious Studies, Professor Robert P. Scharlemann, read parts of the work and made pertinent suggestions for improving it.

In my seminars on Sunnī and Shīʿī theology at the University of Virginia, graduate and undergraduate students have often inspired me by seeking clarification of certain issues. I would be failing in my duty if I did not mention Drs. Farhang Rajaee, John Kelsay, and Vernon Schubel. And I must also acknowledge the indirect contribution of Donald Davison and David Gardiner in helping me formulate and discuss, with clarity and accuracy, issues related to Islamic theology.

At the University of Virginia Word Processing Center I received enormous assistance from Mrs. Gail N. Moore and Sibyl Hale, who patiently taught me the use of the system and accommodated my various revisions and editions of the text. I am grateful to them both.

At Oxford University Press, Cynthia Read gave enormous support and Catherine Clements provided expert advice on many occasions to enable smooth publication of the book.

Finally, I am indebted to my wife, Fatima, and sons Alireza and Muhammadreza, who have made great sacrifices in allowing me to do my research in Iran in 1983 and in Jordan in 1986. Life in Mashhad, Iran, in 1983 was full of challenges and hardship; Amman demanded a different kind of adjustment. Under all those circumstances, they have supported my work and have encouraged me to engage in my research on the Just Ruler in Imamite political jurisprudence. In addition, my son Alireza spent many long hours carefully reading out to me major parts of the book for proofreading purposes. It was a great pleasure to hear him read the text.

I end this preface with my preferred invocation. It bespeaks my vision:

> O God, we ask You for [the establishment of] honorable government through which You will consolidate submission to Your will (*islām*) and its peoples, and subdue hypocrisy and its peoples. And make us in that [government] those who will call others to Your obedience and lead them on Your path. And bless us in that [government] with good of this and the next world.

Charlottesville, Virginia   A.A.S.
Muharram 1, 1408 / August 26, 1987

CONTENTS

Introduction

In the study of Shīʿite history two doctrines have played an important part in producing a coherent body of political and legal jurisprudence—namely, the justice of God (al-ʿadl) and the leadership of righteous individuals (al-imāma) to uphold and promulgate the rule of justice and equity. In the highly politicized world of early Islam, numerous ideas and conceptions about God's purpose on earth and leadership of human society floated freely. The swift conquest of vast territories and the ongoing process of supervising the conquests and administering the affairs of the conquered peoples demanded not only strong and astute leadership, but also the creation of a system that would provide stability and prosperity. Central to this social, political, and economic activity was the promise of Islamic revelation: the creation of a just and equitable public order embodying the will of God.

The promise was based on the belief that God is just and truthful. Divine justice demanded that God do what was best for humanity; and divine truthfulness generated the confidence that God's promise would be fulfilled. The proof that God was doing what had been promised was provided by the sending of the Prophet to guide humanity toward the creation of the Islamic world order. The connection between the divine guidance and the creation of the Islamic world order, as a consequence, marked the inevitable interdependency between the religious and the political in Islam. The result in the intellectual realm was the creation of political jurisprudence (al-fiqh al-siyāsī) guided by the tenets of religious doctrine (al-iʿtiqādāt), even when the doctrine was not adequately formulated or clarified, at least in the early stages.

The death of the Prophet marked the first major crisis in the political history of Islam: circumstances demanded that the Muslims explain the situation that seemed to point toward the breach of divine promise. Tension was felt in the awareness of the lack of an objective actualization of the Islamic ideal in the external world. It was precisely at this time that the notion of the Just Ruler (al-sultān al-ʿādil), who would assume political power to bring about the just order, came to be accentuated. The entire question of qualified leadership to further the divine plan and to enable God's religion to succeed must be seen from the perspective of the Islamic promise

3

of the creation of an ethically just order on earth. More importantly, it was also at this time that the Shī'a refused to acknowledge and regard as legitimate the rule of those whom they considered usurpers of a position of leadership that rightfully belonged to 'Alī b. Abī Tālib and his descendants.

The period that followed this first crisis and tension between the ideal and the real gradually marked the growth of discontent among the people, which led to revolutions and rebellions as well as to discussions and deliberations. This is reflected in the early Islamic *fiqh* (theology cum jurisprudence) literature that emerged toward the end of the second/eighth century. This literature wove together the various threads of Islamic belief and practice. No Islamic legal practice, especially in the realm of socio-political interaction, can become comprehensible unless due attention is paid to the doctrinal underpinnings of early Muslim groupings, because the formation of Islamic law took place under religious and ethical ideas.[1] More significant in the rulings of the early jurists was their consideration of whether legal or political injunctions affected the legitimacy of one or the other leader favored by each faction. Thus, even when a particular ruling went against explicit textual evidence provided by the Qur'ān, the overriding consideration for these early jurists was the preservation and legitimation of the authority in power, a consideration that came to be justified under the rubric of *al-masālih al-'āmma* (the general welfare of the Muslim community). The development of the Sunna (the Prophetic customs and practice) in early Islam amply confirms this observation.[2]

By the time of the second crisis, which affected the political jurisprudence of the Shī'ites—namely, the end of the manifest leadership of the Imams through the occultation of the twelfth Imam—Islamic jurisprudence had separated hands with theology, at least formally, although its underpinnings were still the doctrines of the justice of God and the leadership of the Imams from among the descendants of the Prophet. The most important issue during this period for the Shī'a was the right guidance that had continuously been available to the community even though the Imams were not invested with political authority and were living under the political power exercised by the de facto governments. This guidance comprised the Sharī'a, which resulted from the meticulous study of the Qur'ān and from other related material generated by the effort of the scholarly elite of the Shī'ite community to expound the legal contents of the Qur'ān.[3]

The Imam's authority (notwithstanding his lack of political power, he still had the right to demand obedience from his followers) in Shī'ism was clearly seen in his ability to interpret divine revelation authoritatively. What was decided by him through interpretation and elaboration was binding on believers. These rulings formed part of the obligations (*al-takālīf al-shar'iyya*) imposed on believers. The interpretation of the divine revelation by the Imam, which in Shī'ism came to be considered part of the revelation, was regarded as the right guidance needed by the people at all times. It was, moreover, the divine guidance that theologically justified the superstructure erected on the two doctrines of Imāmī Shī'ism: the justice of God and the designation of the Imam, free from error and sinful deviations, in order to make God's will known to humanity. In response to the crisis created by the occultation of the Imam, the Shī'ites developed their own legal and political jurisprudence in which a prominent place was given to the faculty of reasoning (*al-'aql*).

In the jurisprudence of the Imamites the priority of reason was in accord with their rational theology, in which reason was prior to both sources of revelation, the Qur'ān and the Sunna. This does not mean that the revelation was not regarded as all-comprehensive; on the contrary, there was recognition of the fact that it was reason that acknowledged the comprehensiveness of the revelation by engaging in its interpretation and discovering all the principles that Imamites needed to know. In addition, there was recognition of a fundamental need of interpretation of the revelation by reason, all the more so when the authority invested with divine knowledge was in occultation. However, it was not for just anyone to undertake the decisive responsibility to guide the community by interpreting the revelation rationally. It certainly needed authorization from a divine source, a sort of designation that could guarantee to Muslims the availability of right guidance based on Islamic revelation. Only such an authorized person could, in the absence of the Imam, assume the authority that accrued to the Imam as the rightful ruler (*al-sultān al-'ādil*) in Islam. But in view of the prolonged occultation of the Imam and the absence of a special designation during this period, no realization of just rulership was possible. This was reflected in the political jurisprudence of the Imamites, where the recognition of the lack of specific designation by the Imam reemphasized the separation between power (*qudra* or *saltana*, which could exact or enforce obedience) and authority (*wilāya*, which reserved the right to demand obedience, depending on legal-rational circumstances) that had existed even during the lifetime of the Imams. Only the investiture of authority and the assuming of political power could establish the rule of justice and equity.

However, delegation of the Imam's authority to an individual who could assume both the authority and the power of the Imam when there was no Imam to monitor the exercise of that authority was dangerous. This danger was perceived by the jurists, who took upon themselves to produce a coherent response to this situation in their works of jurisprudence in which the Imamite doctrine that the Imam is the only Just Ruler was reasserted. Pending the return of the Hidden Imam, the possibility of absolute claim to political power (*qudra*) and authority (*wilāya*) resembling that of the Imam himself was ruled out. Nevertheless, the rational need to exercise authority in order to manage the affairs of the community was recognized and authoritatively legalized. The establishment of the Shī'ī dynasties during the occultation did not change the basic doctrine of the Imamite leadership. But it made it possible to conceive, however temporary and fallible it might be, a just Shī'ī authority on the basis of rational and traditional argument, grounded in the basic Islamic principle of creating a public order that would "enjoin good and forbid evil." In this acceptance of the historical reality and its gradual legalization, one can discern the efforts of Imamite jurists in response to the oft-repeated question of the Shī'ites: When would the Imam come forward to establish justice and equity on earth, as promised in Islamic revelation?

It is important to bear in mind that Imamite jurists were responding to this question individually in their works on jurisprudence, and there was a lack of any definite organization or a strict uniformity of responsa among them. The Imamite jurists continued to be private individuals, like their Sunni counterparts. Because of their peity and learning, they were accorded reverence by the people and recognition as leaders in guiding the Imamite community. Imamite juridical writings reflect the jurists' individual and independent exertion of reasoning to formulate appropriate responses to the socio-political realities of the Islamic public order. Moreover, these works reflect the tensions within the Imamite school created not only by the occultation of the Imam, but also by intellectual interaction between Imamite and Sunnī scholars. The occultation of the Imam made it possible for them to be pragmatic and realistic in their contacts with contemporaneous de facto governments and in the formulation of their opinions about them, more so if the de facto rulers happened to be professing Imamites. And their intellectual contacts with Sunnī scholars, especially during the formative period of Imamite jurisprudence, made it imperative for them to equip themselves with methodological terminology to rationalize the concrete situation in which the Shīʿites found themselves.

It is for this reason that each work of jurisprudence is abundantly documented by quotations from the Qurʾān and the Prophetic practice as well as critical evaluation of the opinions of those jurists who preceded them in their formulation of a particular legal decision. However, in this process of discussing the documentation for a ruling, the concrete case in point, which may have initially prompted such an investigation, became concealed. In order to reconstruct a concrete case from the evidence provided in a ruling, one has to labor through the normative jurisprudence as a source for the study of the concrete situations in the Muslim polity at a given time in history when the source was actually produced.

The present study deals with the concept of the Just Ruler in Twelver Shīʿīsm in the light of the political and legal jurisprudence worked out by Imamite scholars from the early days of the Shīʿī Imams to the present time. It deals with the development of the authority of Imamite jurists, who were also thoroughly grounded in Imamite theology.

The study was inspired by the Islamic revolution of 1978–79 in Iran, where the historical idea of the authority of a jurist in matters of jurisprudence attained full-fledged political realization. In the 1980s, several works, most of them written by social scientists, have treated the concept of the "guardianship of the jurisconsult," largely in the light of the present-day political experience of the Shīʿa in Iran.[4] In these works, scholars have provided students of modern Iran with a wealth of material dealing with the complex socio-economic history, on the one hand, and the intricate realtionship between the Iranian state and society, on the other. Of particular relevance to this study is Said Amir Arjomand's work, *The Shadow of God and the Hidden Imam*. The author has not limited himself to the socio-political factors that had relevance to the 1978–79 events in Iran; he has also provided us, to some extent, with important analyses of religious factors that culminated in the legitimation of the power of the temporal as well as the religious authorities in premodern Iranian history.

In the present work I have filled a crucial gap in the existing literature on the development of the Shīʿī juridical authority as it emerges from the study of the political jurisprudence produced during the different periods of the Twelver Imamite history. The concept of "guardianship" (*wilāya*) in general, and the "guardianship" of a jurist (*wilāyat al-faqīh*) in particular, has its genesis in the early history of Imamite jurisprudence. The second element in the concept is, furthermore, peculiar to Shīʿī Imamite jurisprudence, because it is only this school of Shīʿī thought that maintains belief in the Hidden Imam who continues to guide his community through his "generally" designated deputies. Careful analysis of

Imamite juridical texts, then, becomes indispensable for understanding the authority that was regarded by the Shī'ites as delegated to qualified jurists among the Shī'a. The complexity of the material and the interrelationship of different parts of the juridical topics becomes evident when one considers the fact that the guardianship of the jurist was not discussed only in the sections of such works as the Book of Trade (*Kitāb al-tijāra* or *al-makāsib*), where the question of discretionary control of the jurist over properties and persons logically came up; the authority of the jurist was also discussed in the chapters on *zakāt* (alms), *al-khums* (the fifth), *al-qadā'* (administration of justice), *al-hudūd* (administration of legal penalties), *al-nikāh* (marriage), *al-amr bi-al-ma'rūf wa al-nahy 'an-al-munkar* (enjoining the good and forbidding the evil), *al-jihād,* and so on. Thus, an investigator has to go through all the chapters of applied jurisprudence to find references to the authority of the jurist.

An adequate understanding of the *fiqh* works is dependent on two other Islamic sciences on which the demonstrative jurisprudence of the Imamites has been based. The Imamites have relied on reasoning in the form of *ijtihād, which estab-* (i)
lished the intrinsicality of a ruling derived from the authority of revelation in that derived from the authority of reasoning because the point is that rational ruling is intrinsic in revelation ruling (not additional to it). The procedure that reasoning must follow to establish this intrinsicality is the subject matter of *usūl al-fiqh* (theoretical basis of Islamic law). *Usūl al-fiqh,* on the one hand, defines and discusses the (2) extent of revelation and the categories of the injunctions that can be derived from it; on the other hand, it sets forth the theory of juristic practice to deduce further laws that cannot be derived explicitly from revelation. In addition, *usūl al-fiqh* shows the way in which jurists asserted their authority as the interpreters of the will of the Imam by exerting their rational faculty in creating exegetical and terminological devices to deduce appropriate responsa. It is probably valid to maintain that, had it not been for the *usūl* methodological devices, the jurists would not have emerged as the "general" deputies of the Hidden Imam merely on the basis of some documentation provided in the communications of the Imams, because the terminology there had to be extrapolated through exegetical method to infer theologico-political implications for the juridical authority in Shī'īsm. Thus, without a correct understanding of the way the jurists apply the *usūl* to deductively infer laws, it is impossible to determine the exact nature of *ijtihād* in particular judicial decisions that affected political jurisprudence at a particular time.

Because documentation is of utmost importance in issuing a judicial decision, jurists have paid much attention to the *'ilm al-hadīth* and the related religious science of *rijāl.* *'Ilm al-hadīth* is the religious science that studies Prophetic communications transmitted by the close associates of the Prophet and the Imams, in order to determine their authenticity or inauthenticity with a view to their use in giving legal decisions. *'Ilm al-rijāl* deals with the chronological study of the transmitters who figure in the chains of transmissions appended to *hadīth* reports. Biographical information on each informant of a tradition was studied to gain insight into his "reliability" or "weakness," and so on. Both kinds of work have been consulted in this study, so as to get better insight into the reasons for certain rulings based on *usūl al-fiqh.*

In addition to the above technical considerations in the examination of the juridical sources, another important consideration is the period when a particular work came into being. The historical perspective was extremely important for such research, because, as far as possible, the reason for examining the primary material was to allow the source to speak for itself, rather than to impose or to verify an investigator's preconceived notions or thesis in them. It was, therefore, necessary to proceed chronologically, from period to period of the Imamite jurisprudence, to trace the development of the notion of the Just Ruler. Rather than placing the Imamite jurists and their works under the dynasties, like those of the Būyids or the Ilkhānids, and so on, I have tried to establish master-disciple connections between them.

As pointed out above, jurists were responding to the real situation individually within the Imamite school. It is for this reason that we find significant differences not only between jurists who were contemporaries and had studied under the same teacher, but between successive views of a single scholar during different periods of his life. This fact necessitates careful study of a given scholar's judicial decisions, taking into consideration the chronology of his works. This is not always possible and any attempt to do so, especially in the absence of such information from the author himself, can lead to the attribution of views that a jurist might have held at one point in his career, but later revised or abandoned. Furthermore, there exists a body of nonjuridical sources in the form of political tracts written by some of the prominent disciples of these teachers during the premodern and modern eras of Imamite history. They treat the question of political authority under the influence of diverse concrete situations. These tracts, however, were not accorded the status of juridically derived opinions in the area of the ongoing debate on the nature of juridical authority in jurisprudence; rather, they were treated as works of the members of the Imamite community maintaining different views and interests regarding the prevailing political situation and its rationalization. It is for this reason that in this study I have not taken into account a given scholar's thought if it was neither discussed by major figures nor made part of the recognized rubrics in the juridical tradition of the Imamite school. In addition to this formalized characteristic of Imamite juridical sources, one must keep in mind the hostility between scholars belonging to the Imāmiyya. Examples would be the thinly disguised criticisms and mutual refutations in the sources, particularly on the issue of wielding authority and assuming the constitutional right of the Hidden Imam. Imamite scholars were also engaged in preparing individuals who would continue their special interests. In many cases the disciples of an eminent jurist were inclined to disturb practice as little as possible; but when it came to theoretical considerations, they sometimes departed drastically from their teacher on principles and methods, which necessarily caused variations in rulings on the same subject.

The Shī'ī dynasties, notwithstanding their favorable attitudes toward the Shī'a, did not, in my opinion, spark the production of Shī'ite political and legal jurisprudence that reflected an a priori idealized world dominated by the Shī'ī doctrines of the justice of God and the Imamate. The concealed concrete situation, as I have pointed out earlier, had to be meticulously reconstructed by reading between the lines of particular sections of political jurisprudence. Wherever relevant, especially in the later period under the Safavids and the Qājārs, I point out the socio-political background of jurists, which might have influenced particular rulings under consideration.

A Brief History of Imamite Political Jurisprudence

We have fragmentary evidence on *fiqh* works from the times of the Imams al-Bāqir and al-Sādiq (eighth century A.D.), among whose disciples we read of the early *fuqahā'* (jurists). It is probably correct, however, to recognize the beginnings of Imamite jurisprudence among the close associates of these two Imams in two centers of Islamic learning, Medina and Kūfa (Iraq). It was also during this period that Qumm began to be frequented by Imamites. In the second chapter, below, we will be dealing with these early *fuqahā'* and their role in the development of the deputyship of the Imams in Medina. Rulings of these two Imams on various topics of Sharī'a were circulating in the form of *hadīth* reports among these close associates at this time and were available to Imamite traditionalists Kulaynī and Ibn Bābūya, leaders of Imamite learning in Rayy and Qumm in the tenth century, when they undertook the systematic compilation of this information in their *hadīth* works. This intellectual development is covered in the second chapter, below, where I present the history of both *'ilm al-hadīth* and *'ilm al-rijāl* in the Imamite religious sciences during the classical period (tenth–thirteenth centuries A.D.).

The period of epitomizing and systematizing the transmitted material needed to guide the socio-political and religious life of the Imamites was followed by the phase of subjecting this material to the strict discipline of the *usūl al-fiqh*. This was the most important period of Imamite jurisprudence, headed by Mufīd and his erudite pupils Sharīf al-Murtadā and Tūsī in the eleventh century in Baghdad. The element of Imamite jurisprudence that favored *ijtihād*—more specifically, *al-ijtihād al-shar'ī* (reasoning based on the textual evidence provided by the Qur'ān and the Sunna of the Prophet and the Imams)—was now firmly incorporated in the *usūl* works. Moreover, the profound training of these jurists in the rational theology of the Mu'tazilites, and their own exposition of the Imamite theology on the basis of those rational principles, caused Imamite jurisprudence to be inseparably joined to the two fundamental doctrines of Imāmī Shī'īsm: the justice of God and the Imamate. In the second chapter, where I discuss the process of compiling "sound belief" and systematizing "sound knowledge," according to the Imamites, I demonstrate this feature of Imamite jurisprudence. I also deal there with the works on jurisprudence written by these scholars within their socio-political context.

At this juncture, I will give a chronological account of those jurists and of their works that have been regarded as the most important of their kind in the field of Imamite jurisprudence in general and political jurisprudence in particular. I hope this chronological account will provide the reader with some sense of the breadth of the works produced by the Imamites to provide a coherent body of political and legal jurisprudence.

The earliest Imamite jurist whose name occurs on several occasions in connection with the authority of a jurist in the works of Mufīd is Ibn Junayd al-Iskāfī (d. 381/991). He lived during the Short Occultation of the last Imam (A.D. 874–941).[5] He was a prominent scholar of *fiqh* and *kalām*. According to Najāshī, he was the leader of the Imamites to whom the *khums* ("the fifth") that belonged to the twelfth Imam was entrusted. Before he died he left it to his daughter in the form of a trust.[6] Mufīd studied with him. Al-Khatīb al-Baghdādī mentions Ibn Junayd as the leader of the Imamites, and places his date of death in 332/943, and that he died in al-Karkh.[7]

Bihbahānī, in his *Sharh al-mafātīh*, commenting on Ibn Junayd, says that he was among the Imamite jurists who are said to have ruled on some issues on the basis of *al-qiyās* (analogical deduction).[8] This comment is meant as a criticism of Ibn Junayd because, unlike the Sunnī school of al-Shāfiʿī, the Imamite *usūl* does not recognize *al-qiyās* as one of the sources of law. According to the Shīʿī scholars of *usūl*, *qiyās* (which can be inductive reasoning) can give rise to diversity of opinion and thereby affect believers' practice adversely. Ibn Junayd wrote a detailed work on *fiqh* entitled *Tahdhīb al-shīʿa*, which survived in an abridged version, made by himself, entitled *al-Mukhtāra fī al-fiqh al-ahmadī*. It was this latter work that was known to Mufīd and others.

Another "ancient" Imamite jurist whose opinions have been cited by later jurists is al-Hasan b. ʿAlī b. Abī ʿAqīl al-ʿUmānī al-Hadhdhāʾ, a contemporary of Ibn Junayd and Kulaynī. According to Najāshī, he was among the first Imamite scholars to compose a work on *usūl al-fiqh*, entitled *al-Mutamassik bi habl āl al-rasūl* (Adhering to the teachings [lit. rope] of the family of the prophet). These two jurists are known as *al-qadīmān*, the "two ancients," in the fourth/tenth century.[9] Ibn Abī ʿAqīl was among the renowned scholars regarded as thoroughly reliable by Ibn Idrīs al-Hillī,[10] although Ibn Idrīs was critical of the opinions of several "ancient" scholars of great renown.

Following these two jurists during the Short Occultation, we have the list of *al-mutaqaddimūn* (the "ancient") Imamite jurists whose works have survived for posterity. Although Kulaynī (d. 329/941) and Ibn Bābūya (d. 381/991) compiled the *hadīth* reports according to the chapters of *fiqh*, they are rightly classified as "ancient" *akhbārī* scholars (that is, those who accepted unquestioningly the authority of "traditions" in religious knowledge) in contrast to those who developed and employed *usūl* based on rational inferences to deduce ordinances. The "traditionists" did not attach the same importance to *usūl* in deducing rulings and their decisions were heavily dependent on the traditional methodology of understanding the text and discussing its implications within the limited bounds of the ostensible sense of the *hadīth* report.

Mufīd heads the list of the "ancient" jurists whose works demonstrate a rational methodology in deriving laws of the Sharīʿa. He is regarded as *al-shaykh* and is so referred to in all the works of Imamite jurisprudence. Thus, for instance, ʿAllāma Hillī, in the introduction to his *Muntahā*, in explaining the abbreviations of the names that appear in his work, says that he uses *al-shaykh* when he refers to "*al-imām* Abū Jaʿfar Muhammad b. al-Hasan al-Tūsī or al-Mufīd." When he uses *al-shaykhān*, he means both Tūsī and Mufīd.[11] The use of *al-imām* for a jurist in Imāmī Shīʿism is interesting. As a matter of fact, ʿAllāma, in speaking about Tūsī in particular, justifies the use of the title *al-imām* by emphasizing that he is *al-imām al-aʿzam* (the greatest Imam) who "deserves the position of the imām" because of his enormous contribution to Imamite learning in general, and Imamite jurisprudence in particular.[12]

Mufīd wrote under relatively favorable circumstances, under the protection of the Shīʿī Būyid dynasty, a fact at times reflected in his opinions regarding the Just Ruler. Imamite political jurisprudence during the Būyid period, as noted earlier, especially the Baghdad school under Mufīd and his students Sharīf al-Murtadā and

Tūsī, laid down principles for determining the extent of the authority of the jurist— principles that have been invoked throughout the whole history of Shī'ī jurisprudence. In addition, they evolved a new set of *usūl* terminology that allowed for flexibility in dealing with de facto governments against the demand of juridical continuity. Their pragmatism in political jurisprudence made room for rational and sometimes radical interpretation through extrapolation of Imamite sources handed down in the compilations of Kulaynī and Ibn Bābūya. Thus, phrases like *sultān 'ādil* (just ruler) were introduced in judicial decisions dealing with political functions reserved for the theologically acknowledged Just Ruler (the Imam), despite his occultation. Such rulings pointed to the pragmatic recognition of a just ruler on the grounds of certain qualities or acts he manifested, and not because of documentary evidence in the form of an explicit text to that effect. Without these juridical precedents pointing toward the existence of a just ruler whose authority was derived from his qualities and deeds, subsequent Imamite jurists would have lacked the requisite evidence in normative jurisprudence to assert their all-comprehensive claim to the Imam's authority pending his return.

Shaykh Tūsī continued developing the rulings of his teacher Mufīd in several of his works, most significantly in his *Nihāya* and *Mabsūt,* which were the culmination of his lifelong research. Both these works demonstrate Tūsī's command of Imamite jurisprudence and his ability to employ the *usūl al-fiqh* in deriving judicial decisions. In fact, he wrote a separate work on *usūl* entitled *al-'Udda fī usūl al-fiqh* ([Intellectual] equipage in the derivation of the law), explaining the fundamentals of the Imamite faith, the sources of Islamic law, and the procedure to derive legal decisions by arguing for the authority of a priori reason in comprehending revelation. In this work Tūsī discussed the *ijtihād* and related subjects in a separate chapter. I dwell upon his contribution in the context of the development of the Imamite *fiqh* under Mufīd and his pupils at length in chapter 2.[13]

At this point, I want to introduce other jurists of what I have designated as the Baghdad school of the Imamite jurisprudence. As I have argued elsewhere, the Baghdad school was the convergence point for both Baghdad and Qumm-Rayy schools of Imamite learning.

Sharīf al-Murtadā was another prominent teacher in the Baghdad school who studied with Mufīd and trained a number of well-known Imamite scholars, including Tūsī. Sharīf al-Murtadā's contribution lay in Imamite theology as much as political jurisprudence. Both his theological and juridical works demonstrate his interest in using rational principles to arrive at religious truth. He wrote one of the earliest works on *usūl al fiqh* entitled *al-Dharī'a fī 'ilm usūl al-sharī'a* (The means to the [acquisition of] knowledge concerning the principles of the Sharī'a). There had existed treatises dealing with different aspects of *usūl,* but *al-Dharī'a* was the first complete work of its kind. In it Sharīf al-Murtadā cited opinions of different scholars dealing with the theoretical basis of the Sharī'a, refuting those like *al-qiyās,* and explaining Imamite objections to those principles utilized by Sunnī jurists in deducing *shar'ī* decisions. He also wrote some works on comparative and applied *fiqh.* Of greater significance are his responsa (*al-rasā'il al-jawābiyya*) to the practical questions concerning the authority of the jurist, working for the unjust government, convening of the Friday service in the absence of the political authority of the Imam, and so on.

Sharīf al-Murtaḍā, who was himself socially and politically active in the cultural life of Baghdad and held official posts in the ʿAbbāsid administration, paradoxically issued a number of judicial decisions with adverse implications for the legitimation of the comprehensive authority of jurists in works written during the rule of Shīʿī de facto governments. Thus, for instance, on being asked about the legality of Friday public worship during the absence of the twelfth Imam, who alone as the political head of the community could convene the service or deputize someone else to do so, instead of introducing the jurist as deputy of the Imam, he ruled that the prayer was invalid during the occultation. This ruling was taken to imply that the jurist could not assume the Hidden Imam's political function in his capacity as his "general" deputy. Some of these rulings were taken up by Tūsī and expanded upon, or even corrected.

The prominence and the lasting influence of the Baghdad school, throughout Imamite history, was due to the two *shaykh*s, Mufīd and Tūsī, and Sharīf al-Murtaḍā, who trained some of the brilliant jurists of this school. Among the students of these masters of Baghdad, the following have left important works:

1. Al-Ḥalabī, Abū Ṣalāḥ Taqī al-Dīn, or Taqī b. al-Najm al-Ḥalabī (d. 447/1055).[14] Ḥalabī was a prominent jurist and a reliable traditionist on whose authority al-Qāḍī Ibn al-Barrāj of Tripoli reported traditions. He studied jurisprudence under Sharīf al-Murtaḍā and Tūsī. According to Shahīd II, he acted as Sharīf al-Murtaḍā's deputy in Aleppo. Besides his commentary on Sharīf al-Murtaḍā's *fiqh* work *al-Dhakhīra*, he wrote his own work entitled *al-Kāfī fī al-fiqh*. It begins with a theological introduction on the doctrine of *taklīf* (imposition of a *sharʿī* obligation on a believer) and exposition in support of its rational justification. The way Ḥalabī introduces his jurisprudence is unique among the works on the subject written during this time. He classifies *taklīf* into three kinds: *ʿibādāt* (acts of worship); *muḥarramāt* (prohibited acts); and *aḥkām* (ordinances). *ʿIbādāt* is further categorized into *mafrūḍ* (obligatory), and *masnūn* (enjoined). Then he makes the case for their performance by arguing that duties (*al-farāʾid*) and their performance constitute *luṭf* (the "grace" of God that draws a person to obedience and away from disobedience) in all that is rationally obligatory (*al-wājib al-ʿaqlī*), helping the individual to refrain from evil acts. After this introduction he goes on to classify the *ʿibādāt* into ten categories, such as: *al-ṣalāt* (daily worship); *ḥuqūq al-amwāl* (that which is incumbent on wealth, i.e., *zakāt*); *al-ṣiyām* (fasting); *al-ḥajj* (pilgrimage); *al-khums* (the fifth); and so on.[15]

2. Sallār al-Daylamī, Ḥamza b. ʿAbd al-ʿAzīz al-Tabaristānī (d. 448 or 463/1056 or 1070).[16] Sallār studied under Mufīd and Sharīf al-Murtaḍā and sometimes taught in the latter's stead. He wrote works on theology and jurisprudence, and is regarded as a reliable authority. He is well known and extensively cited for his work in *fiqh* entitled *al-Marāsim fī al-fiqh* (The ordinances in jurisprudence), which is also known as *al-Risāla*. Apparently following the inferences of Sharīf al-Murtaḍā in some of his treatises, Sallār is among those "ancient" jurists who do not concede comprehensive authority to jurists in exercising complete discretion in the affairs of the Shīʿa.[17] As a consequence, as we shall see in chapter 5, Sallār is the first Imamite jurist to declare the Friday prayer prohibited during the occultation of the twelfth Imam. The reason for this ruling, as derived from Sharīf al-Murtaḍā's judicial decision, was that no one (not even a well-qualified *faqīh* [jurist]), could assume the comprehensive authority of the Imam, which included the convening and leading of the Friday prayer.[18] This ruling was in complete

contradiction to the ruling adopted by both Mufīd and Tūsī, his contemporaries, and shows Sallār's independence of mind in the application of principles and method to derive judicial decisions.

3. Ibn al-Barrāj, al-Qāḍī ʿAbd al-ʿAzīz Ibn Nahrīr (d. 481/1088). Ibn Barrāj was another prominent jurist who had studied in the Baghdad school of Imamite jurisprudence under Mufīd, Sharīf al-Murtaḍā, and Tūsī. He was Tūsī's representative in Syria and held an official position as a judge in Tripoli for some thirty years. According to Ibn Shahrāshūb, he was born and brought up in Egypt, and traveled to Baghdad to study with the jurists. In this connection, it was Sharīf al-Murtaḍā who supported him financially, fixing his grant at eight dinars per month.[19] Ibn Barrāj transmitted legal rulings on the authority of Muhammad b. ʿAlī b. ʿUthmān al-Karājikī and Halabī. Shaykh ʿAbd al-Jabbār al-Mufīd al-Rāzī, a jurist from Rayy, related rulings on Ibn Barrāj's authority. His *fiqh* works include *Jawāhir al-fiqh* (The gems of jurisprudence), *al-Muʿjaz, al-Kāmil,* and so on.[20]

4. Ibn Hamza al-Tūsī (d. second half of the sixth century). Ibn Hamza's full name, according to Qummī, was ʿImād al-Dīn Muhammad b. ʿAlī b. Muhammad.[21] We do not know the date of his birth or his death. He was among the students of those who had studied with Tūsī, or, according to Khwānsārī, he was among the students of Shaykh Abū ʿAlī, Tūsī's son.[22] Besides being a jurist and the author of a well-known work on jurisprudence, *al-Wasīla fī al-fiqh* (The guide in jurisprudence), he was a preacher who wrote popular works on miracles, such as *al-Thāqib fī al-manāqib* (The penetrating scrutiny of virtuous deeds).[23]

5. Ibn Zuhra, Hamza b. ʿAlī al-Husaynī al-Halabī (d. 588/1192). Also known as Abū al-Makārim, Ibn Zuhra was a famous Imamite scholar belonging to the renowned Aleppo family of scholars, collectively known as Banū Zuhra. He died at the age of seventy-four in Aleppo and is buried at the foot of Mt. Jawshan in *mashhad al-saqt* (a shrine where al-Husayn b. ʿAlī's wife's miscarried infant is buried). Among those who related judicial decisions on his authority is Ibn Idrīs al-Hillī, the leader of the Hilla school of Imamite jurisprudence. Ibn Zuhra's comprehensive work on *fiqh,* which includes a statement on principles (*usūl*) and their application (*furūʿ*), and which I refer to in this study, is entitled *Ghunyat al-nuzūʿ ilā ʿilmay al-usūl wa al-furūʿ* (The indispensable endeavor toward the two sciences of principles and derivatives [of jurisprudence]).

6. Ibn Idrīs, Muhammad b. Ahmad al-Hillī (d. 598/1201). Ibn Idrīs was the leader of the group of Imamite jurists who settled in al-Hilla, a town between Kūfa and Baghdad in Iraq. Hilla remained a center of Shīʿī learning for a long time. After the destruction of Baghdad by the Mongols, Hilla became a refuge for Shīʿī scholars, and as such it formed a sort of extension of the Baghdad school, and yet was sufficiently distinct to be designated as the Hilla school. Indeed, the Hilla jurists made the most impressive contribution to the history of Imamite jurisprudence, following that of the Baghdad school under Mufīd and Tūsī. Whereas the Baghdad jurists composed their works under the relative political security provided by the Shīʿī Būyids, the Hilla jurists wrote under much tension caused by the anti-Shīʿī attitude of the Seljūqs in sixth/twelfth century and hostility between the Shīʿites and the Sunnites. This is occasionally reflected in opinions given by these scholars, which seem to have been prompted by the need of precautionary dissimulation (*taqiyya*) of the true belief in certain political matters.

Ibn Idrīs wrote an extremely important work on *fiqh* entitled *al-Sarāʾir al-hāwī li-tahrīr al-fatāwā* (The comprehensive digest of legal opinions). Its importance lies in his independent reasoning in deriving opinions on several important issues connected with the concept of the just ruler. His name has always appeared along with that of Sallār al-Daylamī in accounts of their opposition to the all-comprehensive authority of a qualified Imamite jurist, including the exercise of the constitutional authority of the Imam during his absence. I have discussed his opinions and his criticism of Tūsī's rulings on the question of the authority of the jurist in several places in this book.[24] Ibn Idrīs strikes one as a sort of free-thinker among the "ancients," who not only criticized earlier jurists for methodological flaws in deducing their decisions, but also ruled against some of the well-established rulings in the "ancient" jurisprudence of the Baghdad school.

The Hilla scholars mark a new chapter in the history of Imamite *fiqh*, as do the later jurists of Isfahān under the Safavids. The Hilla scholars mark the development of a new methodology in deducing laws, and in testing whether the derivation of such laws, if in dispute among scholars, was based on the proper application of principles of jurisprudence. Thus, in *Sarā'ir*, Ibn Idrīs has a long introduction on methodology, which includes a detailed explanation, for instance, of how and when a "single" or "virtually unique" tradition (*khabar al-wāhid*) can be accepted for use as evidence in deriving a judicial decision. Like Sharīf al-Murtadā before him, he was opposed to the use of a "single" tradition as evidence, except that he favored the use of a "single" *mutawātir* tradition (frequently reported in an unbroken "chain of transmission") or a "single" tradition whose transmission on the authority of the Imam could not be doubted. One can discern a process of sharpening of intellectual tools that might have become dull over the years in the composition of juridical works. This process reached a climax in the jurisprudence of al-Muhaqqiq al-Awwal and his most eminent student, 'Allāma Hillī.

7. Al-Muhaqqiq al-Awwal or al-Hillī, Abū al-Qāsim Ja'far b. al-Hasan (d. 676/1277). Muhaqqiq was the author of the most widely commented work of *fiqh* entitled *Sharā'i' al-islām*. He marks the end of the line of the "ancient" jurists, and the quality of his *Sharā'i'* carved for him the title *al-muhaqqiq,* the investigator, in demonstrative jurisprudence. His fame as a researcher attracted many scholars to his lectures in Hilla, among whom the well-known Imāmī savant Khwāja Nasīr al-Dīn Tūsī, also known as al-Muhaqqiq al-Tūsī, is mentioned by Ibn Fahd al-Hillī.

Ibn Fahd, in his commentary on Muhaqqiq's other *fiqh* work, *al-Mukhtasar al-nāfi',* recounts the occasion that prompted Muhaqqiq to compose a treatise for Khwāja on how to investigate the problem of the orientation of prayer in Iraq. When Khwāja Tūsī entered Muhaqqiq's lecture, Muhaqqiq paused to pay his respects to Khwāja, who urged him to continue his lecture. The lecture was about the commendability of moving toward the left (*al-tayāsur*) in the orientation of prayer (*al-qibla*) for worshipers in Iraq. Khwāja was opposed to *al-tayāsur* in reference to Iraq. When Khwāja heard Muhaqqiq's opinion, he said that there was no proof to support such commendability. If *al-tayāsur* involved turning away from the *qibla,* then it is prohibited; and if it meant turning toward it, then it is obligatory. Muhaqqiq replied that *al-tayāsur* was from the *qibla* to the *qibla,* meaning it was turning within the legally recognized possibility to attain the proper direction of *qibla* in reference to Iraq. As such it was recommendable.[25]

The uniqueness of Muhaqqiq's juridical methodology lies in first presenting his investigation in the form that was adopted by subsequent Imamite jurists. He revised the chapter rubrics of the previous works and brought into focus many important questions that were not classified and categorized with full rigor in the works of the "ancients." He divided his *Sharā'i'* into four parts. A *shar'ī* ordinance (*al-hukm al-shar'ī*), according to this classification, is set up either with the explicit intention of drawing close to God (*al-qurbā*) or not; when it is set up with that intention, then it pertains to *al-'ibādāt* (actions done by virtue of human-divine relationship). As for the other type of ordinance, it either needs verbal agreement in the form of "making an offer" and "accepting an offer," from two persons or from one on either side; or it does not need that procedure. If it does, then it pertains to *al-'uqūd* (contracts), thus needing verbal agreement on the part of one person, then it is termed *al-'īqā'āt* (one-sided dispositions); if it does not need any verbal statement, then it is a case of *al-ahkām* (ordinances).

In his *Sharā'i'* Muhaqqiq presents the gist of Tūsī's *Nihāya*, which had been based on, and deductively inferred from, the traditions of the Imamites; but he also includes the essential parts of Tūsī's *Mabsūt* and *Khilāf*, which contained references to the opinions held by Sunnite jurists. In addition, he critically evaluated and included in his *Sharā'i'* what was acceptable (1) in *Sarā'ir* of Ibn Idrīs, who had earlier investigated *Nihāya* critically; (2) in the *Muhadhdhab* of Ibn Barrāj, who had examined the rulings in *Mabsūt* and *Khilāf* most meticulously; and (3) in the works of other jurists who had studied the "ancient" jurists objectively.[26] According to the testimony of 'Āmilī, Muhaqqiq was indeed the Imamite jurist who originated careful examination of the juridical decisions of previous jurists, and it was from him that 'Allāma learned and perfected this methodology in jurisprudence.[27]

Muhaqqiq authored several works on *fiqh* and *usūl*, which are all well known and important in their contribution to Imamite jurisprudence. In this study I have depended heavily on his most important work, *Sharā'i'*. His concise opinions in this work on the authority of the jurist and the Just Ruler during the occultation have occasioned detailed comments in subsequent works on *fiqh*. The various commentaries on *Sharā'i'* afford a researcher an excellent opportunity to trace the development of opinions through their elucidation and elaboration by later jurists. No other work on Imamite applied jurisprudence (according to al-Tihrānī, who listed numerous commentaries on *Sharā'i'* in his *Dharī'a*) has occasioned so many commentaries by so many leading Imamite scholars of demonstrative jurisprudence.[28] Moreover, the commentaries of different jurists have dwelt upon different aspects of *Sharā'i'*. For this study I have found most informative Shahīd II's *Masālik* and Najafī's *Jawāhir,* in which opinions of Muhaqqiq on the subject of the Just Ruler, and whether the jurist is purported to be that just ruler, are discussed in great detail.

It is important to keep in mind that Muhaqqiq's legal opinions on the question of the Just Ruler at times reflect the period of relative peace between the Shī'ī and Sunnī communities under the Ilkhānid rulers. There were several factors that contributed to this easing of tension between the two communities. The most important one seems to have been the hostile attitude of the Mongols toward Sunnī Islam dictated in part by the decades of their rivalry with Mamluk Egypt. The Mongols also saw the (Sunnī) 'Abbāsid Caliphate as a hostile power, and their own religious tolerance and political skill led them to support dissident groups. Moreover, after the capture of Baghdad under the guidance of Khwāja Tūsī, enmity against Sunnism resulted in a tolerance of the Shī'a,[29] which in turn resulted in a revival of the atmosphere that had prevailed under the Shīʿite Būyids. The Ilkhānids, especially after their conversion to Islam, seem to have paid much respect to the Shīʿī jurists and their opinions on certain issues, as appears in the case of ʿAllāma who represented the Imamite views in the court of the Ilkhānid Khudābanda (703–16/1304–16).

8. Al-ʿAllāma al-Hillī, al-Hasan b. Yūsuf b. al-Mutahhar (d. 726/1325). ʿAllāma was the greatest Imamite jurist in the eighth/fourteenth century. His contribution is comparable to that of Tūsī at the end of the classical period. He had studied rational sciences with Khwāja Tūsī, and jurisprudence with Muhaqqiq. His firm grounding in both traditional and rational Islamic learning allowed him to produce numerous books and treatises on almost every branch of the Islamic sciences. Like Tūsī, whom he regarded as *al-imām al-aʿzam* (the greatest Imam), and whose intellectual efforts to strike a balance between proofs derived from reason and those from revelation were well known to him, ʿAllāma demonstrated his genius in establishing the intrinsicality between reasoning based on the authority of revelation and direct reason in his works. ʿAllāma regarded *hadīth* criticism as an important aspect of juridical reasoning, especially because it was known that not all the traditions recorded in the early Imamite sources could be regarded as reliable.

ʿAllāma's encyclopedic breadth of knowledge can be attested by references to his two comprehensive works on jurisprudence, *Mukhtalaf al-shīʿa fī ahkām al-sharīʿa* (The differences of opinion among the Shīʿa on the ordinances of the Sharīʿa) and *Tadhkirat al-fuqahāʾ* (An account of the jurists). Whereas *Mukhtalaf* covers in minute detail the problem of differences of opinion among Shīʿa jurists and the way these differences can be resolved (or at least minimized) through reinterpretation of the documentation, *Tadhkira* is a detailed comparative study of all schools of Islamic law. These were carefully studied by ʿAllāma, as is indicated by the listed contents of this work. It was this breadth of knowledge that earned him the epithet *al-ʿallāma* (the most learned), and gave him the unique position of the spokesman for Shīʿī views in official circles. Thus, when the Ilkhānid Khudābanda asked for a jurist from Iraq to solve a problem he was facing in the matter of divorce and marriage, the choice fell on ʿAllāma. All this confirms the unique position he occupied in Imamite learning.[30] It is significant that beginning with ʿAllāma, Imamite jurists form the line of the "moderns" (*al-mutaʾakkhirūn*), which continues up to the present time.

9. Fakhr al-Muhaqqiqīn, Muhammad b. al-Hasan b. Yūsuf (d. 771/1369). Fakhr Muhaqqiqīn was ʿAllāma's son, and his significance in Imamite jurisprudence lies in continuing his father's work in Hilla after the latter had migrated to Iran around 705/1305. He completed some of his father's unfinished works on jurisprudence and wrote a commentary on ʿAllāma's *Qawāʿid* entitled *Īdāh al-fawāʾid fī sharh ishkālāt al-qawāʿid* (Elucidation of the benefits: on the elucidation of the difficulties of *al-Qawāʿid*). More importantly, he continued the tradition of the Hilla school in training the great figures of Imamite *fiqh*, such as al-Shahīd al-Awwal.

10. Al-Shahīd al-Awwal (or Shahīd I, as I have designated him in this work), al-Shaykh Jamāl al-Dīn Makkī b. Shams al-Dīn Muhammad al-Dimashqī al-Jazīnī (d. 786/1384). Shahīd I is among those Imamite jurists who, after studying jurisprudence and philosophy under Fakhr al-Muhaqqiqīn and other Imamite scholars, proceeded to acquire the most cosmopolitan education of his time, which included

illuminationist mysticism as much as secular Arabic literature and stylistics. He sojourned in Mecca, Medina, Baghdad, Cairo, Jerusalem, and Hebron to study with well-known figures of the Sunnī tradition and jurisprudence. These places were the centers of Sunnī Islamic learning in the eighth/fourteenth century, just as Hilla and Karbalā' were centers of Shī'ī theology and jurisprudence. In the "license to teach" (*ijāza*), which Fakhr al-Muhaqqiqīn gave to Shahīd I and which is inscribed on the reverse of *Qawā'id,* the author refers to Shahīd I as "my master, the imam, the great scholar. . . ." He then says: "Muhammad b. Makkī . . . has permission to transmit and disseminate this book and all that my father ('Allāma) wrote. . . ."[31] This is an explicit recognition of Shahīd I's mastery of jurisprudence, and his ability to teach it.

Shahīd I has left numerous works on Imamite jurisprudence, and I have used his major works extensively to study his judicial decisions on the authority of the jurist and the concept of the Just Ruler. He was in contact with the Shī'ī dynasty of the Sarbidārān of Khurāsān (738–783/1337–1381), whose ruler, 'Alī b. Mu'ayyad (came to power in 766/1364), was regarded as a just ruler (*sultān 'ādil*). 'Alī b. Mu'ayyad had undertaken to spread Shī'īsm and was considered one of the excellent rulers of this dynasty for having improved the social and economic conditions of his subjects. He died in 795/1392, nine years after the execution of Shahīd I in Damascus in 786/1384.

This relationship between a Shī'ī ruler, who was highly regarded by his subjects, and Shahīd I is implicitly indicated in the legal decisions that appear in Shahīd's various works on jurisprudence, more particularly in *al-Lum'a al-dimashqiyya* (The Damascene brilliance), a treatise composed upon the request of 'Alī b. Mu'ayyad for the guidance of the Shī'a in Khurāsān. *Lum'a* has remained a fundamental text for all those who aspire to become jurists in the Imamite legal system.

A significant feature of the Imamite jurisprudence produced in the environment of Jabal 'Āmil is the absence of polemics with other schools of Islamic law. Shahīd I's willingness not only to study but also to teach Sunnī *fiqh* reflects his broad-mindedness and his marked inclination toward maintaining Muslim unity. It was probably for this reason that he declined the invitation of 'Alī b. Mu'ayyad to live in Khurāsān and chose instead to reside in Damascus. His home there was frequented by diverse Muslims, including government officials, regardless of their legal affiliation to one or the other Sunnī schools of law. He advised them on various issues.[32] His views on the performance of the Friday worship (see chapter 5, below) clearly show his comprehension of this important worship in the socio-political lives of all Muslims.

Among jurists who studied under Shahīd I, al-Fāḍil al-Miqdād al-Sīwurī's name appears frequently in my sources. Sīwurī died in the year 826/1422. His work *Kanz al-'irfān fī fiqh al-qur'ān* (The treasure of [insightful] knowledge on jurisprudence [extracted from] the Qur'ān) is an important, indeed unique, work in the elucidation of the legal passages of the Qur'ān. It is worth pointing out that the question of the jurist's authority is not taken up in *Kanz,* which corroborates my observation that the entire question of a just ruler in Islam is the result of the socio-political experience of Muslims in history in fulfilling the Qur'ānic vision of creating an adequately just public order on earth. However, once the issue of creating the ideal order became dependent on the existence of a just ruler, the Imam, authors of works on jurisprudence went on discussing it as a consequence of Islamic revelation.

11. Ibn Fahd al-Hillī, Ahmad b. Shams al-Dīn Muhammad (d. 841/1437). He was a student of Shahīd I. Shahīd I had conferred on Ibn Fahd "permission" to teach his works.[33] Ibn Fahd is reported to have written a treatise for one of his disciples by the name of al-Sayyid Muhammad b. al-Mūsawī (d. 870/1495), in which he predicted the emergence of Shāh Ismā'īl Safavī:

> As predicted by Amīr al-Mu'minīn 'Alī on the day of Siffīn [36/657] after 'Ammār b. Yāsir was killed. [At that time] he warned of certain massacres (*al-malāhim*) that would occur with the rise of Chengiz. Among the *malāhim* was the emergence of Shāh Ismā'īl.[34]

In his treatise Ibn Fahd had required obedience to Shāh Ismā'īl because "this *sultān* will have arisen with truth and triumph."[35]

Ibn Fahd's most important work in Imamite jurisprudence is his commentary on Muhaqqiq's *al-Mukhtasar al-Nāfi'* entitled *al-Muhadhdhab al-bāri' ilā sharh al-nāfi'* (The outstanding revised study, aiming to provide a commentary on *al-Nāfi'*).[36] Ibn Fahd here takes up differences of opinion among jurists and comments on them by pointing out errors in judgment on a particular issue and citing the sound opinion in that connection. The work is in the tradition of the great masters of Baghdad and Hilla. What strikes an investigator as the hallmark of the *ijtihād* of Ibn Fahd is the inferences he makes about the position of a jurist in the community. His rulings on this question leave no doubt that he saw the Imamite jurist as the functional imam of the Imamite community during the occultation.

Among those who related on Ibn Fahd's authority are included al-Shaykh 'Alī b. Hilāl al-Jazā'irī, who trained one of the outstanding jurists of the early Safavid era, al-Muhaqqiq al-Karakī, or al-Muhaqqiq al-Thānī (so called in relation to al-Muhaqqiq al-Awwal al-Hillī).

12. Al-Muhaqqiq al-Karakī, 'Alī b. 'Abd al-'Āl (d. 937 or 941/1530 or 1534). Karakī was a well-known Imamite jurist who witnessed the emergence of the Shī'ī *sultān*, Shāh Ismā'īl Safavī in 907/1501, and the consolidation of the Imamite state in Iran under the reign of Tahmāsp (930–84/1524–76). This period of Safavid history is regarded as the period of consolidation of Shī'īsm during which jurists like Karakī took advantage of the protection afforded by the ruler to consolidate their own position with the people, on the one hand, and to deepen their learning and scholarship, on the other.[37] Consequently, his rulings in the matter of a just ruler, and the role of an increasingly powerful jurist, as well as other socio-political issues, mark an unprecedented level of *ijtihād* based on the early "opinions" in the political jurisprudence of the Imamites. This *ijtihād* was partly due to his education in the tradition of the Shī'ī scholars of Jabal 'Āmil. Karakī, like Shahīd I and Shahīd II, had acquired broad legal and administrative knowledge by studying at different centers of Sunnī learning before going to Iraq, and subsequently migrating to Iran.[38]

Karakī's commentary on 'Allāma's *Qawā'id*, entitled *Jāmi' al-maqāsid fī sharh al-qawā'id* (Uniter of the endeavors in elucidation of *al-Qawā'id*), contains opinions that reflect his own political experience in the administration of the Imamite state as a chief *mujtahid* appointed by Shāh Tahmāsp as much as the consolidated position of Shī'īsm under the Safavid rulers. The tone of Imamite jurisprudence, more particularly its political aspect, where the question of the exercise of authority and the delegation of the Imam's discretionary control are treated, becomes more aggressive during this period, which will culminate in the ultimate institutionalization of the comprehensive authority of the jurist in modern times. Imamite jurisprudence moved away from the period of *taqiyya* (precautionary dissimulation), which had been maintained throughout, even under the Būyid *sultān*s), until the establishment of the Safavid state. Under the Safavid rulers jurists for the first time explicitly regarded it possible for the jurist to become *sultān 'ādil* (just ruler), which had been merely implied in the works of the "ancient" jurists when they ruled it permissible for an Imamite jurist to exercise certain authority with theologico-political connotations. The legal decisions of Karakī and other Safavid jurists clearly reflect an elaboration through *ijtihād* of the authority of Imamite jurists in the context of a Shī'ī state.

13. Al-Shahīd al-Thānī (or Shahīd II), Zayn al-Dīn al-'Āmilī al- Jabā'ī (executed in Istanbul in 966/1558). Shahīd II was the contemporary of Muhaqqiq Karakī, but lived in Jabal 'Āmil. Shahīd II's career and contribution to Imamite jurisprudence in many ways resembled those of Shahīd I, whose academic work he continued by writing a detailed commentary on his *Lum'a*, and whose social work he continued by maintaining a good relationship with all Muslims of various Sunnī schools, mastering their major works and teaching them in his comparative lectures. It is significant that Shahīd II was introduced to the illuminationist philosophy of al-Suhrawardī. He learned it in Cairo, where he studied mystical philosophy with al-Shaykh Muhammad 'Alī al-Gīlānī and al-Shaykh Shams al-Dīn Muhammad b. Makkī.[39] In order to trace the development of rulings in regard to the authority of the jurist, I have made extensive use of Shahīd II's commentary on Shahīd I's *Lum'a*, entitled *Rawdat al-bahiyya* (The brilliant garden), and on Muhaqqiq's *Sharā'i'*, entitled *Masālik al-afhām* (Methods of comprehension).

The *ijtihād* of jurists like 'Allāma, Shahīd I and Shahīd II, based on *usūl al-fiqh*, refined the rough edges of political jurisprudence as it had continued up to this time. But toward the end of the Safavid era, under a prominent traditionist (*muhaddith*), Mullā Muhammad Amīn b. Muhammad Sharīf al-Astarābādī (d. 1033/1624), who had studied with Shahīd II's son, Jamāl al-Dīn al-'Āmilī, the role of *akhbār* (traditions) in the deducing of legal decisions was revived.

Debate over the relative fundamentality of *usūl* and *akhbār* in the derivation of judicial decisions had its roots in the theological debate about the priority of reason over revelation. From the early days of Imamite history, even during the Imamate of al-Bāqir and al-Sādiq, there were discussions between the close associates of the Imams about *al-'aql*, symbolizing the authority and even the priority of human reasoning, and *al-āthār* or *al-akhbār*, symbolizing the authority of revelation. During the presence of the Imams, the final decision on any disputable matter undoubtedly lay with the Imams, whose interpretations of the Qur'ān and the Sunna were authoritatively binding because of the belief in their infallibility. However, when the twelfth Imam went into occultation, that certainty in the interpretation of revelation was unattainable, all the more so inasmuch as normal human reasoning was fallible and not protected by the grace of God. In Imamite jurisprudence even *ijmā'* (consensus) had to be protected by the inclusion of the infallible Imam's opinion in it; otherwise that *ijmā'* had no legal validity for deducing law. The authority of the Imam's utterances, whether in the case of *akhbār* or *ijmā'*, was so central to the decision-making process in jurisprudence that even when *ijtihād* was admitted as a valid intellectual process in deducing judicial decisions, it was, as pointed out above, *al-ijtihād al-shar'ī* (reasoning based on the revelation), and not *al-ijtihād al-'aqlī* (reasoning based on the intellect) that was regarded valid. This latter *ijtihād* was equated with *qiyās*—namely, inductive reasoning—whereas the former type of *ijtihād,* in Imāmī Shī'īsm, was deductive reasoning to establish a judicial decision by indicating its conformity with the divine will.

With the development of *usūl al-fiqh,* especially the section that dealt with the accredition and approbation of traditions through critical examination of the documentation appended to them (i.e., textual and contextual analysis), the authority of the traditions compiled by early authorities like Kulaynī and Ibn Bābūya could not be maintained unquestioningly. In the fourth/tenth century, Qumm and Rayy scholars of Imamite learning were inclined to accept the authority of *akhbār* uncritically. But Baghdad scholars, who had a thorough grounding in the rational side of transmitted sciences, like theology, were inclined to conduct rational investigation to ascertain the authenticity of traditions. The tension between the upholders of one or the other attitude toward compiled traditions attributed to the Imams, whether expressed or not, always remained latent in Imamite jurisprudence, because any challenge to the authority of the traditions was not without implications for the recognition of the Imam's authority as the sole interpreter of revelation.[40]

Although the Akhbārī opposition to the well-established Usūlī methodology found support among some prominent jurists, including Muhammad Bāqir Majilisī (d. 1110/1698), for a considerable time, there was no way for Imamite jurisprudence to revert completely to the Akhbārī position, which tended to be more literalist and rigid in its juridical rulings. More significantly, total abandonment of the Usūlī methodology would have led to the imposing of limitations on furthering the authority of the jurist as a person who possessed qualifications to interpret the will of the Hidden Imam. On the other hand, conceding to the inherent limitations in *usūl* did not in any major way alter rulings derived through *ijtihād* based on the textual evidence preferred by the Akhbārī jurists. In the later works of Akhbārī *fiqh,* especially *al-Hadā'iq al-nādira* of Yusuf b. Ahmad al-Bahrānī (1186/1773), one can hardly distinguish major differences of opinion in the form of disputed questions between his work and that of an Usūlī jurist, although the use of *hadīth* in *Hadā'iq* is overwhelming.

14. Al-Sabzawārī, Muhammad Bāqir b. Muhammad Mu'min (d. 1090/1679). Also known as al-Muhaqqiq al-Sabzawārī, he was another prominent jurist, who held an official position of *shaykh al-islām* in the Safavid administration. He had studied under Muhammad Taqī al-Majlisī (d. 1070/1660). Sabzawārī's two works on demonstrative jurisprudence, *Kifāyat al-akhām* (Exhaustive treatment of the ordinances) and *Dhakhīrat al-ma'ād fī sharh al-irshād* (The provision for the day of resurrection on the elucidation of *al-irshad* [of 'Allāma]), have been cited frequently by later jurists.

The later part of the Safavid era saw much growth, both in the number and in the power of Imamite jurists. It was notably the schools of Isfahān and Najaf, and Karbalā', that trained *fuqahā'* in both the Usūlī and Akhbārī methodologies. However, during the period following the downfall of the Safavid dynasty and throughout the interregnum before the Qājār rule was consolidated in Iran, until the later part of eighteenth century, the Akhbārī influence on Imamite jurisprudence prevailed. It was Muhammad Bāqir b. Muhammad Akmal al-Wahīd al-Bihbahānī (d.

1208/1793), the outstanding *mujtahid* of the later Safavid and early Qājār periods, who almost completely uprooted Akhbārī influence from the holy cities of Iraq. It is reasonable to say that from Bihbahānī's time onward, the Usūlī methodology became established as the sole authoritative method of Imamite jurisprudence,[41] although the *hadīth*-related sciences continued to develop as required by *al-ijtihād al-sharʿī*. Moreover, as a consequence and implication of the Usūlī victory, functions of a *mujtahid* became defined with greater clarity at this time. Bihbahānī's dominant position as a *mujtahid* inspired and prepared a large number of his students who attained great influence in Iran during the Qājār period.[42]

A number of Usūlī jurists had studied, directly or indirectly, under teachers who favored Akhbārī methodology, as indicated by the fact that Bihbahānī studied under his father, Mullā Muhammad Akmal, who was Muhammad Bāqir Majlisī's student. Majlisī's close relationship with Safavid monarchs and his vast authority under them is reflected in his various rulings that deal with the authority of the *mujtahid* in an Imāmī state. Although Majlisī has not left any systematic work on jurisprudence, he has covered the chapters of Imamite *fiqh* in his multivolume *Bihār al-anwār*, recording all the *hadīth* reports, as was done in an earlier age by Ibn Bābūya. His political jurisprudence, with which we are concerned in this study, comes close to the political attitude of Sunnī jurists when dealing with those actually invested with power, the de facto *sultān*s. Thus, he believed, like Ghazālī or Ibn Taymiyya, that tyrannical rulers and their oppressive officials should not be opposed; rather, he advises that one should employ *taqiyya* in their regard, to avoid exposing oneself to pointless harm.[43] His opinions favoring the all-comprehensive authority of the jurist have been criticized by some later authorities as springing from the love of power; but they were also reinforced by others, as witness Bihbahānī's rulings in this connection as handed down by his students.

Imamite political jurisprudence was bound to take into consideration the new exigencies created by the Imamite monarchs of the Safavid and Qājār dynasties who had also declared the Imamite school of jurisprudence the official source of law in their dominions. It is probably correct that Imamite jurists during this period saw a rare opportunity to implement the ideals of the Imamite system in the public order. To that end, both the interpreters of the system (that is, the jurists) and the method of interpreting the system itself had to be consolidated. Although the jurist, as a deputy of the Imam, needed to be buttressed with discretionary control over the affairs of the community (that is, *wilāyat al-tasarruf*), the legal system, which was being threatened by the rigidity and literalism of the Akhbārī trend, had to be salvaged. In achieving both these goals, Bihbahānī and his prominent students, like Kāshif al-Ghiṭā' and Najafī, played a crucial role. Indeed, Bihbahānī is regarded as a *mujaddid* (reviver) of the Imamite community in the twelfth/eighteenth century for his very timely contribution in reasserting the fundamentality of *usūl* in the interpretation of law.[44] In addition, his efforts in establishing an Imamite jurist's candidacy to exercise power in an Imāmī state, and even regarding his rule as preferable to the rule of an unjust authority, resulted in the subsequent consolidation of the guiding role of a *mujtahid* in the Imamite community.

15. Kāshif al-Ghiṭā', al-Shaykh Jaʿfar al-Kabīr (d. 1228/1813). Kāshif al-Ghiṭā' was the student of Bihbahānī and al-Sayyid Mahdī b. Murtadā Bahr al-ʿUlūm (d. 1212/1797) in Karbalā' and Najaf. Kāshif al-Ghiṭā' had maintained close relations with both the Ottoman authorities in Baghdad and the Qājār monarch of Iran, Fath

ʿAlī Shāh (1212–50/1797–1834). His public life and his acceptance of an official position on behalf of the Sunnī Ottoman governor of Baghdad to bring about a peaceful settlement with the Persian army advancing on that city are all reflected in his rulings, in which he saw the leading role a *mujtahid* could play in the state affairs of a Muslim community.[45] His own involvement in leading a defensive *jihād* during the siege of Najaf by the Wahhābīs in 1220/1805 was dictated by his judicial decision in which a *mujtahid,* qualified to issue legal opinions in his position as the deputy of the Hidden Imam, can give permission to the monarch to lead a *jihād* against the enemies of Islam, or even himself do so. Furthermore, he saw the possibility of having a just ruler, pending the return of the Hidden Imam, as is indicated in the dedication to Fath ʿAlī Shāh of his major *fiqh* work, *Kashf al-ghitāʾ ʿan mubhamāt al-sharīʿat al-ghurra* (Unveiling the cover from the ambiguities in the most illustrious Sharīʿa).

Kāshif al-Ghitāʾ was a leading jurist of Najaf in the line of *uṣūl al-fiqh.* His use of *ijtihād* in his ruling on *jihād,* and other issues related to the declaration of *jihād,* is evident in his treatise entitled *Jihādiyya.* It is explicit argumentation on the right of an Imamite jurist to authorize a qualified person to lead a *jihād,* which implies that a jurist, in the absence of the actual legitimate authority of the Imam, can legitimize the authority of an Imamite by supervising the public order. It was this comprehension of the role of the general deputy of the Imam that allowed Kāshif al-Ghitāʾ to authorize Fath ʿAlī Shāh to undertake a *jihād* as a *sultān* against the Russians.[46]

16. Al-Najafī, al-Shaykh Muhammad Hasan (d. 1266/1849). Najafī was Shaykh Jaʿfar al-Kabīr's student in the tradition of Bihbahānī's Uṣūlī methodology. Najafī's monumental work, *Jawāhir al-kalām fī sharh sharāʾiʿ al-islām* (The gems of discourse in elucidation of *Sharāʾiʿ al-islām* [of al-Muhaqqiq al-Hillī]), in forty-two volumes, has overshadowed the important works of many of his contemporaries— for example, Mullā Ahmad b. Muhammad Mahdī al-Naraqī al-Kāshānī (d. 1245/1829), and Shaykh ʿAlī Kāshif al-Ghitāʾ (d. 1254/1838).[47] Najafī's commentary on *Sharāʾiʿ* has not been surpassed in its depth of discussion and encyclopedic information on the opinions of all the Imamite jurists up to his own time, actually citing their works carefully and even at times the oral communications of his contemporaries, including his teacher Bihbahānī. *Jawāhir* needs to be studied independently both to write the history of Imamite jurisprudence in its application, and also to analyze the principles of juristic practice used in the demonstrative jurisprudence of the Imamites as it culminated in Najafī's intellectual endeavors.

Najafī's position as a *marjaʿ al-taqlīd* (the most learned authority acknowledged among the Imamite jurists of his time, whose rulings were followed by all the Shīʿa), gave the position of a *mujtahid* among the Imamites an unprecedented recognition. Najafī's prestigious and well-deserved recognition among the Shīʿa in general, and among the scholars in particular, was instrumental in the acknowledgment of his contemporary, the great *mujtahid* Shaykh al-Ansārī, as the leader of the Imamite community and as their sole *marjaʿ al-taqlīd* in Najaf around 1266/1849, before Najafī's death. By this time Najaf, after the death of Tūsī in the eleventh century, revived as the center of Shīʿī learning, more particularly of jurisprudence, under the leadership of Ansārī.

17. Al-Ansārī, al-Shaykh Murtadā (d. 1281/1864). Ansārī was a younger contemporary of Najafī, whose lectures he attended for a while in Najaf. He is regarded as the "seal of the *mujtahid*s," because of his lasting impact on the methodol-

ogy of Imamite jurisprudence and his reaffirming the theoretical potentialities of the functions that a *mujtahid* could undertake as the functional imam of the Shī'a. He had studied with all the major figures of Imamite jurisprudence in Iraq and Iran before assuming leadership of the Imamites in 1849. I will discuss his political jurisprudence, more specifically his treatment of the comprehensive authority of a jurist, in greater detail in chapter 5. At this juncture, in the context of this brief history of the development of Imamite political jurisprudence, it must be stressed that it is his contribution to *usūl al-fiqh* for which he has been designated the "seal" of those who employ *ijtihād*—that is, the *mujtahids*.

Ansārī has left several works, of which *al-Rasā'il* on *usūl* and *al-Makāsib* on applied *fiqh* enjoy the unparalleled prestige among jurists. In the latter work, Ansārī has demonstrated his intellectual perceptivity in applying the principles of legal theory in deducing laws. The most original aspect of *ijtihād* in *Makāsib* is the continued reference to *al-'urf* in making judicial decisions. Ansārī's use of the concept of *al-'urf* in the *mu'āmalāt* (transactions) section of *fiqh* is significant. This is because he argues in several places, in the context of the social relations, that it is important to analyse terms and understand their conventional and customary significations before any legal decision can be made. Ansārī does not use the concept of *al-'urf* in the meaning of "social conventions and common practices" known to Muslim jurists from the early days of Islamic jurisprudence and at times equated with *al-ijmā'* (consensus).[48] Thus, *'urf* in Ansārī's usage corresponds to the meaning of "ordinary language," which is the subject of discussion in analytical philosophy (clarification and solution of problems).

Ansārī demonstrates his application of *al-'urf* in his analysis of the phrase *ulū al-'amr* (those invested with authority). The phrase is discussed in the context of the discretionary control that can be exercised by various persons, including "those invested with authority," in the managing of the affairs of a Muslim community. He explains the use of the term in revelation and then embarks upon the sense in which Muslims ordinarily understand the phrase. Following the discussion he says that the apparent sense derived from the term in its conventional usage is that "it is a person to whom it is obligatory to refer in all those matters for which there is no single person responsible." He then concludes that *'urf* would expect the *ulū al-'amr* to assume political responsibility as part of the general matters for which they are responsible. As such, anyone who is invested with authority to become *ulu al-'amr* can be said to possess discretionary control over the property and lives of the people.[49]

Ansārī continued in Majlisī's tradition insofar as his attitude to rulers was concerned. His political quiescence is reflected in his treatment of the functions of Imamite jurists in the Imamite state, which concedes no political authority to them in their position as the deputies of the Imam. Nevertheless, his rulings regarding the qualifications of a jurist who can become *marja' al-taqlīd* (accepting the authority of the most learned jurist in the matters of the Sharī'a)—for example, the fact of such a one's being the most learned (*a'lamiyya*)—made the theoretical as well as functional position of Imamite religious leadership a highly influential and centralized office, acknowledged as such by the Shī'a through the requirement of *taqlīd*. Among those who reaped the fruits of Ansārī's *ijtihād* in the matter of *marja' al-taqlīd* are included his famous pupils al-Sayyid Muhammad Hasan al-Shīrāzī (d. 1312/1894), leader of the Imamites during the crisis that followed the Tobacco

Concession in 1890;[50] and al-Ākhund Muhammad Kāzim al-Khurāsānī (d. 1329/ 1911), the leading *mujtahid* supporting the Iranian constitutionalist movement.

The constitutional revolution of Iran in 1906, in which prominent jurists played a leading part with men who had received a modern education, brought the political jurisprudence of the Imamites into focus, albeit retrospectively.[51] For the jurists, the revolution meant the implementation of Islamic law in all its aspects throughout Iran. In the light of the tyranny of secular rulers and the oppression suffered by the people at the hands of government, Ākhund Khurāsānī, al-Mīrza Muhammad Husayn al-Nā'inī (d. 1355/1936), and other eminent figures provided necessary guidance in legitimizing the development of modern political institutions, especially the consultative assembly with the responsibility of legislating for the modern state under the guidance of the leading *mujtahids*. But in the process of modernizing the political institutions of the Imamite state, both the role of Imamite jurisprudence and that of its exponents become drastically limited to personal and religious law, especially when the revolution failed to unite disparate intellectual elements to carry out its fundamental aim—according to the jurists, that of implementing Islamic law. And there had been a conflict between traditionally educated jurists and supporters of the constitutionalist movement who had been given a modern education. They disagreed on the implementation of Islamic law, with the result that the jurists became suspicious of the religious sincerity of liberal thinkers.[52] Consequently, Ansārī's refined methodology in deducing laws, especially in the area of socio-political relations, could not be tested for its applicability in the modern situation. Moreover, the position of the *marja' al-taqlīd* became more and more confined to strictly spiritual matters, with the result that by the time the Imamite religious leadership and community experienced another revolution, in 1978–79, both the leadership and the general body of the Shī'a had been conditioned to believe that, doctrinally as well as functionally, jurists could not assume any political leadership, especially in the modern context.

Some prominent jurists undertook to rationalize constitutional movement by composing political tracts in support of parliamentary legislation and legitimation of democratic government in terms of Islamic revelation. But they did so without discussing it as a juridical elaboration of the authority of the jurist (*wilāyat al-faqīh*), who could assume the authority of the Imam in occultation. Thus, major works on jurisprudence produced during the postconstitutional period not only ignored political jurisprudence discussed by the Qājār '*ulamā*', they also failed to take into account modern developments in society, such as the new financial structures that replaced medieval and premodern financial institutions. These works continued to discuss normative jurisprudence without any concern for real situations.

As I pointed out earlier, the peculiar characteristic of Islamic jurisprudence is its individualistic approach to its subject matter. It is for this reason that we cannot speak of the uniform opinions of the Imamite jurists who have lived and taught at the three main centers of Imamite learning—Najaf, Qumm, and Mashhad. What unite them are doctrinal considerations and juridical methodology as laid down by Ansārī. Otherwise, when it comes to making a judicial decision, an Imamite jurist acts individually and independently, for he alone is to be held responsible for his inferentially derived opinion. The Iranian constitutional revolution, adumbrated by the rapid socio-political development in the context of modern intellectual currents in the Islamic world, in the years following the Second World War, brought about significant change in the general education of jurists who began to address some of the problems faced by a modern Shī'a in managing public life. The Iranian Shī'a seem to have exerted an indirect pressure on the *marja' al-taqlīd* to tackle some of the issues that needed to be resolved if modern Iran was to remain faithful to the application of religion in its socio-political life.

The guidance that came from Shī'ī centers of religious learning during this period, whether from Najaf or Qumm, demonstrates that the legal and political jurisprudence of the Imamites was responding to the methodology of Ansārī, more specifically its use of *al-'urf,* as a major source of authoritative guidance to the Shī'a in the modern world. Consequently, the juridical rulings of the *marja' al-taqlīd* (including al-Sayyid Abū al-Hasan al-Mūsawī al-Isfahānī [d. 1945], al-Sayyid Husayn al-Burūjirdī [d. 1961], al-Sayyid Muhsin al-Hakīm [d. 1970], al-Sayyid Muhammad al-Hādī al-Mīlānī [d. 1975], al-Sayyid Muhammad Bāqir al-Sadr [executed 1980], al-Sayyid Abū al-Qāsim al-Khū'ī, and al-Sayyid Rūh Allāh al-Mūsawī al-Khumaynī), as they deal with various modern topics—such as the creation of modern Islamic financial and political institutions—mark a new era in Imamite jurisprudence.[53]

It is relevant to note at the end of this brief outline of Imamite *fiqh* that for the first time in history Imamite jurists have begun to discuss their judicial decisions in the context of *uṣūl al-fiqh* in the modern Consultative Assembly of Iran. In the past that body was less receptive to the idea of implementing Imamite-applied *fiqh* in the socio-political life of the Shī'a of Iran.

The question that has often come up for discussion with some jurists is the sensitive issue of the human prerogative to legislate for Muslim society. In fact, during the constitutional revolution, some jurists regarded legislation as an innovation (*bid'a*) and even as legislation against Islam itself. There is tension between what is proclaimed to be divine law and hence perfect, in no need of further revision or review, and what is legislated by human beings in the form of positive law, in continuous need of revision. Can Imamite jurists or the members of the Consultative Assembly in a Shī'a state arrogate the right of legislation for the community, a right reserved to the divine Lawgiver?

Debate on this issue among Imamite scholars is deeply entrenched in Imamite theology, and it involves the metaphysical dimension of Imamite cosmology. God, who is perfect and just, has a purpose in creation and in sending the Prophet with a perfect system to create Islamic public order. That being the case, the question of revision of the system does not arise, because, according to the Qur'ān, God has declared: "This day I have perfected your religion for you" (5:5). Consequently, there is no need for further legislation in the Islamic state: all that is needed for human society has already been legislated by the divine Lawgiver. On the other hand, it is acknowledged that human beings need to perfect their rational faculty in order to discover the inevitable conformity between divine laws and that which reason demands a priori. The more human reasoning attains maturity, the more it will discover and comprehend the divine wisdom in legislating for human society through the divine agency of the Prophet. Thus, the effort of human reasoning must be directed toward discovering and extracting laws by deduction from the principles laid down in the divine sources of guidance, and not toward legislation by inductive reasoning. This is one of the reasons why the de facto "legislative" assembly is called a *consultative* assembly in line with the Qur'ānic injunction, "their affair being counsel between them" (42:38).

In the final analysis, the debate centers on the classical theme in Muslim theology: reason versus revelation. There is no doubt that in Imamite theology and, by extension, in Imamite jurisprudence, reason is recognized a priori; but the authority of revelation is even more crucial in resolving the tension experienced by human society—namely, the discrepancy between the ideal and the real. It is important to remember in the Imamite context that the perfect rule of justice and equity will be established only by the Imam al-Mahdī, and it will also be through divine intervention when human society will have failed time and again to create such order on earth. Until then, reason dictates and revelation confirms that human society ought to *strive,* as the Qur'ān puts it succinctly: "There is nothing for the human being except to strive" (53:39).

The present study has attempted to cover this vast material in the political jurisprudence of the Imamites, to trace the authority of the jurist, and to answer an important question regarding Imamite political authority during the occultation of the Imam: Can there be a just ruler in the absence of the Just Ruler as such? In searching for an answer to this question, I proceeded to investigate the basic components of the Imamite worldview, which, as I pointed out earlier, is dominated by two distinctively Shī'ī beliefs: the justice of God and the Imamate of the rightful successors of the Prophet. Both are part of the *usūl al-dīn* (fundamental doctrines of the faith) in Shī'ism. The interrelationship and interdependency of these two doctrines are indisputable in Imamite theology, and they have provided the basic religious focus for the Shī'ī community so as to orientate authoritatively its whole perspective on life. The basic religious focus is the Imam of the Age, who represents God on earth. Given the occultation of the Imam, the Shī'a focused its religious belief on the deputy of the Imam as the spokesman of the Imam and the "gate" (*bāb*) between him and his followers.

This assumption of the leadership of the Imamite community by the prominent followers of the Imamite school, due to the prolonged occultation of the Imam, was a key element in the development of its authority. This authority was based on the principles of necessity and need, which were rationally derived by taking into consideration the necessity to lighten the burden of a people and the need to safeguard its religious and social life by providing it with sure guidance. Thus, the search for the authority of the deputy of the Imam was not conducted in theological texts, because no delegation of the Imamate was conceivable without divine designation; it is found in juridical works, where practical guidance could appear by considering all situations and all individuals whose welfare pertained to the justice of God.

With all this in mind, I undertook to examine the juridical works produced throughout Imamite history, limiting my investigation to the representative works in each period of its unfolding. I proceeded to examine minutely Tūsī's *Nihāya* as the first complete work of Imamite jurisprudence, looking for the appropriate chapters on *fiqh* where one should investigate the Shī'ī concern for rightful authority during occultation. Not surprisingly, the political jurisprudence was located in the sections that deal with the affairs of this world, the *mu'āmalāt;* at the same time, the part that deals with religious observances (*'ibādāt*), although not dealing with the question of authority directly, contains rulings that have implications for the constitutional authority of the Imam. These are the sections dealing with the Friday prayer and the obligation of *jihād* and everything related to it. In the *mu'āmalāt* section, discussion of the authority of the Imamite jurist as a deputy of the Imam, or as a qualified individual who could supervise the affairs of the community, is found in the sections on the administration of justice, including the institution of legal penalties; family law, including marriage, divorce, and whatever is related to them; and transactions, embracing property, contracts, and similar matters. The authority of a qualified jurist ostensibly manifests itself in the administration of justice, which, in Imamite jurisprudence, is included in the section dealing with "enjoining the good and forbidding the evil." Moreover, the discussion related to this religious-moral obligation and its rational underpinnings directs an investigator to locate the tangible Imamite authority in the absence of the properly designated Imam.

It was always tempting to deny the pragmatic value of decisions made under this general rubric of "enjoining the good and forbidding the evil," and to venture into an interpretation of suspected self-aggrandizing motives of an individual jurist in upholding a particular decision that enhanced the position of the Imamite jurist. But given the religious and ethical ideas upon which Imamite political jurisprudence was constructed, it was more plausible to advance the view that the twin doctrines of the justice of God and the necessity of the Imamate underlay those rulings in political jurisprudence that could be explained by the help of *usūl al-fiqh*. The relevant discussion on *usūl* is found in the section on reasoning as a source of law where are treated the objective nature of good and evil, and the continued moral-religious responsibility of mankind to enjoin good and forbid evil.

The next step in the research was to try to ensure that the sources were interpreted as they were originally comprehended by those who wrote and taught them. Problems in the interpretation of juridical texts are enormous, and no library research can substitute for personal contact with Imamite jurists who can expound on a ruling authoritatively on the basis of their broad education in the related sciences of *hadīth, rijāl* (dealing with the transmitters of *hadīth*), *kalām* (dialectical theology) and Qur'ānic exegesis. Recent studies attempting to interpret some of the questions raised in this work have shown inadequate technical comprehension of the texts, apparently relying on modern dictionaries without due attention to the historical context of the technical language of jurisprudence. Although trained in the traditional sciences and methodology myself, I deemed it indispensable to spend further time in Iran after the 1978 revolution in order to hear lectures of jurists on political jurisprudence in the context of the changed circumstances. What I had learned in the religious centers of Iraq and Iran before the revolution and what I acquired in the postrevolution period in Iran are now part of this study. Hence, I can say with confidence that the sources here speak for themselves through their expert exponents. This in no way exonerates the work from any shortcoming in the conveying of ideas in English, which is an exercise in "interpreting the interpretation" from one language to another.

In light of the above explanation of my approach, I have organized my research into five chapters as follows:

Chapter 1, "The Deputyship of the Shī'ite Imams," discusses the early development of the authority of the close associates of the Imams up to the occurrence of the Complete Occultation of the twelfth Imam in A.D. 941. There were two important factors that determined the deputyship and leadership of the prominent associates (*rijāl*) of the Imams: their "sound belief" (*īmān sahīh*), and "sound knowledge" (*'ilm sahīh*). These two factors would remain constant for the assumption of the deputyship throughout the later development of the authority of the jurists. Moreover, these two factors, in addition to a third factor of "sound character" (*'adāla*), would give jurists a position of such weight in the Imamite community as to exercise the political right of the Hidden Imam in his absence. The development of the *rijāl*'s deputyship could be regarded as an intrinsic part of the Imamite worldview that required the presence of a religious figure to lead the community to the Islamic ideal—the notion of the Just Ruler.

Chapter 2, "The Imamite Jurists, Leaders of the Imāmiyya," traces the development of the authority of the jurist following the Complete Occultation in A.D. 941. It also delineates the history of the compilation of the criteria for "sound knowledge" and "sound belief" in Imāmī Shī'īsm—the sources of authority for the Imamite jurist in the community.

Chapter 3, "The Imamite Theory of Political Authority," discusses the doctrine of the Imamate and the *wilāya* (authority) of the Imam to set the stage for discussion of the *wilāya* of the jurist in political jurisprudence. The chapter defines and elucidates important phrases like *al-sultān al-'ādil* (the just ruler) and its opposite, *al-sultān al-jā'ir,* in jurisprudence, as well as the ambiguity that sometimes surrounded the identity of a just ruler in the judicial decisions of the Imamites during occultation. The implied sense of generality in the usage of *al-sultān al-'ādil* allowed an interpretation that identified both the Imamite ruler and a qualified Imamite jurist as a just ruler of the Imamites pending the return of the twelfth Imam.

Chapter 4, "The Deputyship of the Jurists in *wilāyat al-qadā*' (Administration of Juridical Authority)," is a detailed discussion of the important section of political jurisprudence dealing with the *wilāya* of the jurist. The *wilāyat al-qadā*' was a prelude to much broader authority that emerged as a consequence of the prolonged occultation of the Imam. The chapter details the doctrinal underpinnings of rulings and the method employed to deduce them in establishing the deputyship of the Imam so that a jurist could assume the *wilāyat al-qadā*'. It also treats the question of the jurist's qualifications where his idealized, Imamlike position became an important factor in his assumption of the authority that in theory could be exercised only by the Imam. This idealization, in addition to the persistent failure of those who were in power to realize divine justice on earth, created conditions for jurists to finally assume *al-wilāyat al-'āmma* (comprehensive authority) over the affairs of a Shī'a community.

Chapter 5, "The Comprehensive *wilāya* of the Jurists," deals with the free discretionary control of jurists as exercised over the affairs of the community, and also with the problem with its legitimation. This chapter marks the culmination of my investigation, and takes up the question of the leadership in convening and performing the Friday prayer, regarded as the constitutional right of the twelfth Imam. If the deputyship of the jurist can be established in the matter of the Friday prayer, then there is a religious ground as well as a moral one for the ruling that legitimizes the comprehensive authority of the jurist. The chapter then discusses the contradictory views of major Imamite scholars in this regard and treats the question of institutionalization of the *wilāya* through a constitutional process in modern times.

1

The Deputyship
of the Shīʿite Imams

The worldview of the faithful in Imāmī Shīʿism is dominated by the question of the leadership of the Muslim community—the Imamate—that guides and maintains authoritative perspectives of the faithful in Shīʿite Islam. The question of the Imamate becomes the focal point of Shīʿī religious belief system, especially when the Imam of the Age is in occultation. It was this question concerning the fundamental religious belief in the necessity of the Imamate that gave rise to the search for representatives of the Imams, representatives who could generate a confidence among both lay and elite Shīʿites to achieve authoritative orientation in their cosmology—their religious belief system:

> The earth shall not remain without there always being a learned authority (ʿālim) from among us, who will distinguish the truth from falsehood.[1]

This prophecy made by the sixth Imamite Imam, Jaʿfar al-Sādiq, was a clear assurance that through the presence of such a "learned authority" religious leadership will continue to provide the necessary guidance in the absence of the infallible Imam.

In the earliest Shīʿī biographical dictionary that deals with the study of those prominent personages (rijāl) who transmitted the teachings of the Imams, the author, Abū ʾAmr Muhammad b. ʿUmar b ʿAbd al-ʿAzīz al-Kashshī (fl. first half of the fourth/tenth century), in his Introduction, attests to the Shīʿī adherence to the Imamite worldview. Kashshī compiled his work during the occultation of the twelfth Imam and, as a consequence, it is not difficult to construe a depiction of continuous leadership emerging from Kashshī's biographical dictionary, insofar as such a leadership of the prominent transmitters of the Imamite teaching forms a link with the divinely appointed Imamate. It was important for the Imamite worldview to be guaranteed that, even when the Imam was inaccessible, his guidance, through the transmission of the Imams' teachings by "learned authorities" of this school, was available. Thus, Kashshī assures his readers on the authority of a tradition by the Prophet himself. The Prophet declared that the religion of God would refute the interpretations of vain interpreters and repudiate the deviations of extremists and false claims of ignorant persons "just as the bellows remove the dross from iron."[2] It will be the presence of such a righteous person or a learned authority in possession of the authentic religious knowledge derived from the Prophetic source that will preserve the authoritative perspective of God's religion on earth. Such a person will be required to carry on the function performed by the divinely designated Imam, because the possession of authentic religious knowledge will make it incumbent upon him to provide guidance to the community at critical times in the future, as declared by the Prophet in the following tradition:

> When innovations will appear in my community, a learned authority will expose them through his knowledge. And if he does not do so, then may God's curse be upon him.[3]

Imamite biographical dictionaries like that of Kashshī indicate the manner in which, from the time of the Prophet onward, the Shīʿī religious perspective survived historical vicissitudes. And they showed how the disciples of the Imams and the subsequent generations of Shīʿī transmitters of the Imamite teaching during the occultation have left their mark on the path to preserve the focal point of the Shīʿī religious experience. The emphasis laid on the central Shīʿī belief regarding the necessity of perpetual divine guidance through the leadership of the Imams required Shīʿī transmitters to establish themselves as the authorities who not only afforded authenticity to the Imamite belief system, but who also assumed the role of functional imams in the absence of the twelfth Imam. In other words, historical vicissitudes, in addition to the transmitters' personal preparation, made it possible for them to take the place of the Hidden Imam and assert their authority as the qualified interpreters and repositories of the Imamite belief system.

In addition, they provided a continuation of religious leadership among the Imamites. This sense of continuity in the religious leadership furnished the crucial justification needed in the recognition of the Shīʿī transmitters as the only authoritative spokesmen for the Imam's will in the community. More importantly, this recognition, as a fundamental component of the Shīʿī worldview, reinforced the position of the transmitters within the community and enhanced their leadership by giving them a free hand in working out the details of the Imam's will in their juridical corpus dealing with methodology and application, and in administering the social and financial affairs of the Shīʿa. In a distinct way, then, the acknowledgment of the leadership of the Shīʿī scholarly elite can be attributed to the extension of the "apostolic succession," which, strictly speaking, was confined to the twelve Imams in theory, but could now include the transmitters in practice.[4]

The important consideration in defining Islamic religious life was the determination of continuity in a precedent handed down from generation to generation in an "unbroken" chain of transmission, because that precedent constituted the only authoritative basis for religious practice. As we shall see in this chapter, it was indeed in the realm of ascertainable religious leadership founded on "sound belief" and "sound knowledge" that one could locate the authority of the Shīʿī religious scholar. Moreover, it was in the "sound character" of the transmitter that the guarantee of his belonging to that chain of "apostolic succession" was sought. In all this process of ascertaining what constituted the "noble paradigm" in giving expression to the Shī'ite worldview, the role played by lay believers cannot be sufficiently documented. Nevertheless, the determination of their role is not impossible from works concerned with establishing the probity of the Shī'ī transmitters (the *rijāl*), who actually depended on the perception and confidence of the adherents of the Imams in their leadership to exercise real control of the judicial and financial affairs of the community.

It is with this development of the Imamite scholars as the functional imams of the Shī'a, pending the return of the last Imam as the Mahdī, that the present study will be concerned. My thesis, briefly stated, is that, according to Imamite theology, the doctrine of the Imamate of the Hidden Imam, who will assume the leadership of the Muslim community as the Just Ruler (*al-imām al-'ādil*) when he rises at the End of Time, states consistently that divine guidance ought to be available at all times in order for the faithful to live in accordance with the laws of God. This justification for the necessity of the Imamate, moreover, necessitates the existence of an authority that could transmit the Imam's elucidation of the divine laws—that is, the precedent for religious practice—uninterruptedly and authoritatively to the Shī'a when the actual Imam is in concealment.

As we shall see throughout this study, it was as a consequence of the belief in the continuation of religious leadership through the extension of "apostolic succession" that gradually a way was prepared for the religiously learned among the Imamites to assume socio-religious leadership of the Imamite community as functional Imams. And if they possessed the Imamlike qualities, these functional imams could be regarded as just rulers and could assume the comprehensive authority of the Imam, the Just Ruler. This development of the authority of the functional imams, as we shall see in this chapter, was not something that awaited the period of the Complete Occultation of the twelfth Imam; rather, it began during the time when the Shī'ī Imams, living under unfavorable political circumstances (eighth-ninth centuries), could not undertake the full responsibility of guiding their adherents, and hence delegated their partial authority to those disciples who proved their loyalty to them. In addition to the above-mentioned underpinnings of the doctrine of the Imamate, the leaders of the Imamite community were responding to historical circumstances when they assumed the role of the functional imams in the interim period before the appearance of the rightful leader of the community, the Mahdī.

Development of the authority of the jurist in Imāmī Shī'īsm was not something extrinsic to Imamite ideology. More accurately, as I shall demonstrate, it was the logical outcome of the prolonged occultation of the twelfth Imam, in addition to the need for a tangible "noble paradigm" that gave expression to the aspirations of the Shī'a, that required the community to respond to historical vicissitudes without compromising the central doctrine of its religious belief system. This response was worked out by Imamite scholars as the spokesmen of the will of the Imam, and of the ideals and conscience of the community, with extreme caution and careful discussion about the nature of the authority of Imamite jurists during the occultation. At no point, as we shall see, could jurists afford to ignore the fundamental underpinnings of the Imamite theory of political authority in propounding the extent of their own authority, which inevitably depended on the authority of the Imams. This observation will become evident in reconstruction of the discrete components dispersed throughout the literature examined for this study in order to trace the development of the role of Imamite jurists in Shī'ite Islam.

The Early Role of the Shī'ite Jurists (*fuqahā'*)

"When innovations will appear . . . a learned authority (*al-'ālim*) will expose them through his knowledge." The Prophetic declaration renders learned authority in the Muslim community responsible for protecting the purity of the Islamic faith. With this responsibility it also gives learned authority a right to lead the people to authentic interpretation of the faith. Clearly, the right to lead the people is derived from the possession of a special kind of knowledge (*al-'ilm*) by means of which the learned authority will refute the deviations of extremists and the false interpretations of ignorant persons "just as the bellows remove the dross from iron."

The possession of *al-'ilm* is important to bear in mind in the study of Imamite religious authority, because, in the absence of political investiture, this was also the determining factor in proving the claim to Imamate when there arose a dispute among several 'Alid descendants regarding the rightful successor of the previous Imam. Even Imams from the *ahl al-bayt* were expected to demonstrate their possession of special knowledge about the interpretation of Islamic revelation and explain the derivation of injunctions from it. Founders of the Islamic legal schools, as indicated in several sources, were tested in their comprehension of Islamic revelation, whether the Qur'ān or the Sunna.[5] *Al-'ilm, then, was both the source of authority and the means to legitimize the claim of authority, more so when authority was not dependent upon political investiture.*

It should be pointed out that *al-'ilm* provided both the Shī'ī Imams as well as the Sunnī founders of the four legal schools (*al-madhāhib al-arba'a*) religious authority independent of any temporal investiture, which corroborates my observation that religious authority in the Muslim community was dependent upon a person's comprehension of Islamic revelation. Islamic revelation provided scholars with valid precedents to determine the essential elements for any religious-legal activity. However, in Shī'īsm, not all knowledge, derived from any source, on Islamic revelation was to be regarded as authoritative. According to Kashshī, relying on the authority of the fifth Imam, Muhammad al-Bāqir (d.113/732), true knowledge—exegesis of the Qur'ān and the *hadīth*—could not be obtained either in the East or in the West; rather, it could be acquired only from the *ahl al-bayt,* the family of the Prophet—that is, the Imams.[6] For the Imamites, determining the source of their knowledge about Islamic revelation has always been extremely important. This importance is underscored by Kashshī in his Introduction where he painstakingly tries to show that the Shī'ī transmitters derived their information from the Prophet and the Imams. In support of his claim he cites a tradition in which Imam al-Sādiq, commenting on a verse from the Qur'ān, says: "And one should see to his knowledge, as to where it has been taken from."[7] In other words, in order to ascertain the validity of any knowledge, one has to ascertain its source.

In the light of discussion about the source of true knowledge, the Imamites maintained that it was the close associates of the *ahl al-bayt* who received true knowledge, and therefore only these devotees of the Imams were in possession of knowledge necessary to protect the religion of God from corruptive innovations. Moreover, the possession of this knowledge raised these associates to the level of the *ahl al-bayt.* Thus, Salmān, the Persian associate of the Prophet and the supporter of ʿAlī b. Abī Tālib, according to Kashshī, was endowed with special knowledge regarding esoteric matters. He apprehended the knowledge of the Prophet and ʿAlī, and perceived their spiritual station. It was for this reason, says Kashshī, that he was declared as belonging to the *ahl al-bayt*,[8] which meant that he had partially inherited the *ʿilm* that the Imams had inherited from the Prophet as his rightful successors. The religious implication of this declaration is what I have identified earlier as the extension of "apostolic succession" to include the close, loyal associates of the Imams to share in their elevated status as the "noble paradigm" for later generations. Their inclusion among the *ahl al-bayt* also denotes the position of accomplishment that any believer could attain through proven loyalty to the Shīʿī ideology. There was also a political implication of such a declaration regarding Salmān or any other close associate of the Imams. Salmān, according to Kashshī, was one of the three associates of the Prophet who did not "apostatize" after the latter's death by paying allegiance to Abū Bakr, the first caliph of the Muslim community. The other two were al-Miqdād b. al-Aswad and Abū Dharr al-Ghifārī.[9] This means that they were loyal supporters of the Shīʿī view of the leadership of the community, the cornerstone of the Shīʿī worldview.

It is significant that "apostasy," which is usually regarded in the Sharīʿa as a religious violation against God, is used by an Imamite author like Kashshī in the context of denying the right to political leadership of Imams among the *ahl al-bayt*. Thus, al-Sādiq is reported to have said that everyone apostatized after al-Husayn's martyrdom except for three persons—namely, Abū Khālid al-Kābulī, Yahyā b. Umm al-Tawīl, and Jubayr b. Matʿam, who remained faithful to the Imam ʿAlī b. al-Husayn, who followed al-Husayn in the Imamate.[10] From Kashshī's account of the early associates of the Prophet and ʿAlī, it is obvious that in their discussion these companions were mainly engaged in political issues pertaining to the caliphate, and the rightful position of ʿAlī as the Imam of the community. This interest in political leadership continued to the time of the Imamate of al-Bāqir when the Qurʾānic exegesis, *hadīth,* jurisprudence, and other related topics gained in importance.[11]

It is plausible that from the time of al-Bāqir and his successor, al-Sādiq, the point of emphasis in the Imamite leadership was the *wilāya* (the authority that reserved the right to demand obedience), not the *saltana* (power that enforced obedience). Consequently, from this time on, the *wilāya* of the Imam was increasingly conceived in his elucidation and elaboration of *al-takālīf al-sharʿiyya* (religious-legal obligations specified in Islamic revelation), with a view to create a normative Islamic legal system that would be enforceable only when the *wilāya* and the *saltana* will be invested in the rightful Imam from among the progeny of the Prophet.

Thus, in the topics raised for discussion by the disciples of the Imams, as recorded by Kashshī, it is possible to trace a shift from political to theological and legal topics. Whereas the disciples of the Imams preceding the period of al-Bāqir and al-Sādiq raised questions regarding the rightful Imamate, the disciples of al-Bāqir and al-Sādiq were interested in questions of *fiqh* in its early signification, which included questions of theological as well as legal concern. Evidently, the question of the Imamate always remained the focal point of Shīʿī inquiry and, in the absence of political discretion of the Imamite Imams, it increasingly gave way to theologization of the Imamate. In this latter intellectual process the Imamate became interwoven with the necessity of divine guidance for the attainment of the purpose for which God created human beings. Accordingly, it became part of the discussion regarding the necessity of prophecy (*al-nubuwwa*) and the more general question of divine justice, which necessitated that God create capacity (*istitā'a*) in humankind through the creation of natural reason or sending prophets for humankind to attain the end for which it had been created.[12]

It was also lack of political discretion of the Imams, subsequent to the Imamate of 'Alī, that left the theoretical discussions about the Imamate without any concern for the real contingencies through which the Sunnī conception of caliphate had to go, making necessary adjustments dictated by historical circumstances. In Sunnīsm, in the absence of a "rightly guided" caliphate, Islamic revelation was raised to the status of *imām* ("example" or "noble paradigm" to be followed) and the ideal of justice in the Muslim community to which members aspired in their endeavors to remain faithful to the Islamic promise of an ethically just society. In Shī'īsm the Imam became the living embodiment and a paradigmatic precedent of this Islamic ideal, whom believers were required to acknowledge for their religious well-being. Such an acknowledgment was regarded as an essential element for the validity of any religious activity, because only the precedent set by the Imam could be admitted as justification for its inclusion in Islamic law. As a result, the acknowledgment of the Imamate, which became part of the Shī'ī principles of faith (*usūl al-dīn*), became the testing ground for determining the source of normative precedents, as also the loyalty of associates who introduced them as evidence in the religious life of the Shī'a on the authority of the Imams.

No deviation in this regard was tolerated, with the consequence that no matter how learned an authority might be, without the acceptance of the Imamate he could not qualify to be included among the *rijāl* (plural of *rajul*, lit. "a man"; tech. "a prominent personage"), the learned class, who formed the chain of transmission that could be traced back to the Imams. In order to become the trusted associate of the Imams, partially inheriting the knowledge of the Imam through instruction, one had to acknowledge the right of the Imam to assume the position of the Just Ruler of the community even when in reality it was impossible for him to undertake such responsibility. That the number of such associates was small is indicated by al-Bāqir's statement that even among the followers of 'Alī b. Abī Tālib, who had assumed the manifest leadership by accepting the caliphate, there were no more than fifty persons who acknowledged him (*ya'rifūnahu*) the way he should have been acknowledged—through his manifest Imamate.[13]

In Imāmī Shī'īsm, becoming an *'ālim*, who would protect Islam from corruptive innovations, required rigorous training and objective testing—namely, acknowledgment of the Imamate of the Imams from *ahl al-bayt*. According to Imam al-Sādiq, an *'ālim* who accepts the Imamate of an unjust ruler (*sultān jā'ir*, a term reserved for a usurper of the rightful Imam's authority), and who appoints this *'ālim* to condone injustice, is a cursed individual.[14] This statement implies that the *'ālim* who does not reject or confront an unjust authority has failed to perform the central theologico-political obligation of "enjoining the good and forbidding the evil." On the other hand, an *'ālim* who confronted an unjust authority and declared him *zalama* (tyrannical) was regarded as the most learned among the people (*a'lam al-nās*) in preserving the teachings of the Prophetic traditions by his proper comprehension of them.

The implications of these utterances attributed to the Imams were crucial in the development of the political attitudes of the Imamite scholarly elite during the period of occultation. The "most learned" *'ālim* could not remain indifferent to the unjust behavior of the de facto governments. This reflected moral concerns expressed in the traditions—namely, "condoning injustice." There was both revelational and rational justification for any responsible Muslim to fulfil the moral duty of confronting injustice in society. Accordingly, for the ʿālim who became the functional imam of the Shīʿa, it was important not only to involve himself intellectually in the resolution of the political problems faced by the Shīʿa, but also to lead the Shīʿa in action by setting a precedent, if he judged the circumstances opportune.

ʿAlī b. al-Husayn, accordng to Kashshī, regarded Saʿīd b. al-Musayyib (d. 94/ 709) as such a leader. Saʿīd is recounted among those early disciples of the fourth Imam ʿAlī b. al-Husyan who had accepted his Imamate. According to Ibn Saʿd, Saʿīd was among the most renowned jurists of Medina and was considered the "learned of the learned," and was also highly esteemed by the ʿAlid Imams for his knowledge of *fiqh*. During the political turmoil of the Umayyad era, he was among those associates of the Imam who, even when not actively engaged in resisting the caliphal authority, was opposed to the *khulafāʾ al-jawr* or *al-zalama* (the tyrannical or unjust caliphs) and regarded their obedience as disobedience to God. On one occasion a person sought Saʿīd's advice as to whether he should hand over a drunkard to the ruling authority for punishment in accordance with the laws of Islam. Saʿīd replied that if possible he should shield such an offender, implying that the ruling authority, before inflicting punishment on the offender, should himself be punished for his violations of the law.[15]

This precedent in the classical period continued to have an influence on similar situations faced by the Shīʿa vis-à-vis "unjust" governments in subsequent eras. In juridical works, as we shall see throughout this study, besides the utterances and the actions of the Imam, whether factual or contrived, solutions offered by their close associates were also admitted as authoritative precedents for the derivation of judicial decisions. In fact, in Shīʿite discussions on the sources of Islamic law such precedents were regarded as "silently approved" (*taqrīr*) by the Imams, and as such, they form part of the *sunna,* in addition to the utterances (*aqwāl*) and the actions (*afʿāl*) of the Imams. It is plausible that in the judicial decisions affecting the political life of the Shīʿa, besides revelation, which included scripture and the Prophetic paradigm, whether related on the authority of the Prophet or the Imams, there was another important and pragmatic source of information that provided the necessary flexibility for dealing with immediate concrete situations—namely, the precedents set by the loyal associates of the Imams, like Saʿīd b. al-Musayyib. These decisions, which are part of the political jurisprudence of the Imamites, were based on traditions that dealt with the political experience of close associates of the Imam who were able to get the Imam's approval, whether factual or contrived, to render licit their involvement in political life under de facto governments, because an associate's action was not deemed a compromise of his belief or his moral obligation toward his brethren in the faith. This manner of documentation in a legal decision corroborates my observation that the close associates of the Imams formed a link in the "apostolic succession" by setting precedents for the derivation of a *sharʿī* law.

From what has been said above, it is apparent that the Prophetic tradition cited by Kashshī regarding the future leadership by learned authority among the Muslim community when the religion of God would need to be defended required that there should always be a class of *rijāl* to undertake this critical responsibility. These *rijāl*, in the Imamite worldview, took the place of the Imam in occultation. They were the prominent followers of the Imams whose status, as Kashshī informs us, was to be reckoned in accordance with the number of communications related by them. Al-Sādiq is reported to have said: "Acknowledge the status of *rijāl* among us in accordance with the number of *riwāyāt* (communications) reported on our authority."[16] Moreover, their influence and prestige among the followers of the Imams was to be measured in proportion to their ability to transmit the Imams' opinions on various legal as well as theological problems with great clarity. As a consequence, these early *rijāl* were also *fuqahāʾ* (theologian-jurists). According to the requirements set by al-Sādiq, a jurist had to first be a *muhaddath*—that is, a veracious person with true opinion, who reports a thing as he was told, with sagacity.[17] In other words, the *rijāl* were the expert jurists who were trained by the Imams to represent them in matters pertaining to the faith, and as such they were the indirect deputies of the Imams in that sphere of their functioning.

In Shīʿite Islam, matters pertaining to the faith evolved gradually. As discerned from Kashshī's chronological arrangement of his biographical items and as discussed above, matters of faith included anything from support of the ʿAlid cause to the question about God's knowledge and practical issues covering canonical obligations imposed by Islamic revelation. As already noted, it is significant that, because the political aspect of the Imamate ended with ʿAlī, in the subsequent period more attention was given to the religious leadership of the Imams, whereas their temporal role was confined to the theory of the Imamate. It is for this reason that in Shīʿism it is almost impossible to comprehend the theory of political leadership without study of the intellectual infrastructure that sustains and organizes the concept of religious leadership. It was the subsequent version of the Imamate that was passed down to the Imams. Although early disciples of the Imams did not conceive the Shīʿī Imamate in two spheres, temporal and religious, with the former being postponed for the future, this division gradually became obvious to them during the Imamate of the fifth and the sixth Imams, al-Bāqir and al-Sādiq. It was probably during this period that the Shīʿites were made aware of the futility of insisting that their Imams assume their full responsibililty as the Just Ruler (*al-sultān al-ʿādil*) of the community. The Shīʿī expectation of their Imam was summed up in their aspiration to see the establishment of just rule on earth by the descendant of the Prophet. This was their belief in the Mahdīism of their Imams and their being the "redresser of wrong," the Qāʾim.

Although the Sunnī Muslims shared this Shīʿī soteriological vision about the future caliph of God, al-Mahdī, the tendency among their scholars was to locate that vision in the divinely ordained community under the rule of divine law. This was indeed the fundamental difference between the Shīʿī and the Sunnī understanding of salvation history. In the Shīʿī view, priority was repeatedly and consistently given directly to the person of the Imam as the successor of the Prophet and indirectly to his paradigmatic precedent as preserved in the juridical literature. In the Sunnī understanding of that history, priority was given to the Prophetic precedents in the Sunna in such a way that it overshadowed the person of the Prophet.[18] However, with the reorientation of the Shīʿites toward a more politically quietistic posture, by postponing the establishment of true Islamic government to the future, the Imamate became more or less a spiritual office, sustaining the Shīʿī aspiration for the creation of the ideal public order, with the potential of assuming temporal authority when the time came for it.

Such a reorientation of the Shī'ites—from a demand to overthrow the unrighteous rule of the caliphs to a dissimulation-oriented life of hope in a future rule of justice—could not have come about without resistance from among the more politically ambitious disciples of the Imams. These disciples are sometimes characterized as *ghulāt*, extremists, in sources dealing with Shī'ite subdivisions. Here again *'ilm* was a crucial factor in giving credence to the claims made by the *ghulāt* associates of the Imam. Their *ghuluww*, extremism, lay in their claim to the *'ilm al-balāya wa al-manāya*, esoteric knowledge about future events, which was believed to have been taught by the Imams to their special associates.[19] That this knowledge had something to do with the establishment of just government by overthrowing the wicked rule of the caliphs is confirmed by the fact that most of the Shī'a who claimed such knowledge were killed mercilessly by government forces.[20] Kashshī reports two such instances in the early period when Maytham al-Tammār and Rushayd al-Hujrī, among the associates of 'Alī, were put to death. In the case of Rushayd, according to his daughter Qanwa', who also was put to death, by the Umayyad governor 'Ubayd Allāh b. Ziyād, Rushayd was known as Rushayd al-Balāya because he had received the knowledge of *al-balāya wa al-manāya* from 'Alī.[21]

Furthermore, such claims by disciples implied the attribution of divine qualities to the Imams. They were declared the actual source of the *'ilm al-balāya wa al-manāya*. The Imams clearly sensed the danger that such attributions posed for them: disruption of the peaceful relationship between the Imams and government authorities who suspected them of instigating political opposition. This danger was felt early on during the Imamate of al-Bāqir and al-Sādiq in Kūfa, which was the nucleus of Shī'a activity in the eighth century. The Imams did not hesitate to disown such *ghulāt* disciples publicly and to state that these disciples had adulterated their communications by adding things that they themselves had never uttered.[22]

There was another crucial reason to denounce the *ghulāt:* these disciples threatened the survival of the Shī'a community within the larger Muslim community. By ascribing to the Imams reports that exempted the Shī'a from performing religiously prescribed duties, such as *salāt*, in return for their acknowledgment of the Imam of the Age, the *ghulāt* endangered the acceptance of the Shī'a as part of the tolerated deviation of the *umma*.

Kashshī provides a striking example of such attribution in his note on al-Mufaddal b. 'Umar, who was among the *ghulāt* followers of Imam al-Sādiq and who later joined the extremist Shī'ite faction of the Khattābiyya. According to Kashshī, a man by the name of Yaḥyā b. 'Abd al-Ḥamīd al-Hammānī recounted the following incident in his book entitled *Fī ithbāt imāmat amīr al-mu'minīn* (The proof of the Imamate of the *Amīr* of the believers ['Ali b. Abī Tālib]):

> I told Sharīk [b. al-A'war al-Hārithī] that a group of persons maintained that Ja'far b. Muhammad [al-Sādiq] was weak in *hadīth*. He said: "Let me explain to you the true state of affairs. Ja'far b. Muhammad was a pious Muslim, and fearful of God. A group of ignorant persons surrounded him, visiting him frequently and saying that "Ja'far b. Muhammad said such and such." They related traditions, all of which were objectionable, fabricated, and falsely attributed to Ja'far. They did this in order to make money out of it from the people. . . . The masses heard all these traditions. Some followed them and destroyed themselves; others refused to accept such [traditions]. Among the former were al-Mufaddal b. 'Umar, Bayān, 'Amr al-Nabtī, and others. They were the ones who

reported that Jaʿfar [al-Sādiq] had told them that acknowledgment (*maʿrifa*) of the Imams suffices and exempts a person from fasting and praying. They also reported on the authority of his father [i.e., al-Bāqir] and his grandfather [i.e., ʿAlī b. al-Husayn], and that the latter had informed them concerning the events preceding the Day of Judgment. . . . Indeed, Jaʿfar [al-Sādiq] never said anything of the sort and he was surely above such utterances.[23]

This citation evidently proves the point that in the process of disciplining the *ghulāt* element, as in the development of the Shīʿa school of thought in general, the *rijāl* were destined to play a major role. The Imams, from the early days, had seen the possibility of continuing their teaching by training these associates and had taken steps to realize this end. With the growing political turmoil and the atrocities committed against the family of the Prophet, the Imams were unable to maintain normal contacts with their followers. Hence, the close associates were deputized to represent them in relations with their followers. Thus was created the institution of the deputyship (*niyāba*) of the Imam in the early days of Shīʿa history.

The Deputyship of the *rijāl* during the Imamate of al-Bāqir and al-Sādiq

The Imam, according to Shīʿī belief, is the leader of the Muslim community and, as such, the only legitimate authority who could and would establish the Islamic rule of justice on earth. But, in view of the inability of the Imams to assume political leadership, the entire question of their establishing the ideal Islamic rule was postponed. Nevertheless, this postponement applied only to the Imams' temporal authority, because they continued to exercise their religious authority as the descendants and true heirs of the Prophet. They had from time to time delegated their religious-juridical authority to their close, trusted associates, who acted as their deputies or agents in places like Kūfa, Rayy, and Qumm. The deputyship of the Imams was a necessary and gradual process intended to provide guidance for the growing Shīʿa community in far-flung areas. The institution of the deputyship became a prominent feature of Shīʿism partly due to the inability of the Imams to receive their followers or to communicate with them in person at all times. However, there are accounts that show that, during the annual pilgrimage to Mecca, the Shīʿa, in large numbers, met with their Imams. But such meetings in the holy city, which had remained a hotbed of uprisings in the past, were looked upon with suspicion by the caliphal authorities. Some ʿAlid leaders were considered a potential challenge to the establishment and had actually emerged as alternate candidates for the caliphate.

Imam al-Sādiq's cousin, Muhammad al-Nafs al-Zakiyya (the Pure Soul), was expected by pious Muslims from Hijāz and other regions of the empire to rise up as the promised Mahdi of the community and establish an adequately just order in the last decade of the Umayyad rule in the eighth century. According to Abū al-Faraj al-Isfahānī, one such group, which included some early Muʿtazilite teachers such as ʿAmr b. ʿUbayd and Wāsil b. ʿAtāʾ, had come to Mecca from Basra. They not only regarded al-Nafs al-Zakiyya to be the most qualified candidate for the caliphate, especially following the political turmoil generated by the deposition and assassination of Walīd II, the Umayyad in 125/743, but had also initially paid allegiance to him. However, when al-Nafs al-Zakiyya's revolt failed to impress them as a serious attempt to launch an alternate public order, they withdrew their support and approached al-Sādiq, to have him consider assuming the leadership.[24] These contacts between prominent ʿAlids and followers who urged them to assume the caliphate were well known to the founders of the ʿAbbāsid movement. They, too, are listed among those who supported al-Nafs al-Zakiyya against the Umayyads before contending for the caliphate themselves. Imam al-Sādiq did not live long after the establishment of the ʿAbbāsid caliphate, although even he was not spared by al-Mansūr, the second ʿAbbāsid caliph, from harassment and forced appearance in the ʿAbbāsid court at Kūfa for questioning. However, beginning with al-Sādiq's successor, Mūsa al-Kāzim (d. 183/800), all Imamite Imams were either imprisoned as politically suspect or kept under surveillance, whereas other ʿAlid leaders who openly rebelled against ʿAbbāsid authority were dealt with severely. It was under these circumstances that the Imams, from the time of al-Sādiq's grandfather, ʿAlī b. al-Husayn (d. 94/713), who had witnessed the tragedy of Karbalāʾ in 61/680, had realized the futility of any attempt to establish an ʿAlid caliphate. Accordingly, they maintained the idea of the postponement and future establishment of the just social order, after the return of the Mahdi, the promised deliverer of the Muslim community.

The idea of the Imam's delegating his religious authority and functioning through his personal representatives (*wukalāʾ*) among the *rijāl* was, indirectly, a consequence of this belief. The delegation of the Imam's partial authority to perpetuate the religious structure of the Imamite community was indispensable because belief in the Imamate, as it came to be part of the fundamentals of religion in Shīʿism, required the Imam to provide the authentic perspective of Islam even when the Imam could not exercise his political discretion or was to remain in occultation until God commanded him to appear. In this sense, the deputyship (*al-niyāba* or *al-wikāla*) of the Imam was theologically justified. It was, as we have seen above, regarded as part of divine providence—to continue God's guidance of the Muslim community, even under the most adverse conditions of occultation of the rightly guided Imam. In addition, the deputyship of the *rijāl* among their close associates was a solution carefully worked out by Imams who were faced with an unfavorable political atmosphere in the eighth and ninth centuries. For the greater part of their lives, they were kept under surveillance or imprisoned as politically suspect by the caliphal authorities.

Appointment of the personal representatives of the Imam in the early days was done directly, following a request from the Shīʿa in distant lands for a religious guide who could teach the Imam's opinions authentically. This may be seen as the beginning of the institution of deputyship (*niyāba*), when the disciples of the Imam who had heard traditions from them were held in high esteem and were followed by the Shīʿa in areas far from Medina. In his long note on Muhammad b. Muslim al-Thaqafī, a prominent disciple of the Imams al-Bāqir and al-Sādiq, Kashshī relates that ʿAbd Allāh b. Abī Yaʿfūr, another disciple of the two Imams from Kūfa, once came to visit Imam al-Sādiq in Medina. During the visit ʿAbd Allāh told the Imam that it was not possible at all times for him to undertake a journey to meet him. Often a follower of the Imam in Kūfa would come and ask him about a matter and he would have no answer. The Imam said: "What stops you from asking Muhammad b. Muslim al-Thaqafī, who heard [*hadīth*] from my father and was highly esteemed by him?"[25] It was in this way that Thaqafī was indirectly appointed by the Imam to represent the Imamite views and spread the teaching of his school among the Shīʿa in Kufa.

From the early days of the caliphate, Kūfa was an important center of Shīʿī tradition. As such Kūfa was always in touch with Medina, the authoritative source of Imamite teaching, whereas Qumm, another center of Shīʿī tradition, remained in close contact with Kūfa. Thus, among the associates of al-Sādiq one finds several transmitters of tradition who used to travel from Kūfa and also from Qumm to "hear" the traditions in Medina.

From several early biographical notes on these prominent associates of the Imams al-Bāqir and al-Sādiq it is evident that the main qualification of these indirectly appointed deputies of the Imams—and there were quite a few of them in various parts of the empire—was their comprehension of the teachings of the Imams. The extent of their comprehension was determined by the fact of their close association with the Imam. This fact was also decisive in their being held in high esteem by the Imams. That these prominent associates were carefully selected on the basis of their knowledge ("sound knowledge") and loyalty ("sound belief") to the Imams is attested to by the oft quoted tradition of al-Sadiq that among his numerous disciples four were the trusted ones of his father and himself—namely, Burayd b. Muʿāwiya al-ʿIjlī, Abū Basīr Layth b. al-Bakhtarī al-Murādī, Muhummad b. Muslilm al-Thaqafī, and Zurāra b. Aʿyan. They were, according to al-Sādiq, the noble and trusted ones of God on matters of divine injunctions concerning what was lawful and unlawful. "Had it not been for them, the Prophetic traditions would have been disrupted and obliterated."[26] It is obvious that the continuation of these disciples was significant in spreading the Shīʿī ideology as taught by the Imams. This fact was recognized repeatedly by the Imams who depended on these *rijāl* to be trained as *fuqahā* (jurists), to carry on the function of guiding the religious life of the Shīʿa.

The matter of trust and thorough reliability in a disciple was of prime importance for Imams when they were faced with the problem of the overzealous among their disciples, who did not hesitate to interpolate the teachings of an Imam and announce their agency and deputyship in esoteric knowledge, which in theory was confined to the Prophet and the Imams.[27] Such a claim was important for disciples with political ambition because, as pointed out above in regard to *ghuluww* in the matter of esoteric knowledge about future events (*ʿilm al-balāya wa al-manāya*), it was the claim to have received this type of knowledge from the Imam that helped them organize a social upheaval under their own leadership.

Abū al-Khattāb Muhammad b. Abī Zaynab Miqlās al-Asadī, the founder of the extremist Shīʿite sect of the Khattābiya, was a close associate of al-Sadiq. Abū al-Khattāb presumably held a significant position as a *dāʿī* (missionary) of al-Sādiq in Kūfa. Abū al-Khattāb was accused by Imam al-Sādiq of divergences from the ritual law, which points to the fact that questions concerning law were referred to him in his capacity as one of the chief agents of al-Sādiq in Kūfa.[28] Shīʿite sources continue to refer to Abū al-Khattāb long after the conflict between him and al-Sādiq, because of his having interpolated the latter's teachings. There is little doubt that in his early career as the *dāʿī* of the Imam he was held in high esteem because of his profound devotion to him. Thus, when later Imam al-Sādiq repudiated him as going too far, the repudiation greatly disturbed some of al-Sādiq's associates. After al-Sādiq's death, Mūsā al-Kāzim, his successor in the Imamate, was asked by one of his associates the reason for this renunciation in spite of the fact that the Imam had earlier ordered them to be friendly to Abū al-Khattāb. Yūnus b. 'Abd al-Rahmān, another prominent Imamite traditionist, was in Iraq, collecting *hadīth* reports from the disciples of Imam al-Sādiq. He was told how the Imam 'Alī al-Ridā had warned against traditions communicated by persons like Abū al-Khattāb and his followers, who had attributed falsehood to al-Sādiq and had interpolated his teachings in their *hadīth*.[29]

In order to discern the cause for Abū al-Khattāb's excommunication by the Imam and the later development of political acquiescence among Imamite Shī'ites, one must take into account the historical background of these events. On the one hand, the period of al-Sādiq coincided with the decline of the Umayyad rule and the approach of the 'Abbāsid revolution masterminded by the extremist Shī'ite faction of Hāshimiyya. On the other hand, al-Sādiq, in his prominent position as the head of 'Alids, had to exercise great restraint and prudence, and reject any extravagant claims made for him by overzealous disciples like Abū al-Khattāb. Without such a consideration it is difficult to understand why the Imam al-Sādiq considered Abū al-Khattāb and his like a continuous source of threat to his mission and his security, and why later Imams continued to repudiate Abū al-Khattāb and his teachings.[30]

Abū al-Khattāb had asserted that al-Sādiq had transferred his authority to him by designating him as his *wasī* and *qayyim* (deputy or executor of his will) and entrusting him with the *ism al-a'zam* (the greatest name), which was supposed to empower its possessor with extraordinary power in conceiving hidden matters.[31] This assertion, which implied sharing the Imam's "apostolic authority," was decisive in giving him special recognition in Kūfa. It is significant that such assertions were made in the Kūfan environment, away from Medina where the Imams lived. Abū al-Khattāb's popularity in Kūfa is clear from the following he had obtained there. Around the year 138/755–76, he and seventy of his followers, who had assembled in the mosque of Kūfa, were attacked by the order of 'Isā b. Mūsā, who was the 'Abbāsid governor for the period ending in 147/764–65. Abū al-Khattāb armed his followers with stones, reeds, and knives, assuring them that these would prevail against the enemies' swords and lances. The assurance proved to be wrong, and after a bitter struggle some thirty of his followers were killed, and he himself was captured and brought before the governor, who had him crucified together with a number of his followers.[32]

The whole episode of the execution of Abū al-Khattāb in Kūfa in 138/756 shows the extent to which any shrewd and ambitious disciple of an Imam could manipulate genuine religious devotions of the Shī'a in the name of the Imam, no matter how emphatically the latter might deny any designation of the disciple as his *wasī* or his possession of hidden knowledge.[33] The position of the Imam's deputyship, however indirect and limited, inherently possessed the potential of gaining influence in the Imamite community, even more so when the Imams, because of difficult political circumstances, could not maintain regular contact with their Shī'a in distant lands. It was during the Imamate of al-Bāqir and al-Sādiq, more particularly the latter, that the leadership and influence of some of these trusted transmitters of the Imam's teachings, the *ruwāt,* contributed to the tradition in which al-Sādiq declared: "Those who become leaders in their religion are destroyed."[34] As discussed above, there seems to be little doubt that attempts were made by some disciples of an Imam, like Abū al-Khattāb, to appropriate to themselves authority in far away places like Kūfa by claiming esoteric knowledge. In elucidating the qualities of a Muslim the Imam al-Sādiq makes reference to this encroachment upon Imamite authority by Abū al-Khattāb and his like. He says: "A Muslim is the one who endeavors to seek the pleasure of God and to serve Him. . . . What does he have to do with leadership? In fact, all Muslims are one solid leader."[35]

Although it is accurate to classify Abū al-Khattāb as an extremist (*ghālī*) among the associates of the Imam al-Sādiq, there were others who did not make any such claim to esoteric knowledge but were certainly well enough grounded in Imamite *hadīth* and *fiqh* to attain a position of leadership among Imamites. Among this latter group of *rijāl* the most prominent and widely recognized among the Shī'a were Muhammad b. Muslim al-Thaqafī and Zurāra b. A'yan. They were recognized as the agents of Imams, also their reliable representatives in matters pertaining to the law. The growing authority of the transmitters of the Imams' teachings, the *ruwāt,* is symbolized by both Zurāra and Muhammad b. Muslim, whose careers as close associates of Imams are important for understanding the position of functional imams among the Imamites.

Zurāra b. A'yan: A Symbol of the Power of the Early Disciples

His full name was 'Abd Rabbih Zurāra b. A'yan (d. 150/767). Zurāra was his title. A'yan, his father, was a slave from Byzantium and belonged to a person from the Banū Shaybān. His master taught him the Qur'ān and then freed him and offered to accept him as one of his family. But A'yan preferred to remain his friend. Zurāra had many sons and brothers, who formed the large and well-known family of the Banū Zurāra. They were all reporters of several traditions and the authors of various original works and compilations. Zurāra reported from the fourth, fifth, and sixth Imams. Najāshī, another Imamite compiler of a biographical dictionary, sums up his note on Zurāra: "he was a reciter [of the Qur'ān], a jurist, a theologian, a poet, and a man of letters. In him were combined excellence of character, piety, and truthfulness in what he reported."[36] Kashshī, another biographer, mentions a tradition reported on the authority of the sixth Imam, who said: "Had it not been for Zurāra, I think the traditions of my father would have disappeared."[37] With all these good reports about him, there are accounts showing that Zurāra, on some important issues, had contradicted al-Sādiq, having reported something different on the authority of the fifth Imam, al-Bāqir.

Before I embark on a discussion about the relationship between Zurara and Imam al-Sādiq, it is important to treat the subject of authentication (*tawthīq*) of transmitted sciences (*manqūlat*) in the Imamite legal system, which depended on the authentication of the *rijāl,* the transmitters of these sciences. This subject has a bearing on our understanding of the relationship between the Imam and his close associates, and between the "ancient" and "modern" *rijāl* of the Imāmiyya.

Rijāl like Zurāra and Muhammad b. Muslim have been regarded as the pillars of the Imamite religious system. Accordingly, any report that might tarnish their reputation as thoroughly reliable transmitters of Imamite teachings has been regarded either "unreliable" or "weak." In this authentication of the prominent *rijāl* can be discerned the Imamite belief regarding the religious guidance coming from religious authorities who cannot commit any error or act of inadvertance that would affect the authoritativeness of what they transmit. Although such religious authorities, who formed legitimate members of the "apostolic line of succession," were the Prophet and the Imams only, at times their qualifications were extended to the *rijāl*, for them to become functional imams and, hence, the paradigmatic precedent for religious life during the occultation. As we shall see in chapter 4, this is most evident in discussion about the qualifications of a jurist who can assume the *wilāyat al-hukm* (the authority to administer justice). But let us turn our attention to the authentication of the *rijāl* and the problems some reports in the *rijāl* books raise about persons like Zurāra.

In Imamite Shī'ism several methods of authentication of transmitted sciences have been recognized from the early days. One of the most basic methods has been to investigate whether one of the *ma'sūmīn* (the infallible personages—i.e., the Prophet or the Imams) has left any objective documentation in the form of a clear text (*nass*) to the effect that a certain transmitted report is related on their authority. This was, by far, the most difficult process of authentication because of the fact of attribution of falsehood to the Imams and interpolation in their communication, as discussed above in relation to *ghuluww* (extremism) in Shī'ism. In an important directive given to Hishām b. al-Hakam, al-Sādiq is reported to have referred to this latter problem. He instructed Hishām, saying:

> "Do not accept any communication as being from us unless it conforms with the Qur'ān and the Sunna [of the Prophet] and with what has been ascertained as being our utterances. . . . Do not accept something as being from us which is against our Lord and the Sunna of our Prophet. Assuredly, when we relate anything we either say: "God said thus . . . " or "the Prophet said thus. . . ."[38]

Consequently, an investigation would take a researcher through critical examination of the text of the communication, so as to verify its authenticity in the light of historical and exegetical literature. At this point, biographical dictionaries provide material that is valuable for both historical as well as exegetical analysis. Study of these dictionaries has given rise to the religious science known as '*ilm al-rijāl*. '*Ilm al-rijāl* deals with the study of transmitters (*al-ruwāt*) who figure in the chains of transmission (*isnād*) appended to *hadīth* reports in order to ascertain their chronology and reliability in what they transmit. This study has received much attention in Imāmī Shī'ism because of its preoccupation with the question of leadership. In normative jurisprudence only a precedent that can be traced to its revelational source—that is, the Prophet and the Imams—has been regarded as a valid basis for a judicial decision. In other words, the authority of a precedent is dependant upon its being ascertained as part of the "apostolic succession" without which it cannot be used as evidence to deductively infer a ruling. Consequently, '*ilm al-rijāl* can be regarded as the branch of juridical sciences that scrutinizes the transmitter's status in "apostolic succession" in order to admit his or her precedent as authoritative. (Women were also included in the *rijāl*, because of the precedent set by the inclusion of Fātima, the daughter of the Prophet, in "apostolic succession" in Shī'ism.) This precedent is in the form of a tradition narrating the Prophetic paradigm. '*Ilm al-rijāl*, then, critically studies information about prominent transmitters, *rijāl*, in order to determine the level of authenticity of an informant, which directly affected the authenticity of a tradition reported by that person. Biographical information on all informants of a tradition is studied to gain insight into their character and to ascertain their moral probity and religious standing in order for them to be declared *thiqa* (reliable), or *da'if* (weak), or *majhūl* (unknown), or *kadhūb* (liar), and so on. It was on the basis of such examination that a *hadīth* report was classified as *sahīh* (authentic), *muwaththaq* (reliable), *hasan* (fairly reliable), *da'īf* (weak), and so on. Hence, the result of the investigations in *'ilm al-rijāl* was of great consequence for the religious practice of the Muslim community.[39]

To indicate the way in which a transmitter was authenticated, Kashshī in his biographical note on 'Imrān b. 'Abd Allāh al-Qummī mentions two reports in which Imam al-Sādiq praises 'Imrān and prays for him and his descendants.[40] This communication from the infallible Imam establishes the reliability of 'Imrān in regard to what he reports. However, on careful examination of the chain of transmission appended to the two reports, one discovers that the *isnād* is defective because it includes "some [unknown] persons from Kūfa" (a breach in continuity of "apostolic succession") whose reliability it is impossible to establish. As a result, what has been transmitted by 'Imrān b. 'Abd Allāh on the authoriy of al-Sādiq cannot constitute a valid basis for its inclusion as a precedent.

In Imāmī Shī'ism from the time of the "ancient" (*al-mutaqaddimūn*) jurists (fourth/tenth century), the use of transmitted sciences by narrators whose reliability could not be established in the *rijāl* works was problematic. Reliance on such information, when necessary, was regarded as having been derived from "biographical suppositions" (*al-zunūn al-rijāliyya*). No legal judgment could become binding if derived from this kind of supposition. It is usually maintained that in Imamite jurisprudence, *al-'aql* (reasoning) plays a dominant role in the form of *ijtihād* (independent reasoning of a jurist). However, even in *ijtihād*, as we shall see in more detail in chapter 4, its role is limited to the establishing of the intrinsicality (*mulāzama*) of a ruling derived on the authority of revelation (*al-sam'*) to that derived on the authority of reasoning (*al-'aql*). For instance, through *ijtihād*, reasoning should be able to establish the intrinsicality between the prohibition regarding fasting on two particular days of festival (those following the fasting month of Ramadān and during the *hajj*) (already established in revelation) and the reason behind this prohibition. Consequently, in Imamite jurisprudence, in order to derive a ruling in the Sharī'a, in most cases there is no other source than the *hadīth* reports related on the authority of the Prophet and the Imams—the *ma'sūmīn*. These reports, before they can be used as a piece of evidence to support a legal ruling, have to be authenticated by a study of the chains of transmission. It is not without cause that one of the fundamental requirements for Imamite jurists is that they should thoroughly equip themselves with a knowledge of *'ilm al-rijāl*, rather than with *al-mantiq* (logic), as one might expect in the exercise of *ijtihād*.[41]

Sometimes the problem of a transmitter's reliability was resolved by authenticating statements provided by early, eminent Imamite jurists. Their statements about the absolute or relative reliability and trustworthiness of transmitters in what they transmitted and what was recorded in the works of other eminent scholars, such as Ibn Bābūya (d. 381/991), Mufīd (d. 414/1023), Najāshī (d. 450/1059), Tūsī (d. 460/1067), and others of their stature, were regarded as authenticating statements, valid in deriving legal decisions. This inclusion of Imamite jurists in the authentication process was an inevitable consequence of the end of the manifest Imamate, followed by the occultation of the last Imam. They, in fact, became part of the religious authority affording the necessary continuity in the "apostolic succession" that, theoretically, at least, was admissible only in the Imamate.

It is safe to assume that the authentication provided by the "ancient" jurists, on which most of the subsequent research was based, remained the standard on the thorough or relative reliability of the *rijāl* and their transmission. I will return to the *rijāl* works and their function in Imamite religious learning once again in chapter 2. Of these "ancient" jurists, Tūsī appears to be the most frequently cited authority in the *ijāzas* (permissions) that were given to disseminate the transmitted sciences. The fact is that Tūsī's works form a link between the early authors of *usūl* (fundamental) works—which consisted only of those traditions that had been heard directly from the Imams and had become sources for the four major compilations of the Imamite traditions[42]—and the later transmitters. Hence, for the latter group, in most cases, there was no other way of authenticating transmitters for their transmitted information except through an investigation of Tūsī's works and the application of personal, independent judgment (*ijtihād*) and insight. This observation is corroborated by the fact that in the case of Kashshī's findings—the earliest and principal work in this discipline and the one that contains the controversial material on eminent associates of the Imams like Zurāra and Muhammad b. Muslim—the original work has not been preserved except in the abridged version by Tūsī. Consequently, any reference to Kashshī in the writings of post-Tūsī Imamites was based on this abridgement.

Kashshī's original work was entitled *Kitāb ma'rifat al-nāqilīn 'an al-a'immat al sādiqīn* (Information on those who reported on the authority of the truthful Imams). The abridged version of this work, entitled *Ikhtiyār ma'rifat al-rijāl* (Select information on the *rijāl*), includes early contradictory narratives about the close associates of the Imams during the formative period of Shī'a thought (second/seventh–third/eight centuries). Kashshī's inclusion of this material and Tūsī's allowing it to be made part of the abridgement indicate that the appraisal of the associates of the Imams was not constrained by the usual concern among religious minorities to provide ideologically acceptable accounts of their early, prominent personages. It was this kind of information that led to the criticism of Kashshī's uncritical methodology by Tūsī's contemporary, Najāshī, who considered Kashshī's *Rijāl* informative but full of errors caused by incorporation of material related by "weak" transmitters (*al-du'afā'*).[43]

This digression was necessary to appreciate Kashshī's *Rijāl* in delineating the relationship between Zurāra and Imam al-Sādiq. As was said earlier, Kashshī recounts traditions showing that, although Zurāra was held in high esteem, he also maintained different views on some important issues and contradicted al-Sādiq by documenting these views on the authority of al-Sādiq's father, the fifth Imam, al-Bāqir. Zurāra had an eminent position as a teacher among the disciples of al-Bāqir and al-Sādiq. This is acknowledged by a close associate of al-Sādiq by the name of Jumayl al-Darrāj, who was also a prominent transmitter of *hadīth* and a teacher. He said, "In the presence of Zurāra we were like children around a teacher in the *kuttāb* (elementary school for the study of the Qur'ān)."[44]

These disparaging traditions in Kashshī about Zurāra, however, did not prevent him from being authenticated as a thoroughly reliable transmitter of Imamite traditions. As a matter of fact, he is among those few associates who are included in the *ahl al-ijmāʿ*—that is, those who had participated with the infallible Imam in reaching a legally authoritative consensus.[45] In later days, as we shall see in chapter 2, the process of authentication of *rijāl* linked contemporaneous Imamite leaders— for example, the Āyat Allāh—to the *ahl al-ijmāʿ*. In a sense, then, the well-authenticated jurist was also an *imām,* recognized by the will of the Imamite community. Zurāra was also acknowledged as the most renowned *faqīh* (jurist) among the close associates of the Imams. His legal opinions, cited on the authority of al-Bāqir, were regarded as authentic and even binding. Both Zurāra and al-Sādiq had remained al-Bāqir's disciples and as such Zurāra's transmission of al-Bāqir's opinions was considered to be as authoritative as al-Sādiq's relating of the same. However, for al-Sādiq, as Kashshī reports, on some occasion it was necessary to remind Zurāra of the difference between his own authority as the Imam of the Shīʿa and Zurāra's position, no matter how eminent, as the close associate of the Imam. This point is explicitly made in a long tradition reported by Kashshī in which al-Sādiq sent a message to Zurāra through the latter's son, ʿAbd Allāh. Al-Sādiq said:

Give your father [Zurāra] my regards and tell him [this]: "I find faults in you [O Zurāra] so as to defend you from my direction [because my finding faults in you will endear you to your enemies]. Assuredly, the people [in general] and the enemies [in particular] hasten [to seek out] anyone whom we have taken as an associate and whose position we have praised. [They do this] so as to harm anyone whom we love and we have taken as an associate. [Moreover] people reproach such a person because of our love for him, and see it appropriate to harm him and even kill him. [On the other hand] they praise all those with whom we find fault. Thus, we criticize you because you are a person who has attained fame because of [your relationship with] us and your devotion to us. In the eyes of the people you are for this reason blameworthy, and not deserving praise, because of your love and devotion for us. Hence, I desired that I should find fault with you in order for them to praise your authority in religion, in spite of [my criticism of] your shortcoming and imperfection. We are in this way repelling their evil from you. [What we are doing in your case resembles that which is revealed in the Qurʾān in the story of Moses and Ezra.] It is as God has said in the Qurʾān: "As for the ship, it belonged to poor people working on the sea, and I wished to mar it, for there was a king behind them who was taking every (good) ship by force" (18:79). This is the revelation from God. I solemnly declare that the ship was marred in order for it to be saved from the king. This was in no way to destroy it with His aid. Indeed, the ship was perfect and it was not easy to swallow the fault in it. Praise be to God. Do understand this allegory [as I have explained it to you], and may God have mercy on you, for you are to me, by God, the most beloved of the people. And I love my father's [al-Bāqir's] associates, living or dead. Surely, you are the most excellent of the ships in this deep, exuberant ocean. But behind you is an oppressive, usurping ruler watching the passing of every solid ship which passes through the ocean of guidance [i.e., the Imam] in order to seize it by force and then take illegal possession of it and its occupants. God's mercy be on you while you live and His mercy and pleasure be on you when you die. Your sons, al-Hasan and al-Husayn, have brought your letter to me. And may God protect them both. Do not let things that my father or I have ordered [you to do] annoy you. And Abū Basīr [al-Murādi] came to you [and reported to you] against what we had instructed him to do. By God, we have not ordered you or him to do anything that was not within our and your ability to shoulder. For all that [we have asked you to carry out] we possess ordinances (*tasārīf*) and explanations, in accordance with the truth. If we are allowed, we will indeed teach you that the truth is in what we have commanded you to do. So return the authority to us and hand it over to us intact; and then wait for our injunctions and be satisfied with them. The one who scatters you [like a shepherd does in order to save his flock] is your protector [i.e., the Imam], toward whom God attracts His creatures. He is more knowledgeable about the welfare of his flock in that which is corrupt in their affair. If he wished, he could differentiate between them so that you are forsaken, then bring them together so that you are protected from their wickedness and the fear of their enmity in [the performance of] actions permitted by God. From such a person will peace and deliverance come to you. It is your obligation to submit and refer to us and await [the realization of] our and your matter and our and your deliverance. Even when our Qā'im rises and our spokesman speaks and commences the teaching of the Qur'ān and the ordinances of religion, the injunctions and the obligations as God revealed them upon Muhammad (peace be upon him and his progeny), that day those who are with deep insights (*ahl al-basā'ir*) among you will begin to object [to his teachings]. This [objection] will be an outright disavowal.[46]

The report indicates that al-Sādiq was faced with a grave problem of limiting the authority of the *ruwāt* in the elucidation of religious precepts. Further evidence of this situation is provided in al-Sādiq's disavowal of traditions reported on the authority of his father by Zurāra, who was probably challenging the Imam's authority in this matter. When al-Sādiq heard something communicated on his father's authority by Zurāra, something in contradiction with what he had related, al-Sādiq would frequently say that what had been communicated was "neither my father's religion nor mine."[47]

In some instances, as reported by Kashshī, Zurāra was in direct opposition to the views expressed by al-Sādiq, as related in the following incident. On one occasion Zurāra came to al-Sādiq and related to him what al-Hakam b. 'Utayba (d. 113 or 114/731 or 732), had reported on the authority of al-Bāqir. Al-Hakam, according to Kashshī, was a Murji'ite jurist who used to relate traditions on the authority of al-Bāqir. However, his reporting was not accurate and, adds Kashshī, al-Hakam used to give opinions on legal matters in contradiction to the opinion of the Imams.

Najāshī recounts an occasion when there arose a difference of opinion among the associates of Imam al-Bāqir, who asked his son to bring out a notebook on *hadīth* written by 'Alī b. Abī Tālib to show it to al-Hakam b. 'Utayba. Furthermore, he had been Zurāra and his brother's teacher before the tradition regarding true knowledge ("true knowledge . . . could be acquired only from the *ahl al-bayt* [the Imams]") was reported to them.[48]

It is in the context of this biographical information on al-Hakam that the following discussion about what al-Hakam had related on the authority of al-Bāqir becomes comprehensible. Al-Hakam, according to Zurāra, had heard Imam al-Bāqir relate the *hadīth* about the evening prayer (*al-maghrib*) which was to be performed *before reaching* al-Muzdalifa, during the annual pilgrimage (*hajj*). This ruling is contrary to the generally held opinion about the evening prayer which has to be performed *in* al-Muzdalifa. Al-Sādiq told Zurāra: "I pondered about this *hadīth*. My father has never said this. Al-Hakam has falsely attributed it to him."

MAGHRIB

Zurāra left the presence of the Imam, saying to himself: "I have not seen al-Hakam attributing falsehood to his father [al-Bāqir]."[49]

On another occasion, Ziyād b. Abī Hilāl, an associate of the Imam from Kūfa, went to al-Sādiq in Medina and recounted what he had heard on the authority of Zurāra regarding the question of *istitāʿa* (human capacity to perform an act), so as to seek confirmation from the Imam as to its being his opinion. The Imam denied the explanation offered by Zurāra as being his own and cursed him for spreading distorted views. Ziyād asked the Imam's permission to inform Zurāra about this, which the Imam gave. On his return to Kūfa, Ziyād went to visit Zurāra and related to him the Imam's opinion on *istitāʿa*, without mentioning the curse. Hearing this, Zurāra said: "Indeed he has presented to me [his definition of] *istitāʿa*, whereas he himself does not know. This master of yours (i.e., the Imam) does not have insight into the speech of the *rijāl* (the eminent transmitters of the teachings of the Imams)."[50]

The implications of these reports in Kashshī are of great consequence. On the one hand, they show that disciples like Zurāra, because of their close association with the Imams and their statements regarding the eminent position of the *rijāl*, were looked upon by the Shīʿa as representatives of the Imams in Medina, who could inform them of the authoritatively binding opinion of the Imam on a given problem, whether legal or theological; on the other hand, Zurāra's comment about the insight of the Imam explicitly shows that the associates, who were in close contact with more than one Imam, found themselves in the sensitive position of having to verify the authenticity of the Imams' utterances. Furthermore, the function of guiding the adherents of the Imams in religious matters, especially in Kūfa, was becoming even more central due to the systematization of the sources of the Law, at somewhat the same time. As a matter of fact, Zurāra and Muhammad b. Muslim al-Thaqafī, according to Kashshī, used to hold discussion on some legal as well as theological issues with Abū Hanīfa, the founder of the (Sunnī) Hanafī school of law.[51]

It was also in such an intellectual environment in Kūfa that Zurāra, probably under the influence of Abū Hanīfa's endorsement of the use of *al-raʾy*, "opinion" (in the particular meaning of "sound, considered opinion," as a source of Islamic Law), began to give precedence to it over *athar* (tradition). At one time Zurāra was asked about a legal matter. He gave his answer, whereupon he was asked if his reply was based on *al-raʾy* or *al-riwāya* (the tradition from the Imam). Zurāra explained how he had derived his decision, and then added: "Is not the one who derives his decision on the basis of *al-raʾy* better than the master of *āthār* (tradition) [who accepts them uncritically]?"[52]

Statements like this have raised questions about the reliability of Kashshī's *Rijāl*, especially among later Imamite jurists, who have regarded such statements as based on "weak" transmission, which should not have been included by Tūsī in the abridged version of the *Rijāl*. The implications of the statement regarding the superiority of the *raʾy* over *āthār* becomes obvious if we bear in mind the Imamite worldview that orients the faithful toward the Imam of the age, who is the sole source of *āthār*, the authoritative religious tradition, not *raʾy*, however sound that "opinion" might be. This reliance on intellectual capability, as we shall see in chapter 4, would give rise to the authority of jurists who would assume the role of functional imams and would exert their independent reasoning (*ijtihād*) in the comprehension of *āthār*.

However, the Imams were still alive and the delegation of their juridical authority to their early associates was, at the most, partial. This is substantiated by occasional refutation of Zurāra's transmission of al-Bāqir's teachings by the Imam al-Sādiq, who was informed about conflicting statements made by Zurāra in Iraq. Zurāra's influence there was immeasurable and al-Sādiq's refutation of him was doubted by some. Zurāra's nephew had once asked the Imam if it was true that he had disavowed Zurāra. The Imam replied that he had not disavowed him; but in order to silence those who visited him and reported on his authority something that was contrary to fact, he would say: "I disclaim to God the one [not necessarily Zurāra, who might be innocent of what they ascribed to him] who says thus."[53]

It is plausible that persons like Zurāra symbolized the growing authority of the transmitters of the teachings of the Imams, the so-called *rijāl*, who had the advantage of learning from several Imams and, in fact, in some cases, like that of Zurāra, were senior in age to the living Imam, an additional factor, it seems, in their growing influence and prestige. Moreover, one should also mention the *bayt al-zurāra*—that is, the Zurāra clan—whose members were prominent traditionists, theologians, and jurists, and who, evidently, added to Zurāra's prestige as the leader of the Imamites in Kūfa. Humrān, 'Abd al-Malik, Bukayr, and 'Abd al-Rahmān were his brothers. They were regarded as being loyal to the Imams al-Bāqir and al-Sādiq. All of them died during the lifetime of al-Sādiq, whereas Zurāra lived until the time of Mūsā al-Kāzim.[54] Zurāra's severals sons were also among the *rijāl*. They reported traditions from al-Sādiq and his successors.[55]

Muhammad b. Muslim al-Thaqafī: The Kūfa Deputy of the Imam

A careful evaluation of another contemporary of Zurāra and prominent disciple of al-Bāqir and al-Sādiq—namely, Muhammad b. Muslim al-Thaqafī al-Kūfī (d. 150/767–68)—further substantiates the contention that it was the role of *rijāl*, as the jurists of the Shī'a and the deputies of the Imam in their juridical authority, that gave them enormous power in administering the social as well as the religious structure of the Shī'a community in places where the Imams were not physically present. This development of the power of the jurists in Shī'ism at this time must be regarded as the beginning of the gradual and lengthy process that evolved more fully in the period subsequent to the Imamates of al-Bāqir and al-Sādiq, when their successors were even less accessible to their followers. With the occultation of the last Imam, as we shall see, the jurists became the sole leaders of the Shī'a, and their guardians in all matters affecting their lives.

We have seen above how Thaqafī was indirectly appointed by al-Sādiq as his personal representative in Kūfa, to whom the Shī'ites could refer in matters pertaining to religion. This appointment had become necessary because 'Abd Allāh b. Abī Ya'fūr and others like him were unable to reach the Imam's presence at all times, and were not equipped with sufficient knowledge to provide guidance to those who needed it in legal matters. Thaqafī, along with Zurāra and other highly esteemed associates of the Imam, had become a leader in Kūfa, which is attested by al-Sādiq's warning that those who became leaders (*al-mutara'isūn*) in their religion would be destroyed, and among these he enumerated Zurāra, Burayd al-'Ijlī, Thaqafī, and Ismā'īl al-Ju'fī.[56]

Thaqafī had spent four years in Medina with al-Bāqir and had transmitted thirty thousand *hadīth* reports from this Imam alone. His status as the traditionist and jurist of the Shī'ites was recognized even by Abū Hanīfa, who dared not challenge his legal opinions based on the utterances of the Imams al-Bāqir and al-Sādiq. An incident is related of a woman who came to Thaqafī, in the middle of the night, to seek a ruling on a problem. The problem was set out by the woman as follows:

> She said, "My newly married daughter was pregnant. She died when she was in labor. But the child is moving in her womb. What should be done?"
>
> Al-Thaqafī: "O you, God's slave-girl, one day I asked a similar question of Muhammad b. 'Alī b. al-Husayn [al-Bāqir], and he ruled that the womb of the dead woman should be torn open and the child should be taken out. So you too should follow this ruling and do the same." Al-Thaqafī then added, "O you, God's slave-girl, I am a man who lives in seclusion [and nobody knows me]. Who referred you to me?"
>
> The woman replied, "May God have mercy on you. I had gone to ask this problem of Abū Hanīfa, the one who employs *al-ra'y* [in legal matters]. But he said he did not know anything about this problem and he told me, 'Go and ask Muhammad b. Muslim al-Thaqafī. He will inform you about the ruling in this matter. When he gives you the ruling, come back to me and tell me what he has said.' "

The following day Thaqafī went to the mosque and heard Abū Hanīfa asking the same question of his disciples. Thaqafī cleared his throat from where he was sitting. Apparently, this was meant to be a warning to Abū Hanīfa not to attribute the answer he had heard from the woman to himself, because Abū Hanīfa is reported to have told him: "May God forgive you. Leave us alone so that we can live in peace."[57]

The above incident, whether factual or contrived, shows that even adversaries of the Shī'a, like Abū Hanīfa, are believed to have held the disciples of the Imam in high esteem because of their grounding in the knowledge of the *sunna* of the *ahl al-bayt* and legal opinions derived therefrom. However, like Zurāra, Thaqafī was also reported to have enraged al-Sādiq by quoting some contradictory opinions in regard to doctrinal matters. Al-Mufaddal b. 'Umar is reported to have heard al-Sādiq cursing Thaqafī for having maintained a corrupt view concerning God's knowledge. He had said that God did not know a thing until it came into existence. This became the cause for al-Sādiq's curse.[58] Nevertheless, Thaqafī does not appear to have disputed al-Sādiq on legal issues, nor does he seem to have contradicted this Imam's opinions by citing variants of reports on the authority of al-Bāqir. From al-Sādiq he had transmitted some sixteen thousand *hadīth* reports.[59]

The Imam had shown great confidence in Thaqafī when at one point he came to his defense against the *qādī* of Kūfa, Ibn Abī Laylā (d. 148/765), a contemporary of Abū Hanīfa. Al-Sādiq was informed that Ibn Abī Laylā had refused to accept Thaqafī as a witness, probably due to his Shī'īsm. The Imam immediately sent him a message of protest, asking how could he reject the witness borne by someone like

Thaqafī who knew the injunctions of God better than he did, and who knew the *sunna* of the Prophet better than he did.[60]

In all likelihood, it was because of this confidence in his disciple that the Imam referred Thaqafī to Ibn Abī Ya'fūr when the latter came with a request for guidance on behalf of the adherents of the Imam in Kūfa. Kūfa, the center of the Shī'a movement from the early days of the caliphate, was also important politically, economically, and intellectually in the eastern sector of Islamic lands. The Shī'a in Kūfa had looked upon the Imams in Medina to provide religious leadership through the appointment of deputies. This kind of leadership, as noted above, had evolved in the reorientation of the Shī'a toward political acquiescence in the face of frequent failure of Shī'a leaders to establish their political power. Consequently, Thaqafī's appointment had a twofold purpose: first, it was to guide the Shī'a in its religious life; secondly, it was to maintain a peaceful relationship with the Muslim community at large, so as to provide a necessary continuity in the teachings of the Imams from Medina. However, after al-Sādiq's death in 148/765, Medina as the intellectual center was weakened by the 'Abbāsid caliphs who, realizing the potential challenge to their authority from that direction, kept all the succeeding Imams incarcerated or under strict surveillance in their own capital in Iraq. However, it is important to bear in mind that the 'Abbāsid weakening of Medina's influence was not, of course, aimed at the Shī'a only. It was part of a much larger policy, discussion of which is beyond the scope of the present work.

Decentralization of Shī'ite Religious Authority and the *rijāl* under the Successors of al-Sādiq (second/eighth–third/ninth Century)

The death of Imam al-Sādiq in the year 148/765 marked the beginning of the decentralization of Shī'ite religious authority. On examining the sources on Islamic sects and their subdivisions, it becomes evident that under the Imamate of al-Sādiq the Shī'ī community had attained an ideological unity. Even when some of the Shī'ī had followed other 'Alid leaders for a time, they had ultimately reverted to the acknowledgment of al-Sādiq's Imamate. It is plausible that it was due to the astute leadership of the Imam al-Sādiq that the Shī'ī community during this period was saved from further factionalism under the leadership of some of the extremist *rijāl* like Abū al-Khattāb. Although the Khattābiyya (followers of Abū al-Khattāb) and other extremist groups (e.g., the Mughīriyya, Mubārakiyya, and so on) were suppressed mercilessly by the newly established 'Abbāsid government, fully aware of their destabilizing messianic aspirations, they reemerged following the death of al-Sādiq. But this time, instead of following one of the close associates of the Imam, no less than three sons of al-Sādiq were proclaimed as Imams by different groups. Most of the extremist groups, including the followers of Abū al-Khattāb, acknowledged the Imamate of Ismā'īl b. Ja'far al-Sādiq, who had predeceased his father. This faction was known as the Ismā'īliyya.

Some Shī'ites had acknowledged 'Abd Allāh, the eldest of the surviving sons of al-Sādiq as the Imam, and this group was known as the Fathiyya, after 'Abd Allāh's surname *aftah al-ra's* (broad-headed) or *aftah al-rijlayn* (broad-footed). The Fathiyya were more numerous than the Ismā'īliyya, especially in Kūfa where they had the leadership of some of the *rijāl,* notably 'Alī b. al-Tāhī al-Khazzāz and others among the Banū al-Zubayr and Banū al-Fadāl.[61] In the development of these Shī'ī factions after the death of al-Sādiq, the disciples of the Imam had acted as the missionaries of new subdivisions and had paved the way for the acknowledgment of their respective leaders. However, the majority of the Shī'a acknowledged the younger son of al-Sādiq, Mūsā al-Kāzim, as the Imam. It is significant that the *rijāl*—the transmitters of the Imam's knowledge—were largely responsible for bringing about the acclamation of the Imam al-Kāzim as the only true successor of al-Sādiq. Among prominent leaders (*shuyūkh*) mentioned by Mufīd, the Imamite historian, are the jurists and the loyal and trusted associates of al-Sādiq, such as al-Mufaddal b. 'Umar, Mu'ādh b. Kathīr, 'Abd al-Rahmān b. al-Hajjāj and many other traditionists who were to clear the way for the designation of al-Kāzim.[62] It was probably from this time onward that the close associates of the Imams were destined to play a crucial role in the acknowledgment of the Imam of the Age (*sāhib al-'amr*), because after al-Sādiq's Imamate no Imam seems to have been recognized as the single leader of all the Shī'ites. Most of the time there appear to have been more than one contender for the Imam's position, in the absence of one centrally recognized leader of the 'Alids. Moreover, in certain cases, like that of the ninth and tenth Imamite Imams, who had become Imams at the age of eight or nine, it was the influential leadership of the *rijāl* that became instrumental in the allegiance paid to them.

Kulaynī's *Kitāb al-hujja* in his *al-Kāfī* deals with the designation of the twelve Imams as the leaders of the Muslims. Careful study of the twelve sections treating the designation of the twelve Imams corroborates the fact that with the Imamate of Mūsā al-Kāzim, when the Imamate was contended for by more than one person, it was necessary to document the claim to the Imamate and record many more traditions, used as textual proof (*al-dalīl al-naqlī*), in support of the Imamate. Consequently, in comparison with the section on the designation of al-Bāqir and al-Sādiq as Imams, which has three and eight traditions, respectively, the section on the Imamate of al-Kāzim has sixteen traditions, and the content of these communications is vindicatory because of the claim by other sons of al-Sādiq. The subsequent section on the Imamate of 'Alī al-Ridā also has sixteen traditions and is the longest section dealing with this subject. On the other hand, the section on the appointment of the tenth Imam, 'Alī al-Hādī (d. 254/868), whose long Imamate, in the absence of any other prominent 'Alid at that time, was more widely acknowledged, has only three traditions recorded as proof of his eligibility for the Imamate.

The significance of these traditions dealing with the designation of the successors of a deceased Imam—traditions transmitted by their prominent disciples—can be measured in the story of what happened after the death of al-Sādiq, when followers came to Medina to identify the rightful successor of the dead Imam. Hishām b. Sālim and other well-known associates of the Imam entered the residence of 'Abd Allāh, the eldest surviving son of al-Sādiq. Many persons had gathered there, assuming that he, as the eldest son, would be the Imam after his father. One of the associates, in order to test his knowledge, asked him a question regarding the amount of *zakāt* (legal alms) that one was obliged to pay. 'Abd Allāh replied that it was five *dirhams* on two hundred—that is, two and a half percent. "How about on one hundred?" asked the disciple. "It is two and a half *dirhams,*" answered 'Abd Allāh. The disciples were surprised to hear that and declared that even the Murji'ites, ostensibly followers of Abū Hanīfa, who were known as *ashāb al-ra'y*, did not maintain such an opinion. 'Abd Allāh confessed that he did not know the opinion of the Murji'ites in this matter.

Hishām b. Sālim and his companions left 'Abd Allāh in a state of disappointment, not knowing to whom to turn in Medina. At that moment an old man in the street signaled him to follow him. Hishām became suspicious and asked his companions to let him go alone with the old man who, as Hishām supposed, could have been an 'Abbāsid spy wanting to find out who the next Imam of the Shī'ites would be so as to get rid of him. Hishām followed him, fearing for his life. The old man led him to al-Kāzim's house, where the old man left him to meet with al-Kāzim. Hishām requested permission to ask questions, as he did when he met with his father, al-Sādiq. Al-Kāzim granted him the permission and Hishām asked him several questions, to which he received satisfactory answers, a proof that al-Kāzim was in fact the Imam. He then asked him why he had concealed the fact of his being the Imam, especially when his father's adherents were in confusion on the matter of his rightful successor. Hishām then sought permission to inform other associates about the rightful Imamate of al-Kāzim. Although permitting him to do so, the Imam cautioned him to inform only those who were loyal and sincere, and only after exacting a promise from them not to publicize his Imamate, for his life was in danger. Hishām left the Imam, rejoined his companions, and explained to them how he had been guided to the rightful Imam. They too came to the Imam and asked questions, whereupon it became evident to them that he was the Imam. Afterward, they began to spread the news about al-Kāzim's Imamate and 'Abd Allāh was left with very few followers, all others having acknowledged the Imamate of al-Kāzim.[63]

The Imamate of Mūsā al-Kāzim coincided with a turbulent period in the early 'Abbāsid dynasty. The early caliphs of this line were occupied with the internal problem of establishing their authority, especially in the absence of well-defined rules of succession, following the revolution based largely on the Shī'ī ideology, which had actually brought them to power. This fact, in addition to socio-political factors, was also the reason why they felt insecure in their claim to be the legitimate successors of the Prophet, when there were 'Alid leaders with far more numerous followers who could and did challenge their claim to legitimacy. It was mainly because of such a challenge from the direction of the Imams that the 'Abbāsid caliphs adopted an inflexible and intolerant attitude toward them, especially following the death of al-Sādiq. Under these circumstances, the sole responsibility for guiding the Shī'a was delegated to the close associates of the Imams.

An additional function carried out by these associates, and which is mentioned more frequently from this time on, was the collection of the *khums*, a tax intended for pious purposes, particularly for descendants of the Prophet, to be administered by the Imam.[64] The *khums* was given to the Imams by their followers at different times. In distant lands the agents of the Imams were responsible for collecting it and bringing it to the Imams. The revenue from *khums* seems to have been a significant source of income for the Imams, who were not, according to the law, entitled to receive other forms of benevolent charity, such as *sadaqāt* (voluntary alms). It was also a mark of their strength and that of their followers.

In his account of the Wāqifiyya (those who had "stopped" with the Imamate of al-Kāzim) and their notion of the messianism and occultation of al-Kāzim, Kashshī reports that at the time when this Imam was imprisoned by the 'Abbāsid Hārūn al-

Rashīd, an amount of thirty thousand dinars for *khums* had been deposited with his two agents in Kūfa. One of these agents was Hayyān al-Sarrāj. The two agents had spent this money in buying houses and trading, and had made a considerable profit. When al-Kāzim died and the news reached them, they denied his death and spread the story that the Imam had not died, and could not die, because he was the promised Mahdī, and he had disappeared.[65] It emerges from this story that the idea of the occultation of al-Kāzim may possibly have been invented by those agents who wanted to benefit from the material wealth that could have been claimed by the succeeding Imam.

The Wāqifite thesis that al-Kāzim had not died and would return as the messianic Imam to establish justice on earth was in part the work of some of the Imam's disciples in Kūfa who were stressing early messianic intervention. On the other hand, the Qat'iyya, who held that his son 'Alī al-Ridā had succeeded al-Kāzim, were headed by disciples who postponed messianic activity to a future time. Moreover, the Wāqifiyya-Qat'iyya dispute during the Imamate of al-Kāzim reflects the impatience of the Shī'ites at the delay in the establishment of 'Alid rule, and the way in which some eminent disciples were engaged in the Wāqifiyya-Qat'iyya doctrinal controversy regarding the Imamate of the dead or the living 'Alid leader. Similar controversy on the Imamate occurred when the eleventh Imamite Imam, Hasan al-'Askarī, died in 260/873–74, and when Wāqifiyya-Qat'iyya factions were formed once again. The former denied the death of the Imam (which carried the implication that the Imamate ended with al-'Askarī), whose return as the messianic Imam was to be awaited; the latter upheld the succession of last Imamite Imam, who went into occultation to return in the future. It is important to note that some close associates and agents of al-'Askarī contributed enormously in giving recognition to the Qat'iyya thesis about the succession of al-'Askarī's son as the twelfth and last Imam.[66]

In this doctrinal dispute among the Shī'ites, at different times during the Imamate of al-Kāzim or his successors, 'Abbāsid officials did not remain neutral. They gave support to one or the other Shī'ite group, whichever served their purpose of weakening the 'Alid leadership in Medina. The harsh treatment of the Imams by the 'Abbāsids was another way of dealing with a possible challenge from the Shī'ite leaders. Thus, for instance, al-Kāzim spent most of his time in the 'Abbāsid prison in Baghdad, where he was poisoned to death in 800. Similarly, Hasan al-'Askarī was imprisoned for some time and then kept under surveillance in Sāmarra, then the 'Abbāsid capital, where he died in 873–74. The 'Abbāsids were aware of the significant contribution of the close associates of the Imams, who were made their deputies and were put in charge of the affairs of the Shī'a community when the Imams were for the most part inaccessible to their followers. Some of these disciples had established a relationship with the caliphal court and the vizierate. Still others, like 'Alī b. Yaqtīn (d. 182/789–90), a close associate of al-Kāzim, held a high positon in the 'Abbāsid administration and had an intimate relationship with the caliphs. The position and prestige of some of these associates of the Imams in the imperial 'Abbāsid government was partly responsible for the concentration of the Shī'ites in Baghdad, which, under their leadership, became a very influential center in subsequent years. The last Imams' deputies, who had become the focus of the Shī'ites, were appointed from the associates who lived in Baghdad. Thus, the eleventh Imam, al-'Askarī, who was imprisoned for some time and then kept under surveillance, appointed his trusted associate ʿUthman al-ʿAmrī as his personal representative, and he lived in the old section of Baghdad where the Shīʿites were concentrated.

There was, however, a noticeable difference between the associates of the early Imams, al-Bāqir and al-Sādiq, who had lived during the tumultuous formative period of the Shīʿī ideolgy, and the disciples of the last ʿAlid Imams, who were faced with the survival of that ideology under ʿAbbāsid oppression. It was presumably this epochal difference that was reflected in their respective roles as the disciples of the Imams. The early associates became responsible for the moderation of extremist elements among the Shīʿa, just as the Imams themselves were engaged in their reorientation regarding the concept of the Imamate. The later associates predominated in upholding belief in the Imamate of the rightful person by insisting that the establishment of the ideal rule was dependent upon the consolidation of the rightful Imam as the only legitimate authority (*al-sultān al-ʿādil*) in the Islamic system of government. Because there were obstacles in the path of the Imam in assuming his responsibility as the legitimate leader of the entire Muslim community, it was the duty of his Shīʿa, even in the absence of the Imam's political rule, to uphold his Imamate and remain alert at all times to respond when the Imam would require the Shīʿa to assist him in launching the revolution in order to establish the rule of justice and equity. This is the notion of Islamic messianism—that is, the idea of an Imam who would come at the appropriate time in the future to fulfill his role as the ideal Just Ruler in order to establish the just Islamic rule on earth. It was probably for this reason that the deputyship of the last Imams, especially the twelfth Hidden Imam, was not destined to terminate as long as the aspiration of the Shīʿites for the establishment of just government remained alive during the occultation. This aspiration was kept alive by these deputies, who upheld belief in the future coming of the messianic Imam. This belief, in addition to the unfavorable historical circumstances faced by the community, became a moderating force among the Twelver Shīʿa community. At least temporarily, it postponed any political action, pending the appearance of the Mahdī.

Following the death of the eleventh Imam in 874, the central organization of the deputyship was the only way the last two Imams could foresee the continuation of their teaching on the leadership of the community with all its implications in the realm of religion and politics. The institution of the deputyship of the prominent *rijāl* also prepared the ground for the eminent scholars of Twelver Shīʿīsm, the *fuqahāʾ* (*mujtahids*), to assume the religious leadership of the Shīʿites in the absence of the manifest Imamate of the twelfth Imam. Such a leadership of the deputies during the lifetime of the Imams, from the time of al-Sādiq on, was implicit (except when a person was actually deputized by the Imam), but with the occurrence of the occultation of the twelfth Imam it became explicit. During the early period of occultation, called the Short Occultation, deputies were appointed by the Imam; however, with the commencement of the Complete Occultation in A.D. 941, direct designation by the Imam was no longer possible. Nevertheless, indirect deputyship ensured the survival of the Twelver school in the vicissitudes of the following centuries. The guardianship of the nascent Twelver community was assumed by these indirect deputies who maintained that community's religious as well as social structure during the formative period of Twelver Shīʿīsm in the tenth/eleventh century.

The Complete Occultation marked the end of the period designated in the

Imamite writings as the period of "special deputyship" (*al-niyābat al-khāssa*) during which the deputies who undertook the leadership of the community were believed to have been directly designated by the Iman in occultation. During the Complete Occultation, which is known as the period of "general deputyship" (*al-niyābat al-ʿāmma*), there were no directly appointed deputies of the Imam. The *ruwāt* or the *rijāl*, as the transmitters of the Imam's teaching, were indirectly designated to assume the leadership of the community and perform the duties of the special agents of the Imam, without any single person officially holding the position of deputyship. In other words, the whole community was made collectively responsible to administer its religious affairs in the absence of the Imam. However, only those who possessed "sound belief" (*īmān sahīh*) and "sound knowledge" (*ʿilm sahīh*) were qualified to assume the role of "general deputyship" of the Imam and become the mediator between the Hidden Imam and his followers. The mediatorship of the general deputy during the Complete Occultation and the function he was required to perform were the logical conclusion of the development of the leadership of the *rijāl* from the time of al-Sādiq onward.

As discussed earlier in this chapter, the leadership of the *rijāl* was a necessary extension of belief in the continuous divine guidance available to humankind through the leadership of the Prophet and the "apostolic succession" of the Imams. Consequently, the view that the persons who were to act as the transmitters of the Imamite teachings were the qualified eminent personages (*rijāl*) was not based only on an interpretation derived from the function of the *rijāl* during the early history of Twelver Shīʿīsm, and the instruction of the twelfth Imam to one of his Shīʿa concerning the *ruwāt* who should be referred to in future cases of difficulty in matters pertaining to religion. Rather, it was an intrinsic part of the Imamite worldview, which required the presence of the religious leader, the Just Ruler (*al-sultān al-ʿādil*), to lead the community to the Islamic ideal. Significantly, such a belief was not limited to Shīʿites only; even among the Sunnites the belief that "God will aid Islam in every age through an Imam [in the sense of a "jurist"]" was very strong, as the Sunnī books on *fiqh* indicate. In one of the most important works on comparative Sunnī law, *Hilyat al-ʿulamāʾ fī maʿrifat madhāhib al-fuqahāʾ*, the Shāfiʿite legal scholar, Sayf al-Dīn Abī Bakr Muhammad b. Ahmad al-Shāshī al-Qaffāl (d. 507/1113), begins his study with a Preface in which he asserts that the religion of God, in every age, will be confirmed by an Imam who will take the place of God's Prophet and will protect God's laws, assist God's religion, and promulgate God's injunctions. The Imam for the Shāfiʿī jurists like al-Qaffāl was undoubtedly one of the "rightly guided" caliphs, as long as that idealized caliphate lasted. However, in the subsequent period, the founders of the Sunnī legal schools occupied that position as the exponents of the Prophetic precedent, and were accordingly proclaimed Imams by Sunnī Muslims.[67]

In subsequent Imamite history, it is possible to indicate the extent of the power of the transmitters as the "general" deputies of the twelfth Imam, especially in works on Imamite jurisprudence, where attempts were made to specify the kinds of authority (*wilāya*) the *rijāl* could assume in the absence of the Imam's all-comprehensive authority (*al-wilāyat al-ʿāmma*). The reason for such attempts at specifying the authority of the *rijāl* was that, during the later phase of the Complete Occultation, especially when the Imamite political authority was conceded by the Būyid *sultans*, the general deputyship of the jurists gave rise to important questions about the extent and the legality of assuming functions with theologico-political implications, functions that in theory were confined to the Imam's constitutional authority.

In the prolongation of the occultation, there had evolved an opinion among some prominent Imamite jurists that functions of a constitutional nature—such as the forming of a government, declaring the *jihād* on certain occasions, or convening and leading the Friday service (*salāt al-jum'a*)—could not be performed by anyone except the Imam or the one appointed by him directly. Inasmuch as the jurists, according to this opinion, were not directly appointed by the twelfth Imam during the Complete Occultation, such responsibilities could not be assumed by them. But the jurists had remained the sole leaders of the Imamite community from the time of the Imam's occultation. Hence, when political authority of the professing Imamite rulers was established, the question about the extent of the jurists' leadership was bound to arise, to accommodate a new contingency in the historical experience of the Imamite community, which had always seen in the jurists the filling of the vacuum left by the occultation. There was no need for such a discussion in the period when there was no temporal authority of the kind provided by the Būyids and later on by the Safavids. With the establishment of Imamite temporal authority, the question that was bound to be treated by major Imamite jurists was that of the nature of the Hidden Imam's deputyship under given political circumstances and the extent of the authority of his "general" deputy. In the next chapter I discuss the consolidation of the authority of the Imamite jurists as the leaders of the Imamites, through the systematization of "sound knowledge" and "sound belief."

2

The Imamite Jurists,
Leaders of the *imāmiyya*

The end of manifest Imamate with the twelfth Imam in the year A.D. 874 marked a new phase in the religious as well as social history of the Twelver Shīʿī community. The concealment of the last Imam necessitated the assumption of the leadership of the community by the prominent followers of the school, who, in addition, were the repositories of Imamite learning. Thus, there were two factors that decided the question of leadership among the Imāmiyya during the occultation: first, "sound belief" (*īmān sahīh*)—that is, upholding the Imamate of the twelve Imamite Imams; and, secondly "sound knowledge" (*ʿilm sahīh*)—that is, the learning acquired from the *ahl al-bayt,* the Imams. As seen in the previous chapter, even in the case of the early associates during the lifetime of the Imams, before the occultation both these factors had been held to be determinants of the *rijāl* in representing the Imams among their adherents.

Through this "sound belief" and "sound knowledge" one can discern a kind of continuity made possible by the *isnād*s (chains of transmission) of transmitted religious sciences (*al-ʿulūm al-naqliyya*). The psychological reason for appending an *isnād* to a *hadīth* report was to link a contemporary reporter to the past eminent figure who gives the reported text reliability. In transmitted sciences (*manqūlāt*) each link in the *isnād* is independent *and* dependent, because the requirement of "sound belief" and "sound knowledge" applies to each person mentioned in the chain individually, and to all collectively. It was probably for this reason that, before accepting any part of the transmitted science, it was felt necessary to establish a chain of *rijāl,* who formed the links through which that part of the knowledge became accessible, and to ascertain a continuity of the religious, reliable individuals who collectively ensured the authenticity of that knowledge.

There also is surely a deeper reason for insisting on this kind of continuity—namely, the survival of the original ideology. In the intellectual environment of Medina, Kūfa, or Baghdad, where the Imamite *rijāl* could engage freely in discussion with their adversaries, there always remained the danger not only of compromise but also of interpolation in the original teachings of the Imams. It is for this reason that in Imamite jurisprudence only a few associates of the early Imams were accepted as forming a chain of persons whose consensus of opinion (*ijmāʿ*) in legal matters could be held lawfully binding.

There also seems to be a theological reason for insistence on the proper *isnād* in the arena of transmitted religious sciences. The office of Imamate in Shīʿīsm was established through the process of *al-nass al-sarīh* (explicit designation). I have identified this process of legitimation of authority as "apostolic succession." For any individual to claim Imamate without a proper designation amounted to a false claim. Hence, it was *nass* that provided the Imamate with necessary legitimacy in assuming the leadership of the *umma*. Moreover, *nass* was the only guarantee that the Imam was infallible, for according to the theory of Imamate, only an infallible Imam could designate another infallible Imam. As a result, *nass* and *ʿisma* (infallibility) were made interdependent to ensure the succession of the rightful Imam. In the theory of Imamate, one can thus perceive the *nass* to be the chain (or more precisely, "chain of gold," as described by ʿAlī al-Ridā, the eighth Imamite Imam) that links the Imams in "apostolic succession" to one another and to the actual source of divine authority, the Prophet. Without such a linkage in the Imamate, one cannot comprehend the insistence of Imamite theologians that a believer must accept either all the twelve or none of the Imams, because it is a line of twelve Imams, linked through the principle of *nass* for the assumption of the leadership of the *umma,* that guarantees preservation and protection of the Islamic message revealed to the Prophet.

The *isnād* in a *hadīth* report comprised a chain of transmitters who linked the tradition to the actual source from which it was acquired. Such a linkage was necessary in order to render the report sound. Furthermore, it created a sense of continuity in the religious leadership, because the *rijāl* had the confidence of the Imams and, as noted in chapter 1, in certain cases, were properly designated as their trustworthy representatives among their followers to disseminate their teachings. However, their proper designation by the infallible Imams did not guarantee their immunity from vitiation; rather, it established their reliability and their righteousness. The latter trait of character is known as *ʿadāla* in Imamite jurisprudence. *ʿAdāla* is a quality required of the *ruwāt* (the transmitters of the Imams' teachings), in order to render their transmission "sound" (*sahīh*). Consequently, *isnād,* like *nass*, was the sanctified process that provided the *rijāl*, who were also the *ruwāt*, with the required legitimacy in assuming the socio-religious leadership of the Imāmiyya, especially during the occultation. In the absence of a proper *isnād*, going back to the original source of authority (the Imams), no Imamite jurist would have been able to establish the authenticity of the substance of his teaching, or the place he was occupying.

The meticulous study of the *isnād*s, from the early times, which is evident from the number of works on *rijāl* compiled in the tenth and eleventh centuries, substantiates the point that *isnād*s were not only considered keys to authentic knowledge, they were also critical to authentic authority. *Isnād*s carried important information about the way a particular tradition was transmitted from person to person, giving names of a chain of persons who received the tradition from each other in succession. More importantly, it gave some idea about who studied under whom and who, possibly, had permission to cite on the authority of whom. As we shall see below, it was the knowledge combined with the source of that knowledge that gave a position of weight and influence to some of the prominent jurists and theologians of the Imamite school to exercise their authority in that community. These early scholars saw themselves as forming part of the *isnād* going back to the infallible Imams, and were acknowledged as such by later Imamite scholars, who depended on their expositions and elaboration of the transmitted religious sciences. Moreover, as noted in chapter 1, they became part of the authentication process of the transmitted religious sciences and their authenticating statements resolved the problem of transmitters' reliability in some particular cases.

The majority of these eminent scholars, whether they belonged to the Baghdad or Qumm schools of Imamite learning, following the occultation of the twelfth Imam in A.D. 874, lived in Baghdad, where they produced their literary corpus and also led the Imamite community there. Hence I will concentrate on their contribution to giving final shape to the religious as well as the socio-political tradition of the Imāmiyya in their capacity as community leaders.

Leadership of the Shī'ites in Baghdad

Much earlier than the Short Occultation (A.D. 874–941), Baghdad (more precisely, its Karkh quarter) was inhabited by the old Shī'ī families of Kūfa and was a center of trade as well as of Shī'īsm. Shī'ite doctors from the early days of the foundation of Baghdad had either visited or resided in this capital of the 'Abbāsid caliphate. The Shī'ites of Baghdad figured prominently as the old mercantile class of Iraq, who were also partly responsible for the protection and patronage of a number of Shī'ī scholars. The Imamite jurists, who lived as a minority in the Sunnite circle and who were faced with challenges from within and without, had to deal with two major problems in the absence of the Imam: first, organizing the transmitted sources of the Imamite school in order to systematize the central doctrines, like the Imamate, which were under attack; secondly, organizing the community so as to ensure its survival under the Sunnite dominance. Both these problems were to engage the intellectual as well as the social energies of the scholars. They were determined to adjust and amend their views on certain issues in the light of the criticisms leveled against them by their adversaries, and to provide practical guidance to their fellow Imamites.

However, the most immediate issue facing the whole community was the question of the discontinuation of the Imamate. Although the last 'Alid Imams had prepared the Shī'ites for this future event, their mere physical presence, however restricted and at times cut off by incarceration, made a difference to the minds of their followers. That difference became evident when the dispute about the deputyship of certain individuals was questioned by the Shī'ites. For the Shī'ites, in the tradition of "apostolic succession" that legitimized authority, it was of critical significance that any individual exercising authority in the community should have been designated by the Imam himself. Such a designation, according to the Imamites, was possible during the Short Occultation, even though there were then more individual Imamites who falsely claimed to be agents of the twelfth Imam than at other times in Imamite history. This fact, as has been fully discussed in my earlier work on the twelfth Imam, probably prompted the discontinuation of the special deputyship of the Hidden Imam, which had provided the Shī'ites with the formidable and astute leadership of some highly esteemed associates of the last Imams in Baghdad.[1]

During the Complete Occultation, when the special deputyship was formally terminated in 941 A.D., the question of the leadership of the community had to be based on a criterion other than the Imam's designation of his deputy. It was at this juncture that remarkable importance was attached to the twofold requirement of "sound belief" and "sound knowledge" in any person who held a leading position in the Imāmiyya community. Both these requirements were usually attested to by a person's teacher, with whom disciples spent most of their lifetime studying and reading the transmitted sciences handed down from authority. In other words, there was a sort of "living *isnād*" established to ascertain a person's trustworthiness, sound belief and knowledge, going back to the authoritative source.

The permission (*ijāza*) that was issued by the teacher to students to cite and teach the transmitted sciences on his authority had a twofold purpose: first, it fulfilled the necessary certification required to prove that the individuals had "heard" the traditions from the source that they claimed to have known; and secondly it established their own reliability in the eyes of the members of the community who could then entrust them with community affairs, including its financial structure, which was based on the *khums*—the fifth.

The *khums* tax (see Appendix), which must have been substantial in the merchant quarter of the Karkh, was the main source of revenue for Shī'ī leaders with which to manage the affairs of the community. It was also a financial source independent of state control, and as such it preserved the independence of Shī'ite leaders from government control, although our sources do mention Shī'ī administrative families (e.g., in the vizierate) that might have been paying *khums* tax to these Shī'ī leaders and who might have been exercising some sort of influence on them. The financial independence of these leaders was also partly responsible for the establishment of Shī'a centers of learning in the Karkh, then to the west Baghdad, and elsewhere in Islamic lands, where great Imamite teachers like Shaykh al-Mufīd, Sharīf al-Murtadā, and Shaykh al-Tā'ifa al-Tūsī taught on a regular basis. Students, who came from different places to learn under these great masters, were supported financially by the *khums*. It was from the center in Baghdad, and other such places of Shī'a learning in Najaf, Qumm, and Rayy, that the Imamites sought their leaders during the Complete Occultation. Moreover, it was from these places as nuclei that the community was organized under the leadership of the Imamite doctors who became their guides and guardians in both religious and socio-political matters.

The guidance that was provided in political matters during this formative period of Imamite history was, in fact, the culmination of the reorientation begun during the Imamate of al-Sādiq. As we shall see in the next chapter, on the Imamite theory of political authority, it was precisely what was formulated during this period, however indirectly, regarding the theologico-political role of the Imamite jurist that was going to culminate in the all-comprehensive authority (*al-wilāyat al-'āmma*) of the jurist as *al-sultān al-'ādil* (the Just Ruler) in the future.

Briefly stated, at this juncture, it is plausible that the occultation of the twelfth Imam was believed to be a temporary divine arrangement during its early period. It was bound to end with the creation of the Islamic rule of justice when the twelfth Imam reappeared. As a corollary to this belief, accommodation with non-Imamite political authority was not particularly anomalous, for even the Imams had lived under similar circumstances. Whether at this time or in the subsequent periods when Imamite temporal authority was in effect, Imamite doctors were not constrained to make any modification in Imamite political theory. As it appears in works of theology and political jurisprudence, it remained intact under their leadership. As a matter of fact, they became the protectors of Imamite political ideology throughout postoccultation Imamite history. Nevertheless, it was at this time in history that Imamite scholars, under the historical development of the temporal authority of the Imamites, saw the possibility of the existence of *al-sultān al-ʿādil* (the just authority) who would rule in accordance with the Imamite interpretation of Islamic revelation, pending the return of the twelfth Imam, al-Mahdī.

The Compilation and Composition of "Sound Belief" (*īmān sahīh*)

The classical theory of the Imamate assumed the unity of religious and political authority under the properly designated Imams. However, reality reflected the absence of political authority in the Imams. The Imams, troubled by the role of the sinful occupants of political power and sensing their inability to replace them, worked toward the autonomy of the Imamate; it should not be conditional upon the Imam's investiture as a political leader, as required, for instance, by the Zaydī Imamate. This process of the separation of religious and political, although never proclaimed as such, was implied in many a statement of the Imams and was practically complete when the Imam Jaʿfar al-Sādiq assumed the Imamate in 113/731–32.

After the termination of the manifest Imamate of the Twelve Imams following the occultation, Imamite theologians disregarded the cleavage in their theoretical pronouncements on the Imamate and, following the example of the Imams, kept themselves unsullied by the injustice inherent in political power as such. Nevertheless, they too continued to perpetuate the autonomy of the Imamites as a religious community independent of the shifting political powers of the day, and secured the acknowledgment of the religious authority of the Imams over the community without requiring them to assume political power, more particularly during the occultation. Under these concrete and contrived circumstances, the power of the Imam came to be measured in terms of his knowledge (*al-ʿilm*), which he had inherited from the Prophetic source, through the proper *nass* (designation). Consequently, in Imamite theology "power" in the sense of "authority," having moral and legal supremacy, with the right of enforcing obedience (*mafrūd al-tāʿa*), came to be recognized in the *ʿilm*, especially when the power to exercise control or command over others remained in the hands of those who are frequently mentioned in our sources as instances of *al-sultān al-jāʾir* (tyrannical or unjust ruler).

I shall discuss the question of the legitimacy of political authority in the absence of rule by Imams in the next chapter. Here it is important to bear in mind that the leadership of the Imamite jurists, who did not fail to make good use of the relatively safe and favorable period of the Shīʿī Būyid rule in Baghdad (which was never declared illegitimate [*al-jāʾir*] rule and which allowed them to refine the Imamite *ʿilm*), provided crucial guidance on living peacefully as a religious community, accepting the political domination of the Imamite ruler who would no longer harm the community's essential concerns in a society with a Sunnī majority. It was also during this period that systematic exposition and vindication of "sound belief" was undertaken by these jurists to consolidate the Shīʿī "power" of *'ilm*.

The intellectual movement in Shīʿīsm (which was the result of practical issues concerning the authority of the Imam, his occultation, and so on, being debated by Imamite and non-Imamite scholars in the tenth century) necessitated the compilation of the transmitted sayings of the ʿAlid Imams. Thus, among the transmitted sciences, *hadīth* received the most attention at this time. In fact, Baghdad attracted numerous Muslim scholars who came to that ʿAbbāsid capital especially to collect traditions. *Hadīth* is the basis of transmitted science, and those who compiled these reports were required to "hear" them and obtain permission (*ijāza*) to record and communicate them on the authority of the transmitters. Hence, Baghdad, as a city frequented by Muslim scholars from all over the Islamic empire, was an ideal place for study of *hadīth* as well as for personal contact with the author. This also facilitated the process of collecting and compiling the *hadīth*. In Baghdad, more than any other center of Islamic learning, the science of *hadīth* was cultivated during this epoch.

Although there prevailed a general view among the early Muslims that in order to obtain knowledge of *hadīth* one should hear from a teacher orally, it seems that among the *ahl al-bayt* emphasis was laid on writing.[2] According to Shīʿite scholars, the first notebook on *hadīth* was the one written by ʿAlī, which was dictated to him by the Prophet. This notebook was with Imam al-Bāqir, who asked his son to bring it out and show it to al-Hakam b. ʿUtayba, on the occasion of a difference of opinion among the associates of the Imam.[3] (Al-Hakam [d. 113 or 114/731 or 732] was a Murjiʾite jurist who related traditions on the authority of al-Bāqir.)

Another old collection of *hadīth* belonging to a Shīʿite is mentioned by a Sunnī biographer, Ibn Hajar (d. 852/1449)—namely, Abū Rāfiʿ al-Qutbī's *Kitāb al-sunan wa al-ahkām wa al-qadāʾ*. Abū Rāfiʿ was the Prophet's client and had died in the early part of ʿAlī's caliphate. He was his treasurer and was among the trusted associates of this Imam.[4] After his death, his two sons, ʿAlī and ʿAbd Allāh (scribes of the Imam ʿAlī), and Sulaym b. Qays al-Hilālī, among ʿAlī's companions, wrote "books" on *hadīth* that were in wide circulation among the Shīʿites.

From this time on, especially during the Imamate of the two Sādiqs, al-Bāqir and al-Sādiq, Shīʿī transmitters and compilers of *hadīth* increased in numbers, so much so that al-Hasan b. ʿAlī al-Washshāʾ, a companion of the eighth Imam, ʿAlī al-Ridā, claimed he had met nine hundred masters of *hadīth* in the mosque of Kūfa, who used to say: "It was related to us by Jaʿfar b. Muhammad al-Sādiq that. . . ."[5]

Mufīd, who died in 413/1022, wrote that from the time of ʿAlī up to the period of al-Hasan al-ʿAskarī (d. 260/873–74), Imamites had composed four hundred works known as *usūl*. *Asl*, a technical term among Imamite *hadīth* scholars, refers to a work comprised of only traditions that had been heard directly from the Imams; *kitāb* includes *hadīth* reports related on the authority of the Imams, but taken from another transmitter. The early compilers of the books on *rijāl* used to separate the authors of works on *usūl* from those who were mere compilers or composers of works on traditions. The first Imamite scholar to have paid attention to this classification among the traditionists was Abū al-Husayn Ahmad b. al-Husayn al-Ghadāʾirī (d. 411/1020). He compiled two books on *rijāl:* one dealing with the compilers, the other with those who wrote *usūl*.[6] Such a distinction was decisive for later authors who searched for these early collections and critiqued their reliability by establishing the fact that immediate links in the *isnād* were contemporaries of each other, and of the Imams, if the collection claimed to be among *usūl* works.

This method of criticism led to an extensive search for biographical information concerning the teachers of *hadīth,* which in turn led to the remarkable works on *rijāl,* a by-product of the intellectual endeavors of the compilers of *hadīth.* Thus, Tūsī, in his book on *rijāl,* following his teacher Ghadāʾirī, took special care in classifying the transmitters as those who had "heard" directly from the Imams (authors of *asl*) and those who had received communications mediately (authors of *kitāb*). In addition, he included these transmitters who lived subsequent to the period of the Imams and who related the traditions on the authority of teachers of the *hadīth.* In this way Tūsī's biographical dictionary marks a new development in the science of *rijāl* in two important directions. First, his classification of personages into those who reported directly or mediately from the Prophet and the Imams required him to arrange the biographical notes chronologically. To facilitate the research for an investigator he also arranged his material alphabetically. Secondly, instead of recording his information in two separate works as Ghadāʾirī had done, Tūsī compiled all his information in one book, clearly distinguishing two types of transmitters under two separate headings—namely, those who communicated traditions directly and those who did so mediately.[7]

The Imamite transmitters used several technical terms in their compilations to indicate the manner in which a transmission occurred. If the transmitter "heard" the communication from the teacher himself, then this was indicated by *haddathanī* (i.e., "he related to me . . ."); if the report was "heard" in a group, then *haddathanā* (i.e., "he related to us . . .") was appended to the report. If the transmitted material was "read" to the transmitter, then it was indicated by *akbaranī* (i.e., "he [usually the teacher] informed me . . .") or *akhbaranā* (i.e., "he informed us [by reading the text]).[8] The teacher of *hadīth* was of great importance in the *isnād*s, because the traditions received from him orally were copied by the students who read them out while others listened. If necessary, the teacher improved what was read and sometimes elaborated on obscure traditions. It was only after this kind of editing that the student who had put the traditions in writing could receive permission to transmit them to others by appending to the *isnād* the formula indicating that he had heard them on the authority of his teacher. Subsequent transmitters then simply copied the traditions and the permission to circulate them was implied in the formula *haddathanī* or *akhbaranī,* which was also used in their further transmissions, so similar to the formula used by the first student, who was the only one who actually heard them from his teacher.

Thus, identifying the teacher and those who had personal contact with him was of great significance in the *isnād*s. Moreover, this was the only way to ascertain that the collection was authentic, because in subsequent years, with the degeneration of the *ijāza* system through the lack of personal contact between transmitters and teachers copied texts were often devoid of the reliability afforded by the early edited collections.

The problem of *hadīth* fabrication and interpolation in the Prophetic traditions was not uncommon, and the traditionists had to reach a consensus whether to accept a particular communication related on the authority of a particular transmitter. The Shī'ī tradition had a psychological advantage over the non-Shī'ī tradition—namely, the fact that the Shī'ites in their exegesis and interpretation of the Qur'ān and the *hadīth* always referred to the Imams, who were believed by them to be infallible and whose authority was final in the matter of religious as well as legal precepts. This simplified the difficulties of variants and interpolations brought to their attention from time to time. However, this advantage lasted only as long as the Imams were accessible. As seen in the previous chapter, the Imams did not hesitate to denounce some of their close associates who either changed the wording of their utterances (thereby changing their meaning), or interpolated them.[9]

The problem of interpolation and fabrication became even more acute when Shī'ites were subdivided into various factions. These subdivisions, in most cases, were headed by some of the *ghulāt* associates of the Imams. This trend of fabrication persisted even during the Short Occultation, when the duty of defending authentic traditions and tenets fell upon al-Husayn b. Rūh al-Nawabakhtī, the third deputy of the Hidden Imam. On one occasion he was asked about certain books, "which had filled the homes of the Imamites," written by apostates like al-Shalmaghānī. Al-Husayn advised the Imāmiyya to accept what was reported on the authority of the Imams and to discard apostates' "personal opinions."[10] Moreover, there were transmitters whose "reliability" was tarnished because they had changed their allegiance by accepting an Imam outside the recognized line of the Imamite Imams. According to *rijāl* scholars, this deviation degraded them to the level of those who "followed corrupted beliefs (*al-madhāhib al-fāsida*)." However, some scholars, like Tūsī, were not willing to declare all these "deviators" as "weak" links in what they transmitted. Despite the fact of their having adhered to one or the other Shī'ite subdivision like the Fathite or the Wāqifite, some of them were still regarded as "trustworthy" in their transmission.[11] The critical investigation of *isnād*s led to thorough research in verifying authorities and their truthfulness and accuracy in transmitting texts, by establishing when and where they lived, and which of them had been personally acquainted with the other.

The Shī'ī *hadīth* from the early days of the caliphate had two important centers—Kūfa and Qumm—but Baghdad was the convergence point for these two branches. Moreover, Kūfa was always in touch with Medina, and Qumm remained in close contact with Kūfa. Thus among the associates of the Imams al-Bāqir and al-Sādiq one finds several prominent Kūfīs and also some Qummīs.[12] Qumm remained an important center for eastern cities of Khurāsān, like Nishāpūr and Samarqand, where there were centers of Shī'ī learning. Through the study of the transmitters of *hadīth* in the compilation of the Qumm school, such as that of Ibn Bābūya, one can detect a third school of Imamite *hadīth,* the Khurāsān school, fairly independent of Qumm. However, the traditionists of all these centers used to meet from time to time and transmit their *hadīth* in Baghdad, where transmitters frequently settled permanently to benefit from the great teachers of the Imamite transmitted sciences by hearing them discourse in person.

Among the renowned Imamite traditionists in the tenth century was Kulaynī, who had migrated to Baghdad in his later years (d. 329/941–42). Kulaynī was the leader of the Imamites in Rayy and Qumm. The latter city, as a center of Imamite traditions, provided the majority of Kulaynī's transmitters, and it is likely that his compilation epitomized all that was taught in Qumm by the end of ninth century. Kulaynī migrated to Baghdad, it seems, to publicize Imamite *hadīth,* which he had systematically compiled under the title *al-Kāfī fī 'ilm al-dīn*. He divided the work, as customary, into various "books,"[13] such as the Book of the Unity of God" (*kitāb al-tawhīd*), the Book on the Proof [of God]" (i.e., prophecy and Imamate) (*kitāb al-hujja*), and so on, and collecting all relevant *hadīth* reports with a certain degree of completeness and systematization. The compilation took some twenty years to finish and was probably completed in Rayy, though there is a report to the effect that he completed it in the presence of the deputies of the last Imam in Baghdad.[14] *Kāfī* was taught and propagated by Nu'mānī and other transmitters to whom Kulaynī had read his work.

It was probably the presence of several Imamite scholars and students of the *hadīth* that attracted Kulaynī to Baghdad. His migration and the spreading of *Kāfī* marked a new era in Imamite intellectual activity—namely, the systematization of Imamite doctrines on firm rational grounds. *Kāfī* marked the culmination of the first phase in the religious history of Imāmī Shī'īsm, which had begun in the time of al-Sādiq. This phase related to the compilation of what comprised "sound belief" during the occultation. It laid down the general framework of the Imamite creed, providing complete *isnād*s to give credence to the communications, the majority of which are reported on the authority of al-Bāqir and al-Sādiq.

From Kulaynī's time onward, which coincided with the end of the special deputyship of the last Imam, *isnād* in *hadīth* seems to have taken on a much wider meaning than was customarily understood by the Imamite traditionists. As discussed in connection with the authentication of transmitted sciences in chapter 1, ever since the Imamate had formally been discontinued by the occurrence of the Complete Occultation (A.D. 941), those among the *rijāl* who appeared in the *isnād* as the *ruwāt* (transmitters) were bound to be looked upon as the authorities who afforded authenticity to the material transmitted, and hence they became those who could and did exercise the Imam's juridical authority in his absence. Thus, *isnād*, from Kulaynī's time on, meant that continuation of religious leadership among the Imamites, which is reflected in the chronological link established between Kulaynī and contemporaneous religious leaders—Mufīd, Tūsī, and others. This fact can be discerned in Tūsī's statement that all the major works composed by Ibn Bābūya and his father (both teachers of *hadīth*) were collected by Mufīd, who then transmitted them to Tūsī. Consequently, Ibn Bābūya, Mufīd, and Tūsī form a "chain of transmission" (in chronological order) with the authenticating *isnād* already appended to the transmitted literature.[15] This *isnād* also assured the Imamites that the Imam's will was uninterruptedly and authoritatively transmitted to his Shī'a, thereby linking the Imamite doctors to one another and to the Imam and, in turn, linking the community to them all.

As mentioned above, such a linkage was the source of legitimacy for the Imamite jurists, both in the early days as well as in the later centuries of the Complete Occultation, to assume the socio-religious leadership of the Imamite community. The *isnād* ensured obedience to the jurists in matters of Islamic law—that is, in their exercise of the juridical authority of the Imam. At the same time, such an obedience appeared to fulfil a religious duty for the religiously minded Imamite Shī'ites. The Imamite jurists who were recognized by the Shī'ites as their religious leaders had the unanimous support of both lay and elite in the form of *ijmā'*, applicable to the Sunnite rulers, who as the holders of power had to be acknowledged through the *ijmāʿ al-umma* (consensus of the community). Such jurists in the lifetime of the Imams formed the *ahl al-ijmāʿ*,[16] whose opinion was further authenticated by the fact that such a jurist had been one of those who had participated with the infallible Imam in reaching a legally authoritative consensus. In later times the process of *isnād* linked contemporary Imamite leaders to the *ahl al-ijmāʿ*. In a sense, then, the foremost jurist was also an *imām,* recognized by the will of the Imamite community in contradistinction with the Imam who, in accordance with the theory of the Imamate, was acknowledged in the position of Imamate by the will of God in the form of "explicit designation" (*al-naṣṣ al-ṣarīh*).

It is not surprising that the prerequisites set for assuming such leadership were far more rigorous in Shīʿism than in Sunnism, where, according to the majority of Sunnite legal thinkers, *ʿadāla* in religious leadership was not necessary, so that worshipers could pray the canonical prayers even led by a *fāsiq* (unrighteous person).[17] To the question of *ʿadāla* I shall return below. It is noteworthy that leadership in Shīʿism, whether in the form of the Imamate of the descendants of the Prophet, or the functional imamate of the jurists, was conceived in terms of ideals of Islamic rule (that is, the unity of political and religious authorities under a single leader), and it was the contrast of these ideals with the conduct of the government in power that led to the development of the oppositional attitude of the Shīʿī Imams and their jurist-deputies who were troubled by the injustice of the de facto holders of political power. The ideal presupposition in Shīʿī leadership was preserved in all the theological and juridical writings during the entire history of Imamite Shīʿism, regardless of the concrete situation brought on by the occultation and the wielding of political power by Imamite rulers. Thus, Tūsī, commenting on Sharīf al-Murtadā's theological work on the Imamate, says that it is incumbent on the Imam to be fully knowledgeable about politics, upon which depends his commanding good and forbidding evil, and that such a prerequisite in the Imam is not necessitated merely by rational argument; rather, even revelation demands it.[18] I shall discuss the Imamite theory of political authority in greater detail in chapter 3.

Kulaynī's contribution greatly enhanced the composition of the first Shīʿite creed, which systematically included all the "sound belief" comprised in the fourth/ tenth century. This creed was composed by Muhammad b. ʿAlī b. Mūsā b. Bābūya (d. 381/991–92), the foremost Imamite traditionist from Rayy. According to a narrative preserved by Tūsī and Najāshī, Ibn Bābūya's father, ʿAlī b. al-Husayn b. Bābūya (d. 329/940–41), who was the jurist and leader of the Shīʿites in Qumm, met the third deputy of the Hidden Imam, Abū al-Qāsim Husayn b. Rūh al-Nawbakhtī (d. 326/937–38) in Baghdad and asked him several questions pertaining to the law. Following this meeting ʿAlī b. al-Husayn wrote a letter to the agent of the Imam, sending it through Muhammad b. ʿAlī al-Aswad, so that the letter might be delivered to the Imam. In the letter he requested the Imam to pray for a son. He received a reply in which was written: "We have prayed to God for it [on your behalf], and you will be blessed with two godly sons."[19] Afterward two sons were born to him–Abū Jaʿfar Muhammad (Ibn Bābūya) and Abū ʿAbd Allāh Husayn. It is reported by the latter that Ibn Bābūya used to pride himself on the circumstances of his birth, saying: "I was born through the prayer of the *Sāhib al-ʾamr*" (i.e., the twelfth Imam).[20] In a rescript received from the Imam, Ibn Bābūya was praised as an excellent jurist whom God turned to the advantage of the community.[21] From this account it is evident that Ibn Bābūya's father was in close contact with the deputies of the Imam, and had also met ʿAlī b. Muhammad al-Sammarī (d. 329/940–41), the last agent of the twelfth Imam in the year 328/939–40.[22]

Ibn Bābūya had visited Baghdad in 355/965 and before that in 352/962–63, when he also traveled to Kūfa, Hamdān, and Mecca.[23] His visit in 355/965 to Baghdad was prompted, presumably, by the defeat of his patron, the Buyid Rukn al-Dawla, and the devastation of Rayy. As was the custom of the Muslim traditionists, Ibn Bābūya apparently gathered around him a number of students in Baghdad, where he seems to have lectured on Imamite *hadīth*. Among these students was Mufīd, the great Imamite theologian of the following generation, who transmitted all his books and traditions to Tūsī.[24] The study of Ibn Bābūya's *isnād*s, in his collection of tradition, reflects his long journeys undertaken in search of material, at a young age, beginning in 339/950.[25] His reputation as both a compiler and a teacher was well established very early in his life, and reference to it was made by some of his contemporaries, attributing his unusual intellectual maturity to the miracle of the last Imam through whose prayer he was known to have been born. Thus, Muhammad b. ʿAlī al-Aswad, through whom Ibn Bābūya's father had sent the letter to the Imam's agent, saw Ibn Bābūya at the age of twenty lecturing on matters of *fiqh*. He was astonished at the grasp Ibn Bābūya had of his subject matter and remarked: "There is no surprise [to see you so well versed in *fiqh*], because you were born through the prayer of the *Sāhib al-ʾamr*."[26]

Ibn Bābūya's contribution to Imamite sciences lies in his work known as *Man lā yahduruh al-faqīh* (a title that approximates the familiar "Every man his own lawyer"), which is a codex comprised of chapters of law, using relevant *hadīth* from traditional material. It was intended to serve practical needs of the Shīʿite community; similar undertakings had produced compendia to administer justice in Sunnite government. It was probably this practical consideration that discouraged Ibn Bābūya from appending all the traditions with long *isnād*s. Moreover, the work also included his own rulings, in which case the opinion was of course cited without an *isnād*. As such, Ibn Bābūya's work resembles, to a certain extent, the work of the Medinan founder of the Sunnī school of law, Mālik, who in his *Muwatta'* was concerned with setting down ritual and religious practice in the form of law, as accepted through the process of *ijmāʿ* of the Medinan associates of the Prophet and their descendants. Ibn Bābūya collected not only the *hadīth* of the Imams, but also the opinions and the *ijmāʿ* of the Imamite jurists up to his own time, by citing both traditions and judicial decisions based on the interpretation of these traditions, as the case might require, to illustrate the law. Inspite of the fact that Ibn Bābūya lists in great detail the various "ways" (*turuq*) of his transmission at the end of his compilation, it is the absence of proper *isnād*s in some cases, and also the inclusion of his own rulings, that raised, for Imamite jurists, the question of the authoritativeness of some sections of the book.

The *hadīth* in *Man lā yahduruh* are of two kinds: the *mursal* and the *musnad*. In the former type a proper *isnād* is not considered an absolute necessity. Imamite jurists have trusted this *mursal* type of tradition on the ground that in these traditions Ibn Bābūya's main concern had been to document his *fatwā* and rule on its authoritativeness wihtout seeking its confirmation by citing the opinion in the form recognized by most traditionists. In other words, trust is put in the person of Ibn Babuya rather than in the documentation as such. This point is connoted in the statement of the renowned traditionist of the Qājār period, al-Mīrza Muhammad Husayn al-Nūrī (d. 1320/1902). In his work known as *al-Mustadrak*, he says:

> Among the Imamite jurists who believe that the *hadīth* of [*Man lā yahduruh*] *al-faqīh* should be preferred to the rest of the four books [of the Imamite traditions], they have done so because of the trememdous memory of al-Sadūq [Ibn Bābūya], his excellence in recording it, and his reliability in transmission.[27]

Nonetheless, scholars have placed Ibn Bābūya's work after that of Kulaynī, because Ibn Bābūya had not intended his work to be a *musannaf* (a work composed by sifting through material) of the kind Kulaynī produced, where material was presented in critically sifted form to facilitate the practical use of accumulated traditions. Ibn Bābūya's work included his own judicial decisions, in addition to the *musnad*s, in which "chains of transmission" (*isnād*s) were incorporated. His method was different from Kulaynī's.

For instance, Ibn Bābūya begins his work with the usual first chapter of *fiqh* works on *tahāra*, ritual purity. The first section deals with waters, clean and unclean. He commences his discussion with citations from the Qur'ān regarding the purity of water that descends from the heavens in the form of rain, and concludes that the origin of all the water is from the heavens and thus it is all clean, whether in an ocean or in a well. Then he proceeds to quote two traditions on the authority of al-Sādiq, omitting the *isnād* and the parallel versions, to prove the soundness of his conclusion that all water is clean, unless one knows of its having become unclean. Thereafter, he explains the ruling on water derived from the established practice of the Imāmiyya, without any detailed reference to the relevant traditions. Ibn Bābūya was mainly concerned with the requirement of religious guidance and thus avoided citing *isnād*s and traditions in the process of giving his *fatwā:* If a person finds water and does not know that it is ritually unclean, then he can perform ablutions with it and drink it. But if he finds in it anything that defiles it, then he should neither perform ablutions with it nor drink it, except in extreme need. Then he can drink from it, but should not perform ablutions. He can perform *tayammum* as a substitute.[28]

Kulaynī, whose codex is written for the purpose of enabling students to find their way with its *hadīth* reports in all the chapters of the *fiqh* and in all its problems, takes care to mention all the relevant *hadīth* reports with their proper *isnād*s, without systematically putting forward his *fatwā*, as Ibn Bābūya does. For example, the *hadīth*, "All water is clean, unless one knows of its having become unclean," which is cited as a *mursal* in Ibn Bābūya's *Man lā yahduruh*, is appended with full *isnād* going back to the Prophet himself. On the question of different waters, such as ocean or well water, Kulaynī cites *hadīth* with a complete chain of transmission:

> Muhammad b. Ya'qūb [al-Kulaynī] from 'Alī b. Ibrāhīm, who heard from Muhammad b. 'Īsā from Yūnus b. 'Abd al-Rahmān from 'Abd Allāh b. Sanan, who said: "I asked him [al-Sādiq] about the sea water if it was [ritually] clean. He said, 'Yes.' "[29]

Ibn Bābūya had the advantage of utilizing the existing sources on Imamite *hadīth* to prepare a complete scheme for the whole *fiqh*, which Tūsī, the author of the third and fourth major works of the Imamite *hadīth*, filled out with necessary and relevant *hadīth* for future generations of Imamite jurists. Ibn Bābūya's influence through his work *Man lā yahduruh* can be assessed by the fact that it was used by the Shīʿites as a reference for their religious lives in many parts of the Islamic world, in the same way in which the *risāla*s, consisting of the judicial decisions of the major *mujtahid*s of the Shīʿa world are in use today. It is plausible that what actually made Ibn Bābūya a great *faqīh* of the Imamites in the eyes of the followers of this school was that he was endowed with both "sound belief" and "sound knowledge," and was, more importantly, acknowledged as *al-faqīh* by the twelfth Imam through a rescript (*tawqīʿ*) describing him as such. This acknowledgment was believed to be sufficient by the Shīʿites for Ibn Bābūya's acclaimed leadership of Imamite learning until his death in 381/991–92.

The movement of epitomizing and systematizing all that was needed to standardize the constituent elements of "sound belief" was begun by Kulaynī and carried on by Ibn Bābūya who, by his numerous writings, assured the survival of the Imamite school in the fourth/tenth century. The subsequent phase of presenting the transmitted material, whether *hadīth* or *ijmāʿ*, subjecting it to the strict discipline of the *usūl al-fiqh* (the principles of jurisprudence), was carried on by Imamite scholars in Baghdad. The most renowned among these scholars were Mufīd and his erudite pupils Sharīf al-Murtadā and Tūsī. The latter's works on Imamite *fiqh*, including his two works on the *hadīth*, have remained unsurpassed up to modern times. With Tūsī culminated the process of compiling "sound belief" and systematizing "sound knowledge."

Muhammad b. Muhammad b. al-Nuʿmān, known as Mufīd and also Ibn al-Muʿallim, was born in Dujayl, some sixty miles from Baghdad, in the year 336 or 338/948 or 950. He acquired his basic training in Qurʾānic and Arabic studies under his father, who was a teacher by profession. Thereafter he went with his father to Baghdad at the age of twelve (around 350/961–62), where he began to study theology under Abū ʿAbd Allāh al-Husayn b. ʿAlī b. Ibrāhīm al-Basrī, known as Juʿal (d. 369/970). According to Ibn al-Nadīm, Juʿal was a legal authority and a theologian belonging to the school of thought of the Basrite Muʿtazilī, Abū Hāshim.[30] This schooling under a Basrite Muʿtazilī clearly bespeaks the intellectual atmosphere of Baghdad, where Sunnī-Shīʿī differences were not obstacles in the acquisition of knowledge.

Among Mufīd's other early teachers was an Imamite theologian, Ghulām Abī al-Jaysh, who recognized Mufīd's high intelligence and recommended that he attend the lectures of another prominent theologian, ʿAli b. ʿĪsā al-Rūmānī (d. 384/994–95), who belonged to the Baghdad school of the Muʿtazila.[31] It was Rūmānī who, according to one version of the story, gave him the title *al-mufīd* (the beneficial), following a discussion on the value of *dirāya* (critical examination and classification of *hadīth* as "authentic," "good," "weak," etc.) over *riwāya* (transmission of *hadīth*, also connoting adherence to the letter of the transmitted report) in the matter of rebellion against a just Imam.[32] It was during these years that Mufīd came into personal contact with many Shīʿī scholars. In the year 355/965–66, he obtained the *ijāza* from Ibn Bābūya to transmit and spread his traditions. As noted above, he transmitted the latter's works to his pupil, Tūsī.

Of all the teachers with whom Tūsī had studied, Mufīd figures most prominently in his writings. Tūsī, in his biography of Mufīd, after listing the latter's works, says: "I heard him discourse on all these books. Some of them were read to him; others I read to him several times while he listened."³³ Mufīd was the highest Imamite authority in theology and jurisprudence, and according to Ibn al-Nadīm's testimony, the leadership of Imamite theologians culminated in him. Among his works on jurisprudence is *al-Muqni'a*, on which Tūsī wrote a commentary entitled *Tahdhīb al-ahkām*. This work, on the one hand, shows the scheme of Mufīd's arrangement of the *fiqh* chapters; and on the other, it demonstrates Tūsī's intellectual development under him because it includes not only traditions dealing with jurisprudence, but also narratives touching upon theological issues that had attained prominence because of Mufīd's interest in dialectical theology (*kalām*).³⁴ Tūsī had probably begun to write the commentary at the age of twenty-five and had completed the section dealing with canonical prayers (*al-salāt*) before Mufīd's death.

Tūsī's preeminence in *fiqh* is demonstrated by his keen interest in the theoretical basis of Islamic law, the *usūl al-fiqh*. Both Mufīd and Sharīf al-Murtadā had composed works on *usūl al-fiqh* to enable their students to derive judicial decisions by critically examining the content of *hadīth* reports. Thus, Mufīd wrote a detailed work on the principles of law entitled *al-Tadhkira bi usūl al-fiqh*, which was preserved in an abridged version in the work of his other pupil, Muhammad b. 'Alī b. 'Uthmān al-Karājikī (d. 429/1037), entitled *Kanz al-fawā'id*.³⁵ Mufīd's *Tadhkira* provided the critique and further elaboration of the *usūl* methodology expounded by the two "ancient" Imamite theologians Ibn Abī al-'Aqīl and Ibn Junayd.

Ibn Abī al-'Aqīl is regarded as one of the earliest jurists, following the period of the manifest Imamate (third/tenth century), to discuss *ijtihād* in the context of deducing a legal decision by testing the evidence provided in the transmitted sources through a proper rational procedure. Ibn Junayd is regarded as the earliest Imamite theologian to introduce *ijtihād* in matters of the Sharī'a. He also mastered the method by which "conjecture" (*al-zann*) was applied in the Sunnī *fiqh* to derive a point of law. This latter fact in his methodology rendered his work unreliable because use of "conjecture" was interpreted by Imamites like Tūsī as equivalent to upholding invalid *qiyās* ("inductive reasoning") in deducing a legal decision.³⁶

More detailed and significant was Sharīf al-Murtadā's *al-Dharī'a ilā usūl al-sharī'a*, in which he discussed the need to confront traditions with reason and with reliable evidence provided by the Qur'ān to ascertain their authenticity. Of all the early theologians, Sharīf al-Murtadā maintained a rather negative attitude to *hadīth* reports and criticized early works on traditions of the Imamites, as well as works of their opponents, for being full of reports containing all kinds of errors and falsehoods.³⁷ He was also opposed to the use of *hadīth* reports that were regarded as "single" or "virtually unique" (*khabar al-wāhid*) because of lack of sufficient "ways" (*turuq*, "chains of transmission") of its transmission, in the derivation of a judicial decision. He went so far as to declare practice based on such traditions invalid, because even if it was established that such a "single" tradition was transmitted by a thoroughly reliable authority, it necessitated "conjecture" regarding its veracity. Legal decisions in Sharī'a, states Sharīf al-Murtadā, must be based on certainty; "conjecture" in this connection does not remove doubt that the tradition was unsound or unusable as evidence. Sharīf al-Murtadā confirms his opinion by claiming that there was consensus among all generations of Imamite scholars regarding the prohibition to rely on "single" traditions in deducing law.³⁸

Tūsī, who had studied legal methodology under Mufīd as well as Sharīf al-Murtaḍā, and was thoroughly informed of the Sunnī views on *usūl al-fiqh,* which are mentioned under appropriate chapters of his work on *usūl,* had to perform a crucial task of reinstating *khabar al-wāhid* as well as *ijmā'* (consensus in which the Imam had participated) as authoritative sources for deriving a judicial ruling through *ijtihād. Ijtihād* in Tūsī's usage signified independent reasoning of a jurist who attempted to deduce a legal decision by the use of rational and revelational proofs in intances in which, without the exertion of rational procedure, there would be no resolution of a problem. In his work *'Uddat al-usūl* he undertook this task.

'Udda demonstrates Tūsī's independent investigation as much as his intellectual maturity in the application of what has been described in later sources as *al-ijtihād al-'aqlī* or *al-ijtihād al-mutlaq* (reasoning in which rational demonstration forms both the major and minor premises). Tūsī argued against Sharīf al-Murtaḍā for rejecting the admission of "single tradition" as authoritative evidence in the derivation of a ruling, and claimed a consensus among Imamites that allowed it. And he spelled out the conditions for the acceptance of a "single" tradition—namely, that it had to be related on the authority of an Imamite transmitter who possessed "sound belief" (i.e., he was a righteous Imamite) and that it had to be from either the Prophet or the Imams. Moreover, and importantly, the report itself had to be irrefutably and accurately transmitted.[39] Accordingly, Tūsī did not accept all the "single" traditions as being authentic and usable in issuing a legal opinion.

The question of *khabar al-wāhid* and its use as evidence in the *fiqh* continued to be debated in *usūl* works. It is possible to divide the Imamite jurists into two groups on this issue: those who supported Tūsī's opinion in the matter of *khabar al-wāhid* and implicitly regarded the authority of traditions in juridical methodology as necessary to ensure uniformity in jurisprudence; and those who reinforced Sharīf al-Murtaḍā's rejection of *khabar al-wāhid* in *usūl,* thereby implicitly maintaining the necessity of relying on rational investigation in determining the correctness of a judicial decision. Among the latter group was the leader of the Hilla school of Imamite jurisprudence, Ibn Idrīs al-Hillī (d. 598/1201), who saw behind Tūsī's convincing opinion an unforeseen danger to Imamite jurisprudence. Tūsī's opinion was convincing because there were many traditions whose authenticity could be established by his proposed cautious method. However, the trend toward authentication of such traditions could also lead to the cessation of *ijtihād* among jurists whose opinions would be determined in some cases by depending on "unsound" traditions rather than rational investigation. There is no doubt, as I have pointed out in the Introduction, that Ibn Idrīs marks the beginning of a new chapter in Imamite jurisprudence by challenging Tūsī's conclusions in the area of methodology.[40]

The problem of errors and falsehood in *hadīth* reports, which had been raised by Mufīd and Sharīf al-Murtaḍā—especially the latter, who maintained the superiority of reason over tradition—was not limited to *khabar al-wāhid.* By the time Tūsī wrote his works, there had arisen much discussion about the variant and sometimes contradictory traditions that were used in the promulgation of law. At times such contradictions had caused dissension within the Imamite community, some of whose members had left the faith on account of differences in some traditions. One such person was Abū al-Husayn al-Harawī al-'Alawī, who, according to Mufīd, as reported to Tūsī, was a very religious person, having at one time full faith in the Imamite doctrines. On one occasion, when a matter became obscure on account of variant traditions, which he could not comprehend, he reverted from this school to the Sunnī school of thought. However, contends Tūsī, such a decision was based on lack of perception on the part of al-ʿAlawī, as well as his following the Imamite school without establishing its veracity by reason, because "the difference in the *furūʿ* (derivatives) does not necessitate relinquishment of the *usūl* (fundamentals), which are proven by reasoning."[41]

Nevertheless, the problem had to be resolved by interpreting (*taʾwīl*) the variant and contradictory traditions. Mufīd's *Muqniʿa* provided a complete schema for the entire *fiqh*, which Tūsī decided to fill out with relevant data, not only from *hadīth*, but the total *sunna* of the Imāmiyya, including their *ijmāʿ* and all the accumulated juridical information up to his day. His *Tahdhīb al-ahkām* (Refinement of the [religious] ordinances) was meant to give concrete shape to the Imamite *fiqh*, which then was represented in Baghdad by Mufīd and his associates. Accordingly, Tūsī took each section of *Muqniʿa*, prefaced it with an explanatory paragraph, and showed what could be deduced from this or that section for practical purposes. This was followed by recording the ruling and providing the supporting documentation, first from the Qurʾān, whether in direct quotation or in its interpretation, as necessary; and then from the *sunna*—that is, *hadīth*—going back to the close associates of the Imams (such *hadīth* were known as *al-maqtūʿ*), particularly those reported in an unbroken line (*al-mutawātir*), with complete *isnād*s or other equally authentic traditions.

Tūsī, in appending the *hadīth* with *isnād*, followed a middle path between Ibn Bābūya (who omitted the chain of transmission) and Kulaynī (who mentioned a complete chain). To compensate for his ommission in some cases, Tūsī listed his teachers and their various "ways" of transmission, whether based on directly or mediately transmitted *hadīth*. Further documentation was extracted either from the *ijmāʿ* of Muslims, whether Shīʿites or not, if there existed a consensus on particular questions among them, or from the Imamites.

Tūsī was not interested only in the documentation of the judicial decision; he was also keenly sensitive to the form and its criticism in presenting all the traditional sources (*manqūlāt*). As a result, he spent much time seeking the confirmation of the *fatwā* by citing well-known parallel versions of *hadīth* on particular issues, resolving the apparent contradiction in other traditions by interpreting them and demonstrating their unreliability, either by pointing out the weak links in their transmission or indicating that practice of the Imamites was contrary to what these corrupt traditions imply. Moreover, when two traditions were in agreement in general, and no preference could be given to one over the other, it was explained that the performance of an act should be in conformity with the reasoning based on the principle of law and that any practice that did not agree with that principle had to be abandoned. Similarly, if there was an ordinance for which there was no specified textual proof, then the promulgation of that ordinance was correlated to what the juridical principle necessitated.

Tūsī was aware of the criticisms leveled against Ibn Bābūya's *Man lā yahduruh*, and therefore intervened subjectively, when his interpretation of a particular tradition could be criticized on the ground of its weak *isnād*, by relating the interpretation of another tradition that carried a similar implication and confirmed the same meaning, either through its ostensible sense or through its purport. At other times he authenticated the tradition in question by authenticating the text of the communication even if the transmitter who appeared in the chain was not fully admitted as reliable, because of his being a Wāqifī or a Fathī Shīʿī. In short, Tūsī made sure that the judicial decisions in *Muqniʿa* were carefully documented and were derived from recognized sources of Islamic law, including traditions deemed unquestionably the major source of juridical practice.[42]

Tūsī's other major work, *al-Istibsār fīmā ikhtalafa min al-akhbār* (Pondering over that which is varying among the traditions), was the consequence of his commentary on *Muqniʿa*. The title of the collection denotes the purpose of its writing, which was to purify the existing *hadīth* material of all contradictions that had attached themselves to the *fiqh* items in the course of time through rational investigation. To achieve this end Tūsī selected 5,511 *hadīth* reports from among 13,590 recorded in *Tahdhīb*, using the rubrics provided in the last latter work, and discussed variant versions in order to resolve their apparent contradiction. Whereas in *Tahdhīb* all the traditions, whether in agreement or in contradiction with the *fatwā*, were discussed and their inconsistency reinterpreted, in *Istibsār* only the apparently contradictory traditions were put to rigorous criticism and their inner incoherence exposed.

Tūsī was primarily concerned with the practical implications of those *hadīth* communications in which, as he says in his Preface, elements had been added or ommitted, causing them to become untrustworthy. In order for any tradition to be incorporated in the *fiqh* work, it had to comply with certain criteria, of which one was a sound *isnād*. But even more crucial than proper *isnād* was the reliability of the informants who were actually responsible for the dissemination of the particular ruling. In addition, it was deemed necessary to examine the inner coherence of an *isnād* not lacking in a chain of reliable authorities, but in need of further examination of the religious affiliation of a reporter to throw light on transmitters' utterances.

This kind of meticulous research regarding the *isnād* in the *hadīth* literature led Tūsī to compose his book *Rijāl*, and an even more important work on Shīʿa authors, *al-Fihrist*. I have noted the particular features of his *Rijāl* above. More importantly, in his *Rijāl* he not only treated the transmitters of *hadīth* as to their reliability or untrustworthiness, but also pointed out their factional affiliation to different subdivisions of the Shīʿa, alluding to their competency or incompetency to be admitted in the *isnād*s of the *hadīth* with legal or doctrinal content. His *Fihrist* listed the authors who had directly heard from the Imams and had, as a consequence, composed works on *usūl*, which were used as source material in the compilation of *hadīth* literature. It also discussed other works composed or compiled by these authors and whether they could be truly ascribed to them.

Furthermore, the *Fihrist* was composed in response to criticisms made by adversaries of the Imāmiyya, who taunted them for not having any *musannafāt*, any renowned scholars in the past whose works could be emulated by posterity. This reason for writing the *Fihrist* is explicitly mentioned by Tūsī's contemporary (another prominent student of Mufīd), Ahmad b. al-ʿAbbās al-Najāshī (d. 450/1058–59), the author of another *Fihrist*, the *Rijāl al-Najāshī*.[43]

Both Tūsī and Najāshī arranged their material according to the name(s) of the author(s). After giving his full name and place of origin, and whether he is praiseworthy or blameworthy in his transmission, they proceed to give the list of his writings, emphasizing the source from which the works were received by the author. As is evident from the Sunnī Ibn al-Nadīm's *Fihrist*, Shīʿī books were in wide circulation in Baghdad in those days, and most of these works had been seen by Ibn al-Nadīm himself. He recorded their names and contents meticulously, and this information was also used by Tūsī to compose his *Fihrist*.

The *Fihrist*s of Tūsī and Najāshī contain an almost complete list of the Shī'a works that were known in Baghdad and include the information that could in large measure be regarded as the sum of knowledge about the *rijāl*. This characteristic of these two works has led Imamite doctors to consider these *Fihrists, together with the Kitāb al-rijāl* of Tūsī, and his abridgement of Muhammad b. 'Umar b. 'Abd al-'Azīz al-Kashshī's (d. 369/979–80) *Ma'rifat al-nāqilīn*, entitled *Ikhtiyār ma'rifat al-rijāl*, as the four principal works on Shī'ī *rijāl*. These four works are required texts for all students of Shī'ī transmitted religious sciences. They are, furthermore, known as the four *usūl* (fundamentals) of '*ilm al-rijāl*, which in the science of *hadīth* establish the *rijāl*, whose veracity and reliability were unquestionable, and those others whose authority might be impugned or doubted in any way.

Of greater consequence in the history of Imamite jurisprudence were Tūsī's works on *fiqh*—namely, *Nihāya*, *Mabsūt*, and *Khilāf*. *Al-Nihāya fī mujarrad al-fiqh wa al-fatāwā* (the ultimate in the compendium of jurisprudence and legal decisions), as the title suggests, was the epitome of Tūsī's detailed investigation of Mufīd's *Muqni'a* in *Tadhīb*. It was written before he undertook his critical research on Imamite traditions in his *Istibsār*: *Nihāya* reflects traditional methodology without critical discussion of the traditions. This is corroborated by Ibn Idrīs's criticism of *Nihāya* as a work following the method based on traditions (*maslak al-riwāya*), not on rational investigation through the application of legal principles to deduce a decision (*maslak al-fatwā*). As such, he regards *Nihāya* a book of traditions rather than of rationally investigated decisions and critically examined transmitted documentation (*dirāya*).[44]

That Tūsī followed the ancient method in *Nihāya* is well attested by his recording of the texts of the traditions in the place of legal opinion in the majority of the chapters of *fiqh*. However, in sections dealing with political jurisprudence that indicated circumstances unprecedented in the periods of the Imams and consequently absent from traditions on the subjects, Tūsī does put forward his inferentially derived opinion. Thus, for instance, opinion dealing with the delegation of certain tasks with theologico-political implications to Imamite jurists in the absence of the Imam or his specially designated deputy result from Tūsī's independent reasoning. From later sources on Imamite jurisprudence, especially before Muhaqqiq al-Hillī (d. 676/1277) wrote his compendium entitled *Sharā'i' al-islām*, we know that *Nihāya* was followed in its general outline in organizing and discussing applied jurisprudence.

More important than *Nihāya* was Tūsī's other work on *fiqh* entitled *al-Mabsūt fī fiqh al-imāmiyya* (The discursive work on Imamite jurisprudence), which is considered his major work in this field. *Mabsūt* is clearly the product of Tūsī's *ijtihād* and his methodology as expounded in his '*Udda*. Chronologically, *Mabsūt* was the last work of Tūsī's intellectual endeavors. Accordingly, it is the essence of his exposition of Imamite jurisprudence. Ibn Idrīs, who, as we have seen, was critical of Tūsī's *Nihāya*, held *Mabsūt* as a demonstrative work from which he adopted opinions on various issues in his own work *Sarā'ir*.[45]

In writing *Mabsūt* Tūsī was responding to the criticisms of his contemporary Sunnī scholars who contemptuously regarded the Shī'ites as *ahl al-hashw*—that is, those who accepted traditions, even those with anthropomorphic descriptions of God, as genuine, and interpreted them literally.[46] Accordingly, Tūsī not only incorporated critically evaluated transmitted material derived from Islamic revelation (Qurʾānic exegesis, Sunna, and *ijmāʿ*) in *Mabsūt,* he also included his profound comprehension of subsidiary proofs and details that were tested as evidence in Sunnī jurisprudence. It is probably correct to say that *Mabsūt* is the first work of its kind in Imamite jurisprudence in which the jurist applied his *ijtihād* in order to either elucidate ambiguous evidence or to infer subsidiary evidence on which depended rational elaboration and discussion for the resolution of a problem. Tūsī achieved his goal without going into details of Imamite and non-Imamite sources which he acknowledges of having examined to write *Mabsūt* and which he incorporated in his work on comparative law.[47]

There was considerable interest in comparative law among the Imamites. This interest was probably sparked by their minority status in the Muslim community, in addition to the polemics against them by their adversaries, to which, as I have noted, Tūsī makes frequent references. Sharīf al-Murtadā had written his *al-Intisār* (The triumph), on comparative law, mainly from the perspective of vindicating Imamite rulings under attack by Sunnī jurists. His other work, *al-Masāʾil al-nāsiriyyāt* (Rulings [drawn from the jurisprudence of] *al-nāsiriyya*), which is a sort of commentary on selected legal opinions of al-Nāsir al-Kabīr (Sharīf al-Murtadā's maternal grandfather),[48] also includes opinions of all the major scholars of Sunnī jurisprudence. Tūsī continued the Imamite interest in both the principles and application of Sunnī jurisprudence and composed his work *al-Khilāf fī al-ahkām* (Difference of opinion in [the matter of Sharʿī] ordinances) in comparative applied law. In this work, which is put together in the order of *fiqh* chapters, he meticulously records the opinions of different schools of law and compares them with Imamite rulings on the points of differences between them.

There were practical considerations in the milieu of Baghdad that prompted such comparative research by Sharīf al-Murtadā and Tūsī into the divergences of legal schools. Primarily it was the interest that the Būyid government had shown in consulting Imamite doctors on the religious aspect of public law, which called for informative works on the exact differences between the Imamite school and other schools of law, and the ways in which these differences could be minimized to bring about uniformity in the administration of the Sharīʿa. Such knowledge was supposed to serve practical needs and, more importantly, help create a sense of unity in the sectarian milieu of Baghdad, to which both Sharīf al-Murtadā and Tūsī make reference in the introductions to their works on comparative law.[49] Even more significantly for the Imamites, it was a juridical issue that was not confined to that period alone, as I have pointed out in the Introduction, to systematically discuss the views of the major jurists of each school, and demonstrate the superiority of the Imamite jurisprudence in its juridical decisions. In the Introduction to *Khilāf,* Tūsī explains the purpose behind its compilation: to demonstrate the correctness of the opinions held by the Imamites.

It was a characteristic of the Shīʿite school to draw strength from the vicissitudes of its history. No other religious minority has made so many combacks in its evolution. As Tūsī demonstrates in his formidable scholarly efforts, the occurrence of the occultation of the last Imam threatened at one point to destroy the Imamite form of Shīʿism completely, and the ʿAbbāsids tried to turn the Shīʿī Imamate into their own satellite. As it turned out, both the school and the institution of religious leadership emerged from this upheaval in better condition than when they entered it. Faith in the Imamate of the Hidden Imam, who worked through his direct or indirect deputies, grew stronger throughout the subsequent period. This endurance can probably be attributed to the marked ability of jurists to select from among themselves those who could direct the community. This is certainly borne out by renowned figures such as Kulaynī, Ibn Bābūya, Mufīd, Sharīf al-Murtadā, and Tūsī, who provided the crucial guidance needed for the survival and then the well-being of the Imamites.

There were two other significant factors that contributed to the consolidation of the Imamite faith. First, there was the enormous amount of literature produced to codify the "sound belief" of the Imāmiyya. Following the occultation, Imamite leaders were faced with challenges from within and without. These challenges initiated an intellectual movement that took upon itself to address the issues raised by Imamites and non-Imamites. That these issues were real and in some sense urgent, because the survival of the Imamite school depended on their resolution, is corroborated by the fact that the Imamite jurists not only refrained from advancing certain fixed ideas about Imamite doctrines dogmatically, but were eager to adjust their views in light of the criticism leveled against them by their adversaries and even their own fellow Imamites. In this respect, Tūsī's scholarly writings marked the culmination of this intellectual process, which had begun with Kulaynī. His writings clearly assimilated not only the opinions of the Shī'ī Imams preserved in *usūl* works, but also those of their major disciples, of whom some were acknowledged as their direct or indirect deputies, up to the year 460/1067.

The second factor was the importance attached to the commemoration of al-Husayn's martyrdom and the pilgrimage to his shrine at Karbalā'. The commemoration of the martyrdom of al-Husayn had an unexpected impact on religious fervor. From the early days of the Shī'ite Imams, the word *majlis* ("gathering") connoted the gathering in which all believers participated in mourning the death of al-Husayn.[50] But *majlis* went beyond its basic purpose of recounting the tragedy that befell the family of the Prophet. It provided a platform that was used to communicate "sound belief" to the populace, which had little or no academic preparation to utilize written sources on the subject. It was also in such gatherings that prominent Imamite jurists lectured or engaged in debate with other schools, thus proving their leadership to the Imāmiyya. As a result, there are numerous works on *majālis* (the plural form) by almost all renowned figures of the Imamite school, the contents of which indicate that they were lectures to a gathering that included lay persons. It is plausible that the *majlis* may have been the only platform of communication with the Shī'ī public. It is remarkable that this institution is still nowadays the major medium through which socio-politico-religious ideas are disseminated in the Shī'ī world, at least at the mass level. Moreover, it was through the *majlis* that pious literature was disseminated, literature that was not in need of rigorous documentation. It recounted the miracles of not only the Imams, but also their deputies. It was mainly this pious literature that afforded recognition of Imamite doctors, at least in the eyes of the Shī'ites, as legitimate guardians of the community pending the return of the Master of the Age, the twelfth Imam. The stories about Ibn Bābūya's having been born by a miracle performed by the Imam, Mufīd's confirmation as the learned authority by the Imam, or the defense of Tūsī's work on *fiqh* by the first Imam, 'Alī—all were told in these gatherings, with the purpose of strengthening the belief of the Shī'ites in the tenet that, even during the absence of the Imam, God's grace had been continuously descending on the community through these pious, eminent scholars.

The great traditionist of the Qājār period, al-Mīrza Husayn al-Nūrī (d. 1320/ 1902), relates a story that he found inscribed on the reverse of an old manuscript of Tūsī's monumental work on *fiqh, al-Nihāya fī mujarrad al-fiqh wa al-fatāwā,* and which he also found written in the handwriting of several jurists in another place:

> It is reported on the authority of the jurist Shaykh Najīb al-Dīn Abū Tālib al-Astarābādī (may God have mercy on him) as follows: I found in the book *al-Nihāya,* in the library of the *madrasa* in Rayy, that it is written: It has been communicated to us by a group of our reliable associates that they were conversing and conferring about the way the book *al-Nihāya* was organized and the contents thereof under various rubrics. Each one of them raised an objection to one or the other of the injunctions mentioned by the Shaykh Tūsī, and said that they were not without flaw. Following this discussion they agreed and decided to go on pilgrimage to the tomb of 'Alī (peace be on him), in Najaf. (This happened when Shaykh Tūsī was still alive.) Tūsī, unlike any other time before this, was on their mind because of the matter of *al-Nihāya.* They concurred that they would fast for three days and then perform the major ritual ablution *(ghusl)* on Thursday night. They would pray and would request their Master, Amīr al-Mu'minīn ['Alī], to answer their prayer, so that he might clarify for them the points they had disputed about [in *al-Nihāya*]. This they did and Amīr al-Mu'minīn ['Alī] presented himself to them in their sleep and said, "No author has composed a book of the *fiqh* of the Āl Muhammad [i.e., the Imams] more reliable and worth emulating and referring to than *al-Nihāya* . . . about which you are at variance [with him]. The book is thus because the author depended on the sincerity of his intention on God in writing it, for which he has the station and nearness in respect to God. Hence, do not be obstinate in the matter of establishing the authenticity of that which the author has ensured in his work. Act upon the rulings thereof and endorse its injunctions. Indeed, he suffered hardship in its careful organization and its arrangement; and he strove to disseminate the true rulings in all places."

When they got up from their sleep each one approached the other and told what he had dreamt the previous night, which pointed to the soundness of *Nihāya* and confidence in its author. They all agreed that each one of them should write down his dream in a manuscript before uttering it. They did so and on comparison they found that the dream was, both in word and in meaning, contradictory of their own ideas. They got up and dispersed, feeling satisfied with the outcome of their writing the dream and proceeded to see Shaykh Tūsī. When the Shaykh saw them he said: "You did not have faith in what I apprised you of the book *Nihāya,* until you heard from the pronouncement of our Master, Amīr al-Mu'minīn (peace be on him)." They were astonished to hear that and asked him how he knew about it. Tūsī said: "Amīr al-Mu'minīn presented himself to me as he did to you. Then he conveyed to me what he had told you." He then described his dream circumstantially. "This is the book on which the Shī'ite jurists belonging to Āl Muhammad have based their opinion."[51]

The above story demonstrates the significance attached to the shrines of the Imams and the ritual of *ziyāra*—visitation of these shrines. It is no coincidence that Shī'ite leaders chose their residence in the shrine cities of Iraq. The shrines attracted a large number of pilgrims who firmly believed that their love and devotion for the martyred Imams, expressed through these visits to their tombs, would win them forgiveness of their sins. Pious Shīʿites looked upon the shrines as the places where they could share in the Imam's sanctity.[52] The concept of *al-mashhad* (the site of the grave of the Shīʿī Imam), with all its pious implications, seems to have been well developed by the ninth century: it was probably the strong emphasis on the *ziyāra* to the shrines of the Imamite Imams that led the ʿAbbāsid caliph, al-Mutawakkil, in 236/850–51, to destroy the tomb of al-Husayn at Karbalā' and prohibit under threat of heavy penalties the visiting of that *mashhad*. The prohibition against performing the pilgrimage to Karbalā', which in Shīʿī piety came to be placed on an equal, if not superior, state to the *hajj*, clearly shows the concern of the Sunnite caliphs about the growing influences of this and other such shrines in the lives of the Shīʿī masses. The shrines had also become important centers of Shīʿī learning and as such attracted students from all over the Islamic world. Tūsī had migrated to Najaf, which contained the grave of ʿAlī. And as is evident from the above story, visitation to the shrine was not only an act of piety for a group of scholars, it was also used to obtain confirmation of "sound knowledge" as taught by eminent representatives of the Imamite school.

Unlike the *hajj*, which had to be performed at a set time, the *ziyāra* of the Imams could be performed at any time, although some special days were recommended for this pious act. As a result, there was a continuous flow of pilgrims to these shrines, which allowed Imamite leaders to communicate with the faithful from all over the Shīʿī world, at all times. The *ziyāra* was not only an act of convenant renewal between the Imams and their Shīʿa, it was also a pious act of meeting with the indirectly appointed deputies of the Imam. For ordinary Shīʿites, in those days as in contemporary times, these scholars were the *nuwwāb al-imām*—the deputies of the Imam, the functional Imams. The Shīʿite historical experience allowed for such elevation of their religious scholars, who by virtue of their "sound belief"(affirmed by ʿadāla [righteousness]) and their "sound knowledge" (affirmed by their literary output) were able to provide Imamites with badly needed leadership, astute and prudent. In the final analysis, one can suggest that the survival of the Imamite community depended on the commemoration of the tragedy of Karbalā' and the *ziyāra* to the shrines of the Imams. Both these pious acts seemed to perpetuate the threefold religious experience of the Imamites and their ultimate conceptualization.

These three experiences were: martyrdom (*shahāda*), occultation (*ghayba*), and precautionary dissimulation (*taqiyya*). The first experience symbolized death in the struggle for justice and truth against oppression and falsehood. The second signified the postponement of the establishment of the adequately just Islamic order, pending the return of the messianic Imam. The third denoted the will of the community to continue to strive for the realization of the ideal, if not by launching the revolution (which had to await the return of the Imam), then at least by preparing the way for such an occurrence in the future by employing precautionary dissimulation to conceal the true intent of the very survival of Shīʿism—namely, the establishment of the rule of the Just Imam (*al-imām al-ʿādil*). Pious literature, whether commemorating the sufferings of the Imams or describing salutations at the time of pilgrimage, expresses the essence of this threefold religious experience of the Imamite community. Furthermore, it is this experience that underlies the significance attached to the guardians of pious literature, who also assumed the guardianship of both the faith and the faithful.

Vindication of "Sound Belief"

The historical development of the *kutub al-arbaʿa* (the four books of the Imamite *hadīth*) was traced for the sake of gaining a better understanding of the position and influence of the authors on *hadīth* in the religious and scholarly life of Imāmī Shīʿism. The high esteem accorded these authors of the *musannafāt* was also due to public opinion, which eventually declared the preeminence of the three Muhammads (Kulaynī, Ibn Bābūya, and Tūsī) as the *fuqahāʾ* of *Āl Muhammad*, especially the last, Muhammad b. al-Hasan al-Tūsī, in whom culminated the historical development of the Imamite school in all its aspects. Leadership of the Imamite doctors within the community was measured in terms of their being the reliable transmitters (*ruwāt*) of the teachings of the Imams, but their prestige and influence outside the community was reckoned in terms of their ability to equip themselves with the sophisticated technique employed by the practitioners of *kalām,* and their efficacy in vindicating Imamite doctrines.

In the beginning—that is, before employing *kalām*—Imamite theologians, like the scholars of other Islamic schools of thought, had relied mainly on *al-dalīl al-samʿī,* the so-called scriptural and traditional proof, in explaining the fundamental principles of the faith and their derivatives. In this respect, the difference in their approach and that of other Islamic schools was in the fact that the Imamite scholars in their exegesis and interpretation of *al-samʿ* always referred to their Imams, whose elucidations were regarded as religious precepts and legal injunctions, binding on the community. By contrast, the Sunnīs referred to their jurists, whose opinions were not always binding. The presence of the Imam, who also came to be acknowledged as an infallible successor of the Prophet, simplified the difficulties that arose in the interpretation of the verses of the Qurʾān or in traditions in which interpolations had occurred.

Splintering of the Shīʿites into various factions, and the persistence of the Sunnites and the Khārijites in confronting the ʿAlid Imams and their close associates with profound criticism of their beliefs, forced the ʿAlid Imams to equip themselves with more sophisticated intellectual tools to meet the challenge. Initially—because discussion between different schools was limited to matters pertaining to canonical obligations, such as the timing of daily prayers, ablutions, and so on—arguments based on *al-dalīl al-samʿī* were hampered by the inherent problem of being open to any interpretation that could be imposed on a particular verse or tradition. Moreover, the traditions, bereft of the authority enjoyed by the Qurʾānic text, were open to interpolation as well as alteration, introduced by the personal likes and dislikes of any transmitter. Mufīd, Sharīf al-Murtadā, and Tūsī, among the Imamite scholars in Baghdad, who did not fail to observe the discrepancies and spurious origin of some part of *al-samʿ*, openly declared that not only had the transmitters not reported the whole truth about what the Prophet had said, but they had tampered with the facts regarding certain crucial matters.

In some of the sermons delivered by ʿAlī b. Abī Tālib one can trace the roots of the arguments based on *al-ʿaql* (intellectual reasoning) and demonstrative proofs in support of his contentions against those who disputed his position as caliph. Although ʿAlī may not have intended to speak in the manner that became firmly rooted in the writings of the Muslim theologians (*mutakallimūn*) it is plausible that it was in discussion about Islamic leadership, which was not absolutely defined in *al-sam*ʿ, that *al-dalīl al-ʿaqlī*—the rational argument (the form of demonstration favored by speculative theologians, *mutakallimūn*)—came to be employed, supplementing the earlier *al-dalīl al-samʿī*. Much later in the history of Islamic theology one finds that these very sermons of ʿAlī and his other utterances were utilized by the Muʿtazilites and the Shīʿites to substantiate the fundamental principles of religion (*usūl al-dīn*), as well as to refute the contentions of their opponents.

During the Imamate of Jaʿfar al-Sādiq (d. 765), on the one hand, the Muʿtazilites had developed the *kalām* arguments to prove their doctrinal positions; and on the other, the Zindīqs and the Daysanites, who had apparently rejected Islamic revelation, and the followers of other foreign creeds, were occupied in spreading their own doctrines, which were based on rational argumentation. With the rise of other Shīʿite subdivisions, such as the Kaysānites, the Zaydites, and the Ismāʿīlites, the central group of the Shīʿites, who formed the Imamite school in the later period, were placed under severe hardships because of the criticisms leveled against them by both Shīʿites and non-Shīʿite groups.[53] In order to survive the assaults on their tenets by other sects, especially the Muʿtazilites, who had employed fresh terminology and demonstration in their argumentation, the Imamites had to come forward to vindicate their doctrinal stance with equally sophisticated and effective tools of reasoning.

This period also witnessed the career of Abū Hanīfa, the founder of the Hanafī legal school, who used analogical deduction (*qiyās*) in *fiqh* against the followers of *hadīth*, whose efforts were directed toward filling every chapter of the *fiqh* with *hadīth* reports and proving that *hadīth* was always a sufficient source for the solution of religious questions of rite and law. Sometimes *qiyās* was given preference over even a rare tradition (*khabar al-wāhid*), and Abū Hanīfa is reported to have stated that his decision was based on his own sound opinion (*al-ra'y*) and that it was the best opinion he had discovered. It was Abū Hanīfa's use of his "sound opinion" in arriving at a legal decision that allowed his disciples, like Abū Yūsuf (d. 182/798–99), to add to these opinions their own reasoning, which sometimes led to opinions differing from those of their teacher, Abū Hanīfa.[54]

During the periods of the Imamate of the fifth and sixth Imams, al-Bāqir and al-Sādiq, religious movements between the schools and religious debates between various schools of thought came into being. This situation required the Imamites, on the one hand, to debate with other Shīʿite groups and refute their doctrines; and on the other, to vindicate their doctrines against the attacks of non-Shīʿites, above all, the Muʿtazilites. This task, in the early days, because of the lack of compiled works of the Imamite traditions and the insignificant number of their scholars, was not easy. Under the circumstances, the close associates of the Imams, who were engaged in the vindication of Imamite tenets, had no choice but to refer to the Imams in Medina, who, they thought, were the most knowledgeable and most pious of their age. Thus, the Imamite theologians, at all times, made the Imams the center of their attention and refuted their adversaries with the guidance of their Imams.

It was in this milieu that the traditions from the Shīʿite Imams, especially al-

Bāqir and al-Sādiq, began to be written down. Consequently, traditions mentioned on the authority of these two Imams are more numerous than those reported from other Imamite Imams. The compilation of these traditions was not completed until a century or so later, and all the subdivisions of the Shīʿa (such as the Ismāʿīlites, the Fathites, and the Wāqifites), especially after the death of al-Sādiq, were able to find support for their particular doctrines by interpreting and even interpolating the utterings of the Imams. At the same time, many traditions were fabricated and falsely attributed to these Imams, particularly to al-Sādiq, during whose lifetime, as seen above, extremist Shīʿites had begun their activities in Kūfa. It is plausible that in the absence of a systematically developed Imamite school, many early Imamite authors, at least in the beginning, were prone to uphold beliefs and doctrines that later scholars considered corrupt; they accordingly engaged in the revision or correction (or even reinterpretation) of these beliefs.

Different Shīʿite subdivisions, which either fabricated traditions or interpolated them in order to make their arguments convincing, resorted to fabricating *isnāds* as well, making them look as if they went back to the Imams. The sifting, systematizing, and arranging of *hadīth* reports by the Imamite school, as discussed above, was partly due to this threat caused by falsified traditions and chains of transmission.

After the emergence of *kalām,* Imamite doctors, in order to refute the contentions of their adversaries and vindicate their own tenets, gradually found themselves under pressure to resort to the *kalām* terminology and method of argumentation, although, in the beginning at least, most of the Shīʿites were not favorably disposed to speculative theology. In fact, they pointed to traditions forbidding argumentation and discussion on matters pertaining to faith.[55] But gradually they began to pay attention to the need to acquire the principles of *kalām,* and a group among the disciples of Imam al-Sādiq became the first generation of Imamite theologians. The Imam encouraged them to debate with their adversaries, refute false doctrines, and substantiate the truthfulness of their own beliefs.

However, in this school, as in the case of other schools in Islam, there arose sharp differences between the upholders of *al-dalīl al-samʿī* and those who based their contentions on *al-dalīl al-ʿaqlī.* This led to mutual refutation and condemnation. But it was precisely the need to vindicate the Imamite faith through *kalām* arguments that increased the importance of the Imamite theologians, who cited the arguments of Imams and statements of ʿAlī to corroborate their method of vindication. The theologians, both Imamite and non-Imamite, in the process of their study of the sermons and maxims of ʿAlī, inferred various points that they proved by rational argumentation and they included them in their exposition of the fundamental principles of religion. This use of *al-dalīl al-ʿaqlī* by ʿAlī was interpreted by them as permission to engage in *kalām* disputation.[56] In fact, they considered ʿAlī to be their teacher in the use of *kalām.*

The first Imamite *mutakallimūn* were not in complete accord with one other on matters of doctrine. This seems to have resulted from the lack of collected works on *kalām* written from the Imamite viewpoint, as well as a lack of consistency in employing technical terms and classifying material under appropriate rubrics. Another factor that contributed to these divergences was the presence of the Zindīqs and some of the newly converted and self-seeking Muslims who had attained some prominence and engaged in debate with the Muʿtazilites to refute their doctrines. They also had a certain leaning toward the Imamite school and were able to blend their own doctrines with those of the Imamites. This resulted in some blasphemous doctrines by Imamite theologians, totally at odds with fundamental principles of the Imāmiyya. The Imams and their close associates had to intervene and refute such false doctrines in order to make those theologians aware to their errroneous inferences and stop them from disseminating false doctrines.

Differences of opinion concerning traditions and communications allegedly received from the Imams created a rift between those Imamites who wanted to uphold the importance of *hadīth* for religious and legal practice, and those who were inclined to give equal supremacy to human reasoning in deciding religious questions. The latter group maintained the fundamentality of *al-ʿaql* over *al-samʿ*, and became separated from the traditionists who believed in the fundamentality of *al-samʿ*. The former group formed the early generation of the Imamite *mutakallimūn*, who participated with the Muʿtazilites in their discussion about the fundamental principles of religion.[57] This intellectual activity led to a considerable literary output, written by various parties in refutation of the doctrines of others.

The two school, the Muʿtazilites and the Imamites, did not differ much in their central theological exposition, especially in view of the close association between some of the Shīʿites and Muʿtazilites. This close relationship was the main reason why their doctrines were later confused. Some Muʿtazilite biographers classified some Shīʿite authors as Muʿtazilite and vice versa; Sunnī biographers have, most of the time, confused Shīʿites with Muʿtazilites, or have regarded them as identical. This is evident in their treatment of the Zaydites, an important subdivision of the Shīʿites.

The founder of the Zaydī school was Zayd b. ʿAlī b. al-Husayn, a pupil of Wāsil b. ʿAtāʾ, a Muʿtazilite. Zayd followed Muʿtazilite theological exposition and used to give Muʿtazilite leaders priority in teaching over the Imams from among the *ahl al-bayt*. It was for this reason that a group of Shīʿites in Kūfa renounced his leadership for having followed the Muʿtazilite thesis regarding the Imamate of an inferior (*al-mafdūl*) Imam, and for having refused to curse the first two permissible caliphs. Even the fifth Imamite Imam, al-Bāqir, discussed this matter with his brother Zayd, and exhorted him not to acquire knowledge from those who had broken their battle allegiance with his great-grandfather, ʿAlī b. Abī Tālib, or from those who did not believe in the legitimate authority of the Imams from the *ahl al-bayt* and who made it a prerequisite in an Imam to rise with sword in hand.[58]

Besides the question of the Imamate, the qualifications of an Imam, and the mode of his appointment, there were other issues of controversy and debate between the Muʿtazilites and the Imamites. These issues included some of the basic doctrines of the Imāmiyya, such as *ghayba* (occultation of the Imam), *rajʿa* (the return of the messianic Imam before the final resurrection), *badāʾ* (alteration in divine ruling), and other such doctrines. Some of the blasphemous beliefs, like *hulūl* (transmigration of souls), *tashbīh* (anthropomorphism), and so on, which were ascribed to the Shīʿites because of their being expressed by some of their early *mutakallimūn*, were vehemently rejected by Imamite theologians.

The most important single point on which the Imamites and the Muʿtazilites held divergent opinions was the question of "clear designation" (*nass jalī*) in the Imamate. The Imamites and other non-Imamite groups argued that the early Shīʿites, prior to the Imamate of al-Sādiq, had not encountered this concept, and that it was during the time of al-Sādiq that Shīʿite *mutakallimūn*, like Ibn al-Rāwandī (d. 245 or 298/ 859 or

910) and Abū 'Isā al-Warrāq (d. 247/861), formulated belief in the "clear designation" of the Imam.[59] But the Imamite theologians refuted this point, saying that belief in this doctrine formed the cornerstone of the earliest creed of the Shī'ites. However, their opponents, merely because they could not find a written document stating this belief prior to the writings of the aforementioned *mutakallimūn,* conjectured that the latter had been the formulators of the doctrine in question. In fact, argued the Imamites, just as belief in the justice of God and "middle position" preexisted their systematic exposition by the Mu'tazilite theologians al-Nazzām and Abū al-Hudhayl, so did the belief in *al-nass;* in any case, a belief does not depend for its commencement on its being written down or documented.[60]

The main dispute between the Imamite theologians and the *mutakallimūn* of all the other schools of theology was on the question of the Imamate. Hence, from the time of Imam al-Sādiq onward, Imamite scholars belonging to the Qat'iyya[61] wrote numerous works on this subject, expounding and elaborating their arguments with both scriptural (*sam'ī*) and rational (*'aqlī*) proofs.

Among the authors of this first generation of Imamite theologians, some stand out prominently in early sources. 'Alī b. Ismā'īl b. Maytham al-Tammār, according to Tūsī, was the first person to expound the doctrine of the Imamate theologically, basing it on Imamite principles.[62] He was the contemporary of another remarkable Imamite *mutakallim,* Hishām b. al-Hakam, who frequented the Barmakid circle, entering into the philosophical and theological debates sponsored by its members.[63] Still other prominent figures were Muhammad b. Nu'mān Mu'min al-Tāq (known to his opponents as Shaytān al-Tāq) and Muhammad b. Khalīl al-Sakkāk. The latter is also reported to have written a book in refutation of the contentions of those who did not maintain the necessity of the principle of *nass* in the Imamate.[64]

The consequence of this literary output on the topic of the Imamate and its spread among theologians was that eventually the Imamate earned a place among the fundamentals treated in *kalām.* As a matter of fact, it became one of the most debated issues among the Imamite *mutakallimūn* of the second generation, including some profound scholars like al-Warrāq, Ibn al-Rāwandī, Hasan b. Mūsā al-Nawbakhtī, and Abū Sahl Ismā'īl b. 'Alī al-Nawbakhtī. These theologians elaborated and elucidated concise summaries of earlier *mutakallimūn* and also undertook to refute more thoroughly the doctrinal position taken by the Mu'tazilites and other schools of thought. The works composed by these authors on the Imamate[65] were in accordance with the principles of the Imamite school and were regarded as the most authentic exposition of its beliefs. It was by the endeavors of these scholars that discussion on the principle of the Imamate became part of the Shī'ī *kalām,* where it began to be treated together with the doctrine of *nubuwwa* (prophecy).

After the generation of al-Warrāq, Ibn al-Rāwandī, the Nawbakhtīs, and their immediate successors, the Shī'ite *kalām* found extensive and detailed exposition. Several later theologians wrote works on *kalām* with different arrangement of topics and in varying styles. Although all of them followed the same principles of religion as propounded by the early doctors of this school, in many detailed instances there were divergencies, and each one wrote according to his personal temperament and bent of mind. It was, in large measure, this rise of divergence of opinion among the *mutakallimūn* that gave traditionists an opportunity to criticize and question the utility and validity of rational demonstration in proving religious truths.

It is plausible that the roots of the Akhbārī-Usūlī controversy in later Imamite history can be traced back to this period when the supporters of *al-dalīl al-sam'ī* argued for the scriptural proof and tried to demonstrate that the *akhbār* (communications reported on the authority of the Imams) were a sufficient source for the religious and legal life of Muslims.[66] On the other hand, the upholders of *al-dalīl al-'aqlī*, who could be regarded as the forerunners of the *usūlīs*, maintained that rational demonstration was crucial in the elucidation of religious precepts and in elaboration of what was considered religiously binding on the community. This was particularly the case in those matters where *al-sam'* could not provide the authentic documentation necessary for incorporation of some of the *akhbār* into legal practice.

As mentioned above, Tūsī had made this observation regarding *akhbār* in which discrepancies and contrariety had occurred. Consequently, in his last work on jurisprudence, *Mabsūt*, he explained his method, which was to follow *ijtihād* (independent reasoning), balancing it with *akhbār*, to issue a judicial decision. The reason for following this method, as he explains in his Introduction to *Mabsūt*, was that Imamite jurists had put traditions together without transmitting the actual wording, so much so that legal decisions based on such traditions failed to include the sense of the words of the tradition. Moreover, these jurists had often failed to understand the purport of the traditions. Thus, Tūsī undertook to rectify the situation by bringing together some eighty works on *fiqh* and applying his methodology of "balancing" *ijtihād* with *akhbār*. As a matter of fact, all Tūsī's works, whether on jurisprudence or on *usūl al-dīn*, indicate his efforts to work out a delicate compromise, blending the two forms of argumentation based on *al-sam'* and *al-'aql*. By employing this method he was able to vindicate the Imamite teachings, studying the vast number of traditions and giving them the theological structure of a doctrine.

The history of Imamite jurisprudence, as well as theology, shows that *al-dalīl al-'aqlī* was found necessary not only to substantiate the necessity of the Imamate and to solve the problems of legal practice, but also to defend the absence of infallible authority during the occultation (for the Imam's presence was regarded, at least theoretically, as decisive in all points of law and doctrine). It was presumably for this reason that Ibn Bābūya, the Imamite traditionist, who ventured to demonstrate Imamite doctrines by relying mainly on *al-dalīl al-sam'ī*, found himself under pressure to meet the challenge of the more profound criticisms made by the Mu'tazilite and Ash'arite theologians. These criticims related to matters that were supposed to constitute *al-dalīl al-'aqlī*, and thus had to be dealt with by using the rational argumentation favored by the *mutakallimūn*. For instance, the relationship between the Imam and *taklīf* (religious duties imposed upon a believer) was a theological concept that had to be established by the intellect, because it was a derivative of the more central dogma of the justice of God. This necessitated that as long as religious obligations were in force, the existence of the Imam (who could elucidate those obligations) was indispensable. Such a conclusion did not depend on scriptural proof; only rational demonstration could establish such a necessity.

The prolonged occultation of the last Imamite Imam clearly seems to have put pressure on scholars who followed Ibn Bābūya to vindicate the Imamite creed theologically. It was, in large measure, for this reason that Mufīd began his studies in *kalām* with prominent Mu'tazilite teachers. His training in *kalām* rendered him one of the leading *mutakallimūn* whose intellectual endeavors were able to utilize developments of the preceding period to give a definitive form to Imamite doctrines. Ibn al-Nadīm, a contemporary of Mufīd, remembered him as one of the greatest *mutakallimūn* of the Imamites, in whom culminates the leadership among his associates of the Imāmiyya in law and theology. Mufīd had trained his disciples in *kalām*. The most brilliant and erudite of them was Sharīf al-Murtadā. He composed one of the chief works on the Imamate, *al-Shāfī fī al-imāma* (The complete statement on the Imamate), in reply to the refutation of the Imamite stand on the question of the Imamate discussed in the sections dealing with the Imamate in *al-Mughnī fī abwāb al-tawḥīd wa al-ʿadl* by al-Qādī ʿAbd al-Jabbār.

Sharīf al-Murtadā brought to fruition Mufīd's work in rational theology and became the most prominent of the Imamite theologians following the death of his teacher Mufīd in A.D. 1014. He died in the year 436/1044–45 at the age of eighty, having spent some sixty years in the public life of Baghdad. He was among the first Shīʿī doctors who converted his home into a place of learning, where students from various parts of the empire used to attend his lectures.[67] Tūsī, as we saw earlier, pursued the legacy of Mufīd and Sharīf al-Murtadā, and excelled in jurisprudence, providing the Imamites with systematized and organized documentation for the derivation of Law.

In the intellectual (and to some extent the paradoxically reactionary) milieu of Baghdad in the first half of the fifth/eleventh century, students attended lectures of the eminent representatives of different schools of thought, regardless of their own personal affiliation to one or the other school. Sharīf al-Murtadā's wide knowledge in different areas of Muslim sciences attracted a large number of students to him. The century was pregnant with challenges for religious communities and, most significantly, it was the preeminence of *al-dalīl al-ʿaqlī* that began to be felt at all levels of scholarly pursuits. It was at this time that Imamite *fiqh*, as discussed above, emerged from being just a collection of opinions of the Imams to become an intellectual process based on personal reasoning, issuing legal opinions on matters particularly pertinent to that age. The other factor that increased the importance of intellectual pursuits, and encouraged Shīʿī scholars like Sharīf al-Murtadā and Tūsī to produce systematic and creative religious writings, was the patronage and protection of theologians by the Būyids. The Būyids had shown great interest in the Muʿtazilī *kalām*. Their patronage of the Imamite theologians helped foster a Muʿtazilite form of *kalām* among the Shīʿites.

Among Sharīf al-Murtadā's works on *kalām,* I have mentioned *Shāfī.* According to Tūsī, who abridged it during the lifetime of its author, it includes all the demonstrations of the Imamite theologians in sustaining their doctrine of the Imamate.[68] The fact that it was abridged during Sharīf al-Murtadā's lifetime indicates that it was an important work on the subject of the Imamate, and indeed it has remained unsurpassed throughout Imamite history.[69] His other important work on *kalām* deals with the infallibility of the prophets and the Imams, *Tanzīh al-anbiyāʾ wa al-aʾimma* (Removal of the Prophets and the Imams from blameworthy associations). In this work he treats the subject of the infallibility of religious leaders in general and the blamelessness of the actions of ʿAlī, his two sons, al-Hasan and al-Husayn, the eighth Shīʿī Imam, ʿAlī al-Ridā, and the last Imam of the Twelvers, in particular.[70]

Sharīf al-Murtadā is rightly known as the *mujaddid* (restorer) of the Imāmiyya school whose scholarly endeavors, in addition to his extremely influential sociopolitical position, helped to bring about both the recognition and the firm establishment of the Imamite school in the ʿAbbāsid capital. Ibn al-Athīr (d. 606/1209–10), in his elucidation of the *hadīth* regarding the appearance of a *mujaddid* at the turn of a century, writes that during that epoch the most important schools of Islamic law were those of al-Shāfiʿī, al-Hanafī, al-Mālikī, al-Hanbālī, and the Imamite school. Then he mentions the names of the *mujaddids* of these schools in each century. For the fourth/tenth century, Ibn al-Athīr erroneously mentions Sharīf al-Murtadā; in that century the leadership of the Imamite school was in the hands of Sharīf al-Murtadā's teacher, Mufīd. This point corroborates the fact that Sharīf al-Murtadā's leadership of the Imamite school was widely recognized and had overshadowed Mufīd's contribution. This popularity of Sharīf al-Murtadā, at least in the eyes of the Shīʿa, was partly due to his renowned genealogy, going back to the seventh Imamite Imam, Mūsa al-Kāzim. For ordinary believers, it was important to remember that Sharīf al-Murtadā, in his position as the spokesman of the Shīʿa, was indirectly confirmed by the Imams.

A most popular story that the Shīʿa loved to hear about Sharīf al-Murtadā and his teacher, Mufīd, was the one narrated time and again to emphasize the greatness of these scholars, with the obvious implication of their being approved by the holy *ahl al-bayt*. (Such approval was far more crucial than any of their expository writings for their being acknowledged as the leaders of the Imāmiyya by the adherents of the Imams.) The story has been preserved by the Muʿtazilī commentator on *Nahj al-balāgha* (a collection of sermons and maxims attributed to ʿAlī), Ibn Abī al-Hadīd (d. 656/1258). He related the story on the authority of one Fakhkhār b. Maʿd al-ʿAlawī al-Mūsawī:

> Al-Mufīd, Abū ʿAbd Allāh Muhammad b. al-Nuʿmān, the Imamite jurist, in a dream saw Fātima, the daughter of the Prophet, come to see him when he was in his mosque in the Karkh. She was accompanied by her two sons al-Hasan and al-Hasayn, who were still little. At that time she handed them to him and said: "Teach them *al-fiqh*." He woke up in the state of astonishment at having seen this dream. When the day progressed, following the night when he saw the dream, he saw Fātima, the daughter of [al-Husayn b. Ahmad b. al-Hasan] al-Nāsir, and around her were her maids and in front of her were her two little sons, Muhammad al-Radī and ʿAlī al-Murtadā. He stood up and greeted her. She said: "O shaykh, these are my two sons. I have brought them to you so that you would teach them *al-fiqh*." Al-Mufīd wept and related the dream to her and took upon himself to teach them *al-fiqh*. God blessed them both and opened the gates of knowledge and virtues that made them famous in the world.[71]

This confirms the ability of Mufīd to teach "sound belief" and "sound knowledge," and also glorifies the roles of Sharīf al Murtadā and his brother Sharīf al-Radī, the compiler of *Nahj al-balāgha,* in the revival of Imamite learning. For the Imamites it was important to be told that their scholars were confirmed in their responsible positions as teachers and guardians of the Imamite creed. One often encounters pious narratives in which, for instance, Mufīd received rescripts (*tawqīʿāt*) from the twelfth Imam confirming him as the jurist and theologian of the Imamites. I have mentioned above the manner in which Tūsī's *Nihāya* was approved by the Imam ʿAlī as an authentic work on jurisprudence of the *ahl al-bayt*.

The purpose of using *kalām* was to vindicate the "sound belief" of the Imamite community under attack by theologians of other Islamic schools. The methodology, in this regard, was the one preferred by the speculative theologians—that is to say, it was based on the ability of human reasoning to elaborate and defend the beliefs in question. But for the Imamites in general, much that was treated in *kalām* remained incomprehensible. They were satisfied with the traditions and the literature that spoke to their feelings as the adherents of the Imams, and found difficulty in grasping the complexities of *kalām,* for which they had little use in their religious experience. Nevertheless, it was the theologians' "sound belief," in addition to their "sound knowledge" acquired through both scriptural as well as rational proof, that made it possible for some of the leading Imamite doctors to assume the leadership of the Imamite community in the absence of the twelfth Imam.

Although assuming such leadership of the community was not problematic *legally,* because of the provision regarding the "general" deputyship of a qualified Shīʿī during the occultation, the jurists had to work out in some detail, however hypothetically, the nature of their leadership in the absence of the Imam. The Imamite jurists were faced with a real situation where, on the one hand, they had to act as the custodians of the Imamite theory of leadership, both spiritual and temporal; and on the other, they could not afford to ignore the historical reality wherein Imamite temporal authority could be established. In the next chapter I will examine the Imamite theory of political authority and show the problem that is commonly shared by all those who have to find ways of accommodating the contingencies of a concrete situation without having to compromise an ideal that appears to be threatened by that real situation.

3

The Imamite Theory
of Political Authority

The cornerstone of the Imamite theory of political authority is the existence of an Imam from among the progeny of Muhammad, clearly designated by the latter to assume the leadership of the Muslim community. Acknowledgment of the lawfully established authority of the Imam falls within the category of the religious obligations (al-takālīf al-shar'iyya) imposed on the adherents of the Imam. Among the Shī'ī Imams, it was during the Imamate of Ja'far al-Sādiq (d. A.D. 765) and his successors that the idea that the 'Alid Imam was the sole legitimate authority—by virtue of his being an infallible leader and authoritative interpreter of Islamic revelation, and therefore qualified to establish the Islamic state—became a distinctive feature of Imāmī Shī'īsm. The Imamites invariably regarded the historical caliphate, with the exception of the period when 'Alī was acclaimed as the caliph, as usurpatory and illegitimate.[1] As a matter of fact, even non-Shī'ite Muslims, who were not actively engaged in resisting the caliphal authority as such, opposed the khulafā' al-jawr or al-zalama (tyrannical or unjust caliphs) and regarded obedience to them as disobedience to God.[2]

In Imāmī Shī'īsm, government belonged to the Imam alone, for he was equally entitled to political leadership and religious authority. The fourth Imamite Imam, 'Alī b. al-Husayn (d. A.D. 712–13), in his invocation on the occasion of the Friday service, the convening of which is considered to be the political right of the Imam in his capacity as the temporal ruler, said: "O my God, this position [of convening and leading the Friday service] belongs to your viceregents (khulafā') and your chosen ones . . . which they [i.e., the Umayyads] have taken away [from us]."[3] The implication of such an utterance was that the function of convening and leading a public form of worship by the caliph was usurpatory, and hence legally invalid; and also that the establishment of caliphal authority without prior designation was itself a political innovation in the same way as assuming the leadership of public worship in religious law. Political innovation gave rise to political dissidence, resulting in active or passive resistance to the rulers of the age.

However, even before the disappearance of the twelfth Imam, following the murder of al-Husayn in Karbalā' (A.D. 680), the Shīʿite leaders reoriented their adherents to adopt political quietism, especially after most of the Shīʿī revolts failed to achieve any concrete results in favor of the family of the Prophet. This reorientation of the Shīʿites toward a policy of quiescence, and the acceptance of existing circumstances until a descendant of al-Husayn could establish the true Islamic polity at the proper time, were achieved by al-Ṣādiq's own example. Significantly, he refused to support the direct political action of his cousins and, in spite of being one of the most respected leaders of the ʿAlids, he also declined the invitation to accept the caliphate when the ʿAbbāsids assumed power.[4]

Al-Ṣādiq's attitude toward politics became the foundation of Imamite political theory during the occultation of the twelfth Imam. The absence of the political discretion of the Imam taught its followers to employ taqiyya (precautionary dissimulation) and await a more favorable time for the overthrow of tyrannical rulers and their replacement by the Imam. Although the Imam was entitled to political authority, his Imamate, after all, was not contingent upon his being invested as the ruler (sulṭān) of the community.

Expediently, under the impact of historical circumstances, leadership of the community was divided into a temporal and a spiritual sphere. The former was regarded as usurped by the ruling dynasty, which in theory required proper designation (nass) by the Prophet, but was tolerated as long as its sphere of action was limited to the execution of the law. The spiritual sphere, although in Islam metaphysically connected to the very nature of God who exercised divine authority and governed the affairs of humanity through a divinely appointed human agency like that of the Prophet, likewise required a clear designation, also by the Prophet. The holder of spiritual authority in Shīʿī Islam was the proof (hujja) of God, empowered to interpret Islamic revelation and elaborate on it without committing error. In this respect, the Imam was like the Prophet, who was endowed with special knowledge and had inherited the knowledge of all the ancient prophets and their legatees (awsiyā'). The Imam is, thus, the link with the way of guidance, and without acknowledging him no person seeking guidance can attain it. This spiritual authority with the right to demand obedience, according to the Imamite teaching, was not contingent upon the Imam's being invested as the ruling authority (sulṭān, who could and did exact or enforce obedience) of the community. As such, it resided in ʿAlī from the day the Prophet died, for he became the Imam and the only true amīr of the faithful through the Prophet's designation.

This leadership would continue to be available in the line of the Imams, explicitly designated by the preceding Imams. It was in this latter sense that the Imamate of the community came to be conceptualized. Therefore, religiously speaking, to ignore or disobey these divinely appointed Imams was tantamount to disbelief (kufr). When the twelfth Imam reappears, the temporal and spiritual authorities will merge in his person; like the Prophet he will uniquely unite the two spheres of the ideal Islamic rule. Thus the idea of Imamate by designation among the ʿAlids, continuing through all political circumstances, was cemented by the expectation concerning the Imamate of the last Imam while he is in occultation. This reaffirmed the Imamite hope regarding true Islamic rule by a legitimate Imam from among the descendants of al-Husayn.

In the development of the political attitude of the Imamites, the threefold religious experience of the community, as mentioned briefly in the last chapter, had a major role to play. These were the martyrdom (*shahāda*), the occultation (*ghayba*) and precautionary dissimulation (*taqiyya*).

The first experience followed the conviction that God, who is the Creator and the absolute Sovereign of the universe, is just. Accordingly, God has the welfare of the individual at heart in the demand that humankind heed the guidance provided by divine revelation to the Prophet. This divinely inspired guidance also requires obedience to the Prophet in his capacity as the head of the Islamic polity. By implication, then, the Imam, who is regarded as the rightful successor of the Prophet, must also be upheld as the true leader of the community to whom obedience is due in his capacity as the head of government. When the Imam, following the death of the Prophet, was denied his right to assume the temporal authority invested in him by divine designation, direct political action was justified to establish the rule of justice—to replace a usurpatory rule by a just and legitimate one. The result of the ensuing struggle to install a legitimate political authority entailed the murder of several Shī'ī leaders, which in the light of the above conviction, was apprehended by succeeding generations as martyrdom suffered in order to uphold justice against oppression.

The second experience, the concealment of the legitimate Imam, reflected the failure of the Shī'ī revolts provoked by atrocities of the ruling house. Moreover, it connoted some sort of divine intervention in saving the life of the Imam by moving him from the realm of visible to invisible existence, and conveyed the idea that the situation was, at the time, beyond the control of those who proposed to overthrow the tyrannical rulers in order to establish the Islamic rule of justice. Belief in the occultation of the Imam and his eventual return at a favorable time helped the Shī'ites to persevere under the difficult circumstances and to hope for some degree of reform pending the return of the Imam al-Mahdī. Such expectation necessarily implied the postponement of the establishment of the thoroughly just Islamic order pending the reappearance of the messianic Imam, who alone could be invested with valid political authority.

The third religious experience of the Imamite community was *taqiyya*, employing precautionary dissimulation of its faith. This underlay the political attitude of all the Imams and their followers subsequent to the martyrdom of al-Husayn in A.D. 680. The Shī'ite leaders encouraged the employment of *taqiyya* and even declared it to be a duty incumbent on their followers, so as to avoid pressing for the establishment of the 'Alid rule and the overthrow of the illegitimate caliphate. In a sense, *taqiyya* signified the will of the Imamite community to continue to strive for the realization of the ideal Islamic polity, if not by launching the revolution contingent upon the appearance of the Imam and his consolidation as the leader of the community, then at least by preparing the way for such an insurrection in the future. In the meantime, the Shī'ites had to avoid expressing their true opinions publicly about the shortcomings evident in the various de facto Muslim governments in such a way as to cause contention and enmity.

These three religious experiences of the Shī'ites during the first three centuries of their history were bound to mold the political outlook of the scholars of this community whenever faced with a new political situation. Rulings on such a situation could be traced back to precedents set by the Imams., But the question of the legitimacy of a political rule established by a professing Imamite during the absence of the political discretion of the Hidden Imam was an issue that had no precedent set during the lifetime of the Imams. Imamite jurists could not give a legal opinion based on a precedent set by the Imamate. As a result, they had to guide the community by issuing a legal opinion based on their independent reasoning (*ijtihād*). *Ijtihād* could be evoked in the absence of documentation in support of a given ruling from the communications transmitted on the authority of the Imams regarding the nature of Imamite political authority during the occultation.

It is precisely this question that I propose to exmaine in this chapter. How did the Imamite jurists cope with this new development in the Imamite theory of political authority? On the one hand, the Shī'ites during the Complete Occultation of the twelfth Imam continued to accommodate themselves to the reality of life under the illegitimate government of the 'Abbāsids; on the other, for over a hundred years, they found themselves living under the Imamite temporal authority of the Būyids (tenth–eleventh centuries). How was this apparently paradoxical situation to be resolved in light of the central tenet of the Imamites—the Imam's sole prerogative to rule over the Muslim community—and the new circumstance created by the political ascendancy of an Imamite dynasty for the first time since the concealment of the last Imam?

The Būyid period is an important interlude in Imamite history because of the ascendancy of Shī'ism and the patronage that was afforded Shī'ī scholars by the dynasty to write their works under relatively favorable conditions and even in a real sense of security. Solutions offered to the sensitive question regarding an Imamite Islamic government, headed by the Imam himself or by his specifically designated deputy, were carefully formulated during this period. They are found in the texts that deal basically with jurisprudence. Imamite scholars did not compose treatises similar to al-Māwardī's *al-Akhām al-sultāniyya* in order to deal with the new development in Imamite temporal authority. In fact, no such attempts were necessary in the light of the Complete Occultation of the Imam, which postponed the entire question of legitimate temporal authority to the end of time. Nevertheless, in the meanwhile, practical guidance for the members of the Imamite community living under these circumstances was deemed necessary.

This necessity prompted scholars to express their opinions regarding Imamite temporal authority in those sections of the applied *fiqh* that required the exercise of the Imam's authority (*wilāya*) in his absence. The options that were issued regarding the *wilāya* of the Imam and his deputy went through further elucidation and elaboration during subsequent periods of Imamite history when Shī'ī dynasties like those of the Safavids and the Qājārs came to power. But there were no deviations from any of the politico-theological tenets already formulated. In fact, once it was crystalized, the doctrine of the Imamate was left untouched throughout Imamite history. However, the welfare of individual Imamites very much required the application of reason—the theological underpinnings of the Imamite school—to analyze and resolve problems affecting the Shī'ī community. In this connection Imamite jurists provided ingenious leadership to the Imamite community by scrupulously guarding Imamite doctrines and allowing for interpretation of the implications of Shī'ī teachings in the applied *fiqh* so as to accommodate political contingencies that affected the everyday lives of believers. These implications obviously included the question of leadership of the Imamites during the absence of the Imam.

Imamite Political Authority during the Occultation

In the light of the main concern of the Imamite jurists to provide their followers with practical guidance relevant to their survival under "unjust" political authorities, it is important to state from the outset that none of the classical texts on the fundamental principles (*usūl al-dīn*) of the Imamite school, as examined for this study, deal directly with the possibility, not even as a fait accompli, of temporal Imamite authority during occultation. Such a discussion would necessarily have involved tampering with the terms of the doctrine of the Imamate, which (as stated earlier) was absolutely ruled out because of the absence of any directly designated deputy (*al-nā'ib al-khāss*) of the twelfth Imam. The "special deputyship" (*al-niyābat al-khāssa*), at least during the Short Occultation (A.D. 873–941), was seen as the ongoing guidance of the theological Imamate, adequately legitimized by the Imam's explicit deputization. With the occurrence of the Complete Occultation (from A.D. 941), the ongoing guidance of the community through the deputization of a specific person was terminated, so that there could not evolve concrete *imāma* of a "special deputy" through the process of *niyāba* (deputization) until the return of the Hidden Imam.

However, the question of the leadership of the Shī'ites in the absence of the Imam was a crucial one. A sense of urgency is reflected in Imamite jurisprudence whenever the question of exercising the Imam's authority without his specifically designated deputy comes up in the treatment of religious obligations requiring the presence of either the Imam or his appointed deputy for that purpose. It was under these circumstances that the Imamite jurists had to deal with the issue of the "general" deputyship of the Imam, which was vested in them as the custodians of Imamite teachings.

As for the Imamite community, it is plausible that the three religious experiences of the Imamites—*shahāda, ghayba,* and *taqiyya*—had more or less convinced the community not to anticipate the rise of the Imam to establish Islamic rule before the return of the messianic Imam at the end of time. The policy of political acquiescence was consistently pursued during the occultation, and Imamite leaders did not fail to reiterate the communications of the Imam al-Sādiq in this regard. They reminded the Shī'ites that, although it was true that the Qā'im among the Imams was capable of overthrowing unjust rule, he would appear only when God would command him to do so. This was doubtless judged to be a way to appease the impatience of the Shī'a at the delay in the fulfillment of the promise of justice. Consequently, the Shī'ites, emulating the example of their Imams, had to find ways to accommodate themselves to the enduring historical fact of living under the illegitimate government of the caliphs.

The Imams had on various occasions issued their opinions on how Shī'ites could act under these circumstances without incurring any blame for having obeyed or worked for an unjust ruler (*sultān jā'ir*). The dicta of the Imams on this subject were collected and compiled in works on jurisprudence, in sections dealing with *makāsib,* "earnings," where lawful and unlawful means of livelihood and gain are discussed.[5] The *makāsib* sections dealing with the subject of the legality of working for the government were written to provide practical guidance to Imamites who, during and after the time of the Imams, were known to have held offices in the Sunnī caliphal government. These sections were written with much care and prudence because of the political implications they carried in the matter of regulating the relationship of Shīʿites with a government they considered illegitimate.

In the Imamite jurisprudence, the chapters on *makāsib,* more particularly those treating the legality of working for the caliphal authority, seem to have been written under *taqiyya,* because it is in these chapters that the Imamite doctors were under pressure to include, however ambiguously, their opinions about Imamite political authority during the occultation. This pressure came to be felt more particularly when the Shīʿī *sultān*s of the Būyid dynasty came to power: before that period, the occultation of the last Imam did not alter the basic political situation, when most of the Imams were not in any case invested with temporal authority. Nevertheless, the Imamite jurists who addressed this new development were very conscious of the theoretical position of the Hidden Imam. Hence, on reading their works, one has to seek their opinion by implicit inference, not expecting to find any explicit dictum regarding accommodation with temporal authority. It is this difficulty in discovering their explicit opinion on the question of the legitimacy of Imamite temporal authority in the period of occultation that has led to erroneous opinions among Western scholars regarding the unrighteousness of *any* government pending the return of the twelfth Imam.

In order to elaborate on my thesis, I will begin by analyzing the twin concepts of *sultān ʿādil* (just ruler) and *sultān jāʾir* (unjust, tyrannical ruler) in Imamite jurisprudence.

Sultān ʿādil and *sultān jāʾir:* Just and Tyrannical Authority

The concept of *saltana*—exercise of legal and moral authority by demanding obedience—is directly connected with the fundamental question of *wilāya,* the faculty of that authority which enables a person to assume authority and exact obedience in practice. *Wilāya* in Islam is intrinsically related to the moral vision of Islamic revelation. Islamic revelation regards public life as an inevitable projection of personal response to the moral challenge of creating a divinely ordained public order on earth. Personal devotion to God implies the responsibility of furthering the realization of a just society, embodying all the manifestations of religious faith in the material as well as spiritual life of humankind.

This responsibility of striving for one's own welfare and that of the society in which one lives derives from the fact that, according to the Qurʾān, humankind has boldly assumed "the trust" that God had offered "unto the heavens and the earth and the hills, but they shrank from bearing it and were afraid of it. And man assumed it. Lo! he has proved a tyrant and a fool" (33:72). This trust, according to the major Imamite exegetes, means the divine sovereignty (*al-wilāyat al-ilāhiyya*), which God offered to all creatures.[6] Only human beings, having assumed the trust, have the potential to attain perfection and perfect their environment. Furthermore, only they are not afraid to bear the burden of this trust, and to accept the consequences of being a "tyrant" and "ignorant," because they only can acquire the opposite attributes—namely, those of being "just" (*ʿādil*) and "knowledgeable" (*ʿālim*). In fact, both "tyranny" and "ignorance" are the primary counterpoise of human responsibility in accepting *al-wilāyat al-ilāhiyya*, especially as it concerns God's providential purpose in allowing imperfect humanity to accept this responsibility. It is indeed through the acceptance of this *wilāya* that human beings acquire both the responsibility for their actions as well as superiority over all other creatures in the world, because *al-wilāyat al-ilāhiyya* enables them to put society into order in accordance with their unique comprehension of religion.

However, the *wilāya* is given to humankind with a clear warning that it will have to rise above "tyranny" and "ignorance" by heeding the call of divine guidance. Human beings, according to the Qur'ān, have been endowed with the cognition needed to further their comprehension of the purpose for which they are created, and the volition to realize it by using their knowledge. It is through divine guidance that human beings are expected to develop the ability to judge their actions and choose what will lead them to prosperity. But this is not an easy task. It involves spiritual and moral development, something that is most challenging in the face of the basic human weaknesses indicated by the Qur'ān:

> Surely man was created fretful,
> When evil visits him, impatient,
> When good visits him, grudging [70:19–20].

This weakness reveals a basic tension that must be resolved if human beings are to attain the purpose for which they are created. It is at this point that divine guidance is sent through the prophets and revealed messages to provide either the sources and principles or the basic norms of the social organization under which a divinely sanctioned public order is to be established. The Prophet thus becomes a representative of the divine authority on earth and exercises that authority in conformity with the divine plan for human conduct.[7]

It is in this Qur'ānic context that the Prophet's role as the head of a state and the founder of a religious order should be understood. The sense in which the Qur'ān speaks about the *wilāya* of the Prophet is necessarily in conformity with the Qur'ānic view of the divine guidance regulating the whole of human life, not just a limited segment of it. As a consequence, the *wilāya* of the Prophet meant not merely that the Muslim community be organized in the context of religious devotion to God as explained by the Prophet, but also that it acknowledge his political leadership as well. Thus, the *wilāya* of the Prophet establishes an authoritative precedent regarding the relationship between religion and political leadership in Islam.

It is on the basis of this concept of *wilāya* in the Qur'ān that one can say that in Islam religious and political authority are one and the same. This *wilāya* is concerned with the whole life of the Muslim community, with the result that it never relinquished its belief in the identity of religion and government as it saw them in the founder of Islam. The Prophet's emergence, the Muslims believed, had a fundamental purpose behind it: to transform the tribal structure of the Arab society at that time into a Muslim *umma*—a religio-socio-political community under the divinely planned *al-wilāyat al-ilāhiyya*.

The social transformation envisioned and initiated by the Prophet was the necessary consequence of this *wilāya*, which had to be acknowledged by society as a whole, not merely by individuals as a logical outcome of their faith in God. Acknowledgment of the *wilāya* of the Prophet, necessary to live a new life based on divine norms, led to the accentuation of a crucial requirement for the fulfillment of the social responsibility of the Muslim community—namely, that the community always needs to acknowledge a leader, divinely designated, who would exercise *wilāya* in order to unite its members in their purpose of creating a just social order under the guidance of Islamic revelation.[8]

In light of the above discussion it seems correct to maintain that leadership is of utmost significance in attaining the purpose of Islam, because it is only through divinely guided leadership that the creation of an ideal society could be realized. The question of divinely guided leadership in the fulfillment of divine planning, under the aegis of *al-wilāyat al-ilāhiyya,* thus, assumes a central position in the Islamic belief system or worldview, in which the Prophet, as the active representative of the transcendental God on earth, is visualized as possessing the divine *wilāya.* If the ultimate objective of Islam was conceived as the creation of an ideal community living under a fitting moral, legal, and social system of Islam on earth, then such an ideal, as advanced by the Qur'ān and shown by the example of the Prophet himself, was dependent on leadership that could assure its realization.

This fact was so important that, both during the Prophet's lifetime and immediately following his death in A.D. 632, the question of Islamic leadership became inextricably interwoven with the creation of an Islamic order. Islamic revelation unquestionably presupposed divine guidance through the divinely appointed mediatorship of the Prophet for the realization of Islamic public order. This mediatorship of the Prophet was the logical consequence of the strict monotheistic nature of Islam, which precluded the possibility that God assume human form, ruling directly over believers and governing their affairs. Thus, a ruler to represent God on earth and to exercise *al-wilāyat al-ilāhiyya* was deemed necessary in order to achieve the utlimate goal of Islam.

Moreover, in the light of the basic human weaknesses indicated in the Qur'ān, there had always been a basic tension between the purpose of creation and the obstacles to its achievement. This tension was to be resolved, according to the Qur'ān, by further acts of guidance through the Prophet, who became the "pattern of model behavior (*uswa*)" or, as indicated earlier in this study, the "paradigmatic precedent" for human beings, showing them how to change their character and bring it into conformity with the divine plan.

Studying the Qur'ān in its entirety, it becomes evident that the question of divine sovereignty—*al-wilāyat al-ilāhiyya*—is the integral element in the creation of the ideal society. It is through such a *wilāya* that the leader is able to provide a set of religious and moral laws and rules by which believers manage their affairs, and through which their public order is governed and should govern itself.

In the Shī'ī understanding of the Qur'ānic injunction in which the concept of *wilāya* occurs, the perspective sketched above on the leadership of the Muslim community assumes a central position. The relevance of the *wilāya* to the question of lawful and legitimate Islamic authority in Shī'ism can be deduced from those sections of Qur'ānic exegesis that deal with the passages on *wilāya.* The following verse of the Qur'ān is regarded by Shī'ite exegetes as the most important reference to the *wilāya:*

Only God is your *walī* [guardian] and His Apostle and those who believe, who perform prayer and pay alms while they bow [5:55].[9]

This passage establishes the "guardianship" of God, the Prophet, and "those who believe." The last phrase ("those who believe"), according to Shīʿī commentators, refers to the Imams whose *wilāya* is established through their designation by the Prophet.[10]

The term *al-walī*, as it occurs in the above context, has been interpreted diversely by Sunnī exegetes. Although there is a consensus among them that the verse was revealed in praise of ʿAlī's piety and devotion, the term *al-walī* has been interpreted as denoting the *muwālāt* ("befriending") of ʿAlī and not necessarily the acceptance of his *wilāya* (authority, in the form of *imāma*).[11] But Shīʿī exegetes have taken the term in another of its primary significations, *al-awlā* and *al-ahaqq* ("more entitled" [to exercise authority]), because *al-awlā* in ordinary usage is often applied to a person who can exercise authority (*al-sultān*) or who has discretionary power in the management of affairs (*al-mālik li al-ʾamr*).[12] Furthermore, *al-walī*, as it occurs in the above passage of the Qurʾān, is unlikely to mean a person invested with *wilāyat al-nusra* (the authority of "backing"), because there are numerous explicit references to that effect in other verses of the Qurʾān where believers are exhorted to back the religion of God by promulgating God's laws, a task in which the Prophet and the community of the believers assist each other.[13] Rather, *al-walī*, as applied to the Prophet, signifies a person who is invested with *wilāyat al-tasarruf*, which means possession of the authority that entitles the *walī* to act in whatever way he judges best, according to his own discretion, as a free agent in the management of the affairs of the community. The *wilāyat al-tasarruf* can be exercised only by one so designated by *al-walī al-mutlaq* (the Absolute Authority—i.e., God) or by one who is explicitly appointed by the Prophet in the position of *al-walī bi-al-niyāba* (authority through deputization). Consequently, the Imam who is designated by the *nass* as *walī* possesses the *wilāyat al-tasarruf* and is recognized as the ruler over the people.

This was the meaning of the term in the early usage of the Shīʿī Imams. In a speech to the Umayyad troops who had come to intercept him on his way to Iraq, al-Husayn b. ʿAlī (d. 61/680) explained to his adversaries the reason why he had refused to pay allegiance to the caliph Yazīd, son of Muʿāwiya:

> We the Family of the Prophet (*ahl al-bayt*) are more entitled (*awlā*) to [exercise]
> the authority (*wilāya*) over you than those [who have taken it for themselves,
> (i.e., the Umayyads)].[14]

Accordingly, *tasarruf* has been regarded as the primary and essential meaning of *wilāya*, especially as it is applied to God, the Prophet, and the Imams in the above passage. However, there exists a substantial differentiation in the way *wilāya* is apprehended in relation to God, the Absolute Authority (*al-walī al-mutlaq*), on the one hand, and the Prophet and the Imams, the authority through deputization (*al-wali bi-al-niyāba*), on the other. When the Qurʾān speaks about God's *wilāya*, it signifies *wilāyat al-takwīnī*—the unconditional *wilāya* "originating" in God, with absolute and all-encompassing authority and discretion over all that God has created. To this *wilāya* is sometimes appended *wilāyat al-nusra*, by means of which God helps believers. Thus the Qurʾān says: "God is the guardian (*walī*) of those who believe . . . unbelievers have no guardian" (47:11).

Moreover, the Qurʾān frequently speaks about God's *wilāya* in relation to

believers, by means of which God manages the affairs of believers—their guidance
to the right path and assistance to them in obeying God's commandments:

> God is the guardian (*walī*) of those who believe. He brings them out of darkness
> into the light [2:257].

But when *wilāya* is used in relation to the Prophet, it is designated as *al-wilāyat
al-iʿtibāriyya*—that is, "relative" authority—dependent upon God's appointing him
and deputizing him in that position (*istināba*); or *al-wilāyat al-tashrīʿiyya*, the
religious-legal authority granted in order for the Prophet to undertake the legislation and execution of the divine plan on earth. Thus, the Qur'ān declares: "The
Prophet has a greater claim (*awlā*) on the faithful then they have on themselves"
(33:6).[15]

The *wilāya* of the Prophet over believers is due to his being the Prophet of
God. As such, the point of reference for his *wilāya* is, in actuality, the *wilāya* of
God. It is for this reason that his *wilāya* is signified as "relative"—that is, accorded
as a mark of trust. In this sense, the Qur'ān speaks of only one kind of *wilāya*—
God's *wilāya*—which is the only fundamental *wilāya*. The *wilāya* of the Prophet and
"those who believe" (i.e., in this context, the Imams) is dependent upon God's will
and permission.[16] It is because this *wilāya* was vested in them that the Prophet and
the Imams had more right than other believers to exercise full authority, handing
down binding decisions on *all* matters pertaining to the welfare of the Muslim
community, and requiring complete obedience to themselves.

The corollary of this *wilāyat al-tasarruf* was the belief that not only was the
Imamate the continuation of the Prophethood, because of the authority vested in
the Imams after the Prophet; it also meant that the Imams were the sole legitimate
authority to lead the community in establishing just public order. The Imams became the *ʿādil* (just authority). In a case where the Imam was prevented from
assuming his rightful authority, the interfering power was rendered illegitimate, and
the ruler "unjust" (*al-jāʾir*) and "unrighteous" (*al-zālim*).

It is in this context of the above elucidation of the concept of *wilāya* that the
following *hadīth* report related by Kulaynī becomes comprehensible:

> Imam Jaʿfar al-Sādiq was asked by someone about the passage of the Qur'ān that
> mentions the Trust (*amāna*) which God offered to mankind. The Imam said:
> "This Trust is the *wilāya* of the *Amīr* of the faithful" [ʿAlī b. Abī Tālib].[17]

The Imam's statement makes it clear that it was the act of accepting or rejecting the *wilāya* of ʿAlī that determined whether one had been faithful to the divine
trust or not. The same act, moreover, determined the "righteousness" or "unrighteousness" of the ruling authority claiming to be legitimate. In Shīʿism, from its
inception, the Imams not only possessed the *wilāya* to establish political authority
on earth; they were also regarded as the *sole* legitimate authority who could and
would establish Islamic government. Imamite works treating of the theory of political authority, as it was systematized in the ninth/tenth centuries A.D., unanimously
maintained that legitimate government could not be established except by the one
who is *maʿsūm*—that is, the infallible Imam invested with the *wilāyat al-tasarruf* to
exercise discretionary control over the affairs of the community. Furthermore, it

was held that the process through which this authority becomes known to the public is explicit designation (*nass*) by the one possessing *al-wilāyat al-i'tibāriyya*—the "relative" authority derived through one's being divinely appointed to that office (e.g., Prophethood or Imamate).

In the case of the Shī'ī Imams, 'Alī b. Abī Tālib is known as *fātih al-wilāya* (the one with whom *wilāya* commenced among the Imams), through the Prophet's designation. From 'Alī the *wilāya* passed on to the succeeding Imams, each properly designated by the preceding Imam, who had to be acknowledged by the community through the process of *bay'a* ("paying allegiance"). Those who paid allegiance to the Imam were required to adhere loyally to the decisions made by him and obey him in all matters. The Imam who has been designated and acclaimed in this manner is regarded as *al-sultān al-'ādil* or *al-haqq* (the "just" or "legitimate" authority).

The characterization of the Imam as *al-sultān al-'ādil* implied that his authority was legitimately invested in him to exercise *wilāyat al-tasarruf* in order to establish the Islamic rule of justice. It is plausible that messianic expectation in the Shī'ī, and even in a non-Shī'ī Islamic context, should be seen in the light of the socio-political function attributed to the Prophet's successor—that is, in his role as *al-sultān al-'ādil*.[18] *Al-sultān al-'ādil*, to use the Shī'ī phrase to express the messianic vision, "would fill the world with justice as it is now filled with injustice." Accordingly, *al-sultān al-'ādil* had to be the one who was most pious and most knowledgeable of God's will in order to be invested as the Imam of the community.

Messianic expectation notwithstanding, the Imam as *al-sultān al-'ādil* was responsible for *tadbīr al-anām*—that is, managing the affairs of humanity. One of the fundamental functions of the *sultān* is to implement the Islamic ideology based on the principal duty of "enjoining good and prohibiting evil" in Muslim society. In fact, according to Tūsī, the just *sultān* is "the one who enjoins good and interdicts evil and acts with justice" (lit. places things in their proper places).[19] Inasmuch as the question of "enjoining the good and forbidding the evil" is directly connected to the well-being of human society, the necessity for *al-sultān al-'ādil* was regarded as being rationally deducible. As long as there remained a need for "enjoining the good and forbidding the evil," there remained a need for the fulfillment of the precondition, deductively derived, to this obligation—namely, the existence of a just government headed by *al-sultān al-'ādil*, who can take upon himself to implement such a duty, even by use of force.[20]

Accordingly, in the Shī'ī theory of political authority, power in the sense of authority, having moral and legal supremacy because of the *wilāyat al-tasarruf*, with the right of enforcing obedience to Islamic ideology, can never be invested in a person without proper *nass;* and no government can become legitimate (*al-hukūmat al-'ādila*) if it is not headed by *al-sultān al-'ādil*, who is explicitly appointed by a legitimate authority like the Prophet. If a government is established without *al-sultān al-'ādil* as its head, it is declared unjust and the ruler is a *sultān jā'ir*. Moreover, because the tyrannical ruler, lacking the necessary *wilāyat al-tasarruf*, has encroached upon the authority of the rightful person, he is also *al-zālim* (the oppressor).

Khulafā' al-jawr or *al-zalama* is the title applied to these rulers under whom, according to the Shī'ites, the world was filled with injustice. Disobedience to these *salātin al-jawr* was regarded as obedience to God. Thus, according to Mas'ūdī, there were pious Muslims, like ʿAwn b. ʿAbd Allāh b. Masʿūd during the Umayyad caliphate, who upheld the principle that anyone who opposes an unjust ruler was not devoid of divine guidance, but the unjust ruler was devoid of it.[21]

It is perhaps significant that the terms *mashrūʿiyya* (legitimacy) or *ghayr mashrūʿiyya* (illegitimacy), to denote these two types of government in Islam, do not appear in the major works of Imamite jurisprudence of the classical age. These terms, however, do appear in the works of the Imamites who wrote during the Qājār period, when the question of the legitimacy of a Shīʿī authority to exercise *wilāyat al-tasarruf* during the occultation of the twelfth Imam was being discussed. Thus, in his discussion about the necessity of government during the occultation, Muhammad Husayn Nāʾinī (d. 1936) uses the concept of "legitimate" (*mashrūʿ*) government in connection with the constitutional authority established by the approval of the righteous *fuqahāʾ* (jurists).[22]

Ambiguity of the Term *al-sultān al-ʿādil* as Used in the Writings of the Jurists

Major texts of Imamite jurisprudence have been composed during the occultation of the twelfth Imam, the only legitimate *sultān ʿādil*, according to Imamite political theory. There is a concurrence among all Imamite jurists that the phrase *al-sultān al-ʿādil* is applicable to the one who has been appropriately designated by God (*al-mansūs min qibal allāh*). He is the one appointed to rule (*li al-riʾāsa*) and he is the *sultān al-zamān* (the ruler of the age) whose *wilāya* is all-encompassing and obedience to whom is obligatory on religious as well as moral grounds.[23] However, on the basis of a well-documented *hadīth* report related on the authority of the Prophet— "The ruler (*al-sultān*) is the guardian (*walī*) of those who do not have a guardian"— it would seem that Imamite jurists have sometimes, even if only indirectly, pointed toward a *sultān ʿādil* who was not necessarily the Imam proper, designated by God.

In *Mabsūt* (his last work on jurisprudence, in which he ruled on the basis of his *ijtihād*), Tūsī discusses rulings regarding inheritance (*al-irth*). At one point in a section dealing with inheritance where there is no heir to claim the property, Tūsī says that this inheritance belongs to the Imam and should be handed over to him for distribution among those whom he chooses. However, even those (like the Sunnites) who hold that such an inheritance belongs to all Muslims, and accordingly should revert to the public treasury, maintain that it should be turned over to the just ruler—*imām ʿādil*—if there is one; otherwise it should be preserved in order to be handed over to *al-imām al-ʿādil* when he appears.[24] This ruling indicates that even though the divinely appointed Imam is the only *al-imām al-ʿādil*, during his occultation there is the possibility of the presence of a just imam who could be made responsible for the distribution of the inheritance in accordance with the rulings of the Sharīʿa.

In another place, on the question of authority to condemn a person to death, Tūsī clarifies the Imamite position by declaring that *al-imām al-maʿsūm*, because of his *ʿisma* (infallibility), is the only person who can do so because he does not order the killing of any person without a valid reason. On the other hand, according to the jurists of other schools of law, says Tūsī, on the basis of their doctrine about the leadership of someone who is not *ma'sūm,* they have permitted any authority to order the killing of a person, where deserved, even if there remains a possibility that such an authority may commit an error of judgment. As for the deputy (*khalīfa*) of the Imam, Tūsī rules that he is permitted to order the killing of a person although the deputy's position is similar to that of the imam among the Sunnites.[25]

The use of the phrase *khalīfat al-imām* is unusual in Imamite jurisprudence and even more unusual is Tūsī's likening of it to Sunnī political authority. Taking into consideration the historical circumstances under which Tūsī wrote his jurisprudence, it is possible to discern the use of *taqiyya* (precautionary dissimulation) when discussing issues pertaining to rightful political authority. For Tūsī, as for other Imamite jurists writing under the Shī'ite Būyids, *taqiyya* was the only way to avoid confrontation with the ruling house on the matter of who could be regarded as the *khalīfat al-imām* in the absence of the Imam of the age. Consequently, in Tūsī's usage relating to constitutional authority, there is inconsistency and also intentional ambiguity, which can be clarified only when interpreted within the general context provided by the entirety of Imamite jurisprudence written during this period. To this end, let me consider the way in which Mufīd, Tūsī's teacher, deals with the question of *sultān 'ādil* before I return to Tūsī's works.

Mufīd deals with the question of political authority indirectly in his jurisprudential works when he treats the question of authority that can undertake to execute certain legal requirements in Islamic polity. One of these requirements occurs in the use of force in implementing the obligation of "enjoining the good and forbidding the evil." Although I shall discuss this obligation in more detail in the next chapter, on the administration of justice, it is important to note here that the general obligation of "enjoining good and forbidding evil" constitutes the broadest and most all-embracing socio-political dimension of Muslim communal life.[26] Accordingly, all the duties that go toward the creation of the ideal Islamic community—such as the administration of justice, the execution of legal penalties, and so on—have always been regarded as stemming from this general obligation of "enjoining good and forbidding evil." In various ways this obligation is connected with the question of the leadership of the community, the leader being directly responsible for providing adequate means of fulfilling this obligation.

It is important to note that the question of "enjoining good and forbidding evil" has been discussed by the Imamites in the *doctrinal* section on the Imamate. The main function of the Imam, according to this theory of Islamic leadership, is to lead the community to establish a just public order. In order to attain this purpose, it was necessary to "enjoin good and forbid evil" with heart, tongue, and hand. Although fulfilling this obligation could be done with heart and tongue by any believer, use of force could be employed only by the Imams and the one perrmitted by them to administer it. In fact, some Imamite scholars, as we shall see, have contended that *'isma* (the divine protection that keeps the knowledge of the Imam free from error) is necessary in the matter of "enjoining" and "forbidding" by force, because use of force may lead to the spilling of blood. The use of force must guarantee that the desirable effect will be produced in society, and such a guarantee is available only in the judgment of *al-imām al-'ādil.* There seems to be consensus among Imamite jurists that any exercise of authority that may involve "harm" or "injury" to someone else in "enjoining the good and forbidding the evil" requires the existence of *al-imām al-'ādil,* who "places things in their proper places" (i.e., acts with justice); or, in his absence, there should be someone who has his "permission" to undertake the implementation of this duty. It is in this context that Mufīd discusses the authority that can "enjoin the good and forbid the evil" with the use of force.

In *Muqni'a,* Mufīd rules that it is incumbent upon believers to enjoin good and prohibit evil if it is possible, and provided it is expedient to do so. Thus, when it is possible to prevent an evil by using one's hand or tongue, then it should be done, provided that doers are free from fear, both for themselves and other Shī'ites. If there is any danger in fulfilling the obligation with one's hand, whether in the present or in the future, it should be limited to heart and tongue. Using one's hand—that is, the use of force—should not lead to killing or injuring. If it leads to the spilling of blood, then it should be undertaken only if this course is the only one left to eliminate evil and, more importantly, if the person doing it is reasonably sure that the desired effect will follow. However, use of force that may lead to killing or injuring must be undertaken only with the permission of the *sultān al-zamān,* appointed to manage the affairs of humanity (*al-mansūb li-tadbīr al-anām*). If permission cannot be obtained from the *sultān al-zamān,* then the spilling of blood or injuring a person should not be undertaken.

Mufīd then goes on to mention the obligation to administer *hudūd* (legal punishments), which too has implications for the socio-political dimension of Muslim communal life. The connection between administration of *hudūd* and "enjoining the good and forbidding the evil" in the interest of Islamic public order is very obvious; and hence the question of the use of force vis-à-vis offenses committed against community interests or public order was more critically raised with regard to offenses punishable by the death penalty. With respect to crimes deserving severe penalties, the matter of pain, harm, injury, or even loss of life to someone else through the administration of legal punishment was so serious that anyone who undertook to execute it had to be made responsible to the legitimate authority acknowledged by the Muslim community. It is in the context of this theologico-political consideration that Mufīd rules that it is only the *sultān al-islām,* the ruler of Islam, designated by God (*al-mansūb min qibal allāh*), the rightly guided Imam (or the one delegated by him for this purpose, such as the *umarā'* or the *hukkām*), who can implement legal punishments involving injury or loss of life. Indeed, says Mufīd, the Imams have delegated their authority in this matter to the jurists (*fuqahā'*) of their Shī'a, who should carefully consider how to fulfil it.

However, for a jurist who is appointed by an authority that has come to power by force (*al-mutaghallib*), the obligation of instituting legal punishments becomes personally (*bi al-'ayn*) incumbent only because of the ruler's manifest authority over him or because responsibility for the people has been vested in him by that authority. When a jurist is appointed by an "overpowering" authority, it is necessary for him to assume the responsibility of instituting legal punishment, to make the laws of Islam effective, to enjoin the good and forbid the evil, and to wage *jihād* against unbelievers and those who deserve to be fought against—the wicked. At the same time, it is necessary that the Shī'a assist him when called upon to do so by him, as long as what he requires of others does not in any way contradict the limits set by the faith or does not lead to disobedience to God by obedience to the one appointed by a tyrannical ruler, a "ruler in error" (*sultān al-dalāl*). If that jurist agrees with oppressors in anything that can be construed as disobedience to God, then it is not permissible for any of the Shī'a to assist him in such a matter. They can help him only in the performance of those tasks in his administration that lead to obedience to God, like the institution of legal punishment and carrying out the requirements of the Sharī'a.[27]

In this section of Mufīd's *Muqni'a*, the word *sultān* is used in two ways, corroborating the observation that there are two kinds of *sultān* in the *fiqh* writings of Mufīd. First, there is the *sultān al-zamān*, who manages the affairs of humanity (*li-tadbīr al-anām*). He can probably be understood as any ruler, not necessarily divinely appointed, during the occultation. In this category of *sultān* there is the possibility that an authority comes to power by force, and accordingly is considered tyrannical or lost in error (*al-dalāl*); but such an authority, though Mufīd does not explicitly state it, can be just because it upholds the norms and principles provided by the Sharī'a. This category of *sultān* will become evident in Tūsī's jurisprudence. Mufīd's ruling regarding the *sultān al-zamān* can include *hākim al-shar'*, the Imamite jurist, for Mufīd rules that a jurist, under a tyrannical ruler, should undertake not only to implement the *hudūd* but also to wage *jihād* against unbelievers and those who deserve to be fought against among the wicked. As we shall see below in our discussion on *jihād*, this undertaking of *jihād* is among the constitutional prerogatives of the rightful Imam when he is present.

The second kind of *sultān* in Mufīd's jurisprudence is the *sultān al-islām*, appointed by God (*al-mansūb min qibal allāh*), and such are the rightful Imams of the Shī'a. In this usage Mufīd has left no doubt that this "ruler of Islam" is none other than the twelfth Imam. It is the *sultān al-islām*, according to Mufīd, who has delegated his authority to the Shī'ī jurists through his general permission for them to undertake the functions that essentially require his own presence. In this sense, then, even when a Shī'ī jurist is appointed by a *sultān al-zamān* who might be unjust but who manages the affairs of the Muslim community, he is actually representing the *sultān al-islām*, on whose behalf he is executing the obligation of "enjoining the good and forbidding the evil" in various forms. This point is made even more explicit when Mufīd states that no jurist should accept a public office under an unjust authority unless he is confident that by doing so he would be furthering the cause of believers and protecting their rights. Mufīd states that this must be seen in the following light: even when he is apparently accepting the office from those who have deviated from the right way (*ahl al-dalāl*), he is actually representing the twelfth Imam (*sāhib al-'amr*), who has made it lawful for him, giving him permission during the occultation to undertake obligations that require the specific delegation of the Imam, not that of the ruler who has usurped authority by force (*al-mutaghallib*).[28]

This apparent ambiguity in the use of *sultān 'ādil* in the jurisprudence of the Baghdad school of Imamite *fiqh* has led some scholars either to regard all these judicial decisions as the consequence of "precautionary dissimulation" (*taqiyya*) of Shī'ite jurists who had to avoid confrontation with the de facto illegitimate authorities under whom they issued their opinions; or to attribute them to some self-aggrandizing motives of jurists who contrived to promote their own position as the *imām* of the Shī'a through various exegetical and terminological stratagems.[29] There is no doubt that "precautionary dissimulation" contributed in some measure to opinions that could also be interpreted as purporting the illegitimacy of *any* delegation of the Imam's constitutional authority to *anyone,* including Imamite jurists.

On the other hand, to attribute to self-aggrandizing motives opinions that concede the exercise of the Imam's constitutional authority to jurists is to deny the pragmatic value of decisions that reflect the situation of the Shī'a, decisions carefully scrutinized by rational evidence provided by the Qur'ānic injunction regarding the religious-moral obligation of "enjoining the good and forbidding the evil." If one were to venture into the interpretation of the suspected motives of a given jurist in maintaining a particular decision favorable to the position of the jurists, the opinions (sometimes in the works of a single scholar) are far too contradictory to allow these judicial decisions to be admitted as evidence of their interest in upholding the authority of a just ruler (*sultān 'ādil*) other than the Imam himself. It is even more difficult to ascertain that the jurists under "precautionary dissimulation" were engaged in either conferring or conceding authority to the de facto governments in their rulings based on doctrinal issues connected with the politico-theological implications of the concrete examples provided by the Imams who had lived under the de facto governments.

It is significant that jurists like Mufīd and those who followed him regarded the duty of "enjoining the good and forbidding the evil" as not only rationally and revelationally prescriptive in guiding the conscience of the Shī'a; they also considered *saltana* (power) as a requirement in the fulfillment of this crucial obligation in the public interest. Consequently, in Tūsī's judicial decision on this politically important obligation, the introduction of *sultān 'ādil* is not without implications for the real existence of a just authority other than that of the Imam.

In Tūsī's jurisprudential works, some of which were in the form of further elucidation, appended with meticulous documentation from the works of Mufīd, there is much room for interpretation of such phrases as *sultān al-waqt li al-ri'āsa* (the ruling authority of the time, to head the community) and *sultān al-zamān al-mansūb min qibal allāh* (the authority appointed by God). For instance, Tūsī defines *al-sultān al-'ādil* as the one who "enjoins the good and forbids the evil and acts with justice (lit. places things in their proper places)." This general definition is apparently applicable to *sultān al-waqt* or *sultān al-zamān,* who can fulfil the function reserved to the just ruler without actually detracting from the exercise of authority in the hands of a divinely appointed *sultān.*

This observation on Tūsī's usage of *al-sultān al-'ādil* is further corroborated by the fact that the definition occurs in a chapter of his *Nihāya* where he is discussing "the legality of working for the *sultān* and earning therefrom."[30] The *sultān* here referred to is evidently not the twelfth Imam; rather, he is a *sultān al-waqt* for whom a member of the Shī'a intends to work. Moreover, it is this *sultān* who can be classified as either just or unjust, depending on his commitment to uphold justice by "enjoining the good and forbidding the evil."

It is reasonable to propose that Tūsī employs the title *al-sultān al-'ādil* for both the infallible Imam and the ruling authority. Although it might seem that Tūsī's usage is inconsistent, he is clearly writing with prudential caution when he discusses ideas that pertain to the question of rightful authority during the absence of the twelfth Imam. Accordingly, his jurisprudence sometimes demonstrates ambiguity in using phrases like *al-sultān al-'ādil* and *sultān al-waqt,* which can be interchangeably applied to the twelfth Imam or the contemporary ruling authority, especially an Imamite authority like that of the Būyids. At other times, when such ambiguity was theologically impossible, especially in those cases where it was regarded unlawful to invest that authority in anyone except the Imamite Imam, a qualifying phrase, like "appointed by God," was added to *sultān al-zamān*. Thus, for instance, in the area of "enjoining the good and forbidding the evil" considered necessary for the well-being of Muslim society, *al-sultān al-'ādil* was used in such a way that anyone who could undertake the obligation could assume the title; whereas, in the area closely related to the detailed apprehension of jurisprudence—that is, knowledge of the will of God—*al-sultān al-'ādil* was necessarily the Imam himself or the one specifically designated by him for that purpose.[31] In this latter sense the Imamite jurists were also regarded as constituting *al-sultān al-'ādil*, who, according to the famous rescript of the twelfth Imam and other documentary evidence I shall examine in the next chapter, were designated as his general deputies and as such had his permission to manage the affairs of the Shī'ī community in his absence.

In later works of *fiqh* the application of *sultān 'ādil* for other than the divinely designated Imam became even more explicit as shown in 'Allāma Hillī's discussion on guardianship. In his *Tadhkira*, in the *Kitāb al-nikāh* (Book of marriage), 'Allāma discusses at length the question of guardianship (*wilāya*) in regard to a woman whose benefactor has died and who is to be given in marriage. In such a case, 'Allāma says, all the jurists (both Imamites and non-Imamites) have agreed that the *sultān* has the *wilāya* to give her in marriage or to keep her in an unmarried state. The *sultān*'s authority in this case is similar to the authority of a father. This is so because "the *sultān* is the guardian of those who do not have a guardian." Moreover, the jurists agreed that the *wilāya* of the *sultān* is of a general type because he receives taxes and protects those who do not have anyone to protect them.

Naturally, this *wilāya* cannot be exercised by an unjust *sultān*.[32] There is no doubt that the 'Allāma is referring to the *sultān* as a ruling authority in general, because in the same section he discusses the rulings of all the Sunnī schools of law in this matter and nowhere does he point out that the *sultān 'ādil* that he is referring to in this section is only the Imam in occultation. In his other work, *Tahrīr*, more particularly in the part called *Kitāb al-jihād*, when he speaks about the precondition for the existence of *al-imām al-'ādil* to wage offensive *jihād*, he clearly states that this Imam is the one appointed by God and invested with the authority to uphold the principles of religion and promulgate the divine laws. This latter observation leaves little doubt that 'Allāma Hillī differentiates between the divinely appointed *al-imām al-'ādil* and the *sultān 'ādil* who receives taxes and protects people.[33]

Al-sultān al-'ādil and the Question of *jihād*

The aim of Islamic ideology, as it emerges from the Qur'ān, is to "command good and forbid evil" (3:104, 110; 9:71). This would constitute a moral obligation taken to be binding upon and available to all, and for the disregarding of which all people may be held accountable.[34] Accordingly, the Qur'ān, under certain conditions, gives the state, when representing the society, the right to control "discord on earth," which means to control the breakdown of social and moral order because of general lawlessness that is created by "taking up arms against God and His messen-

ger" (9:33–34), that is, the established Islamic order. The Qur'ān's repeated demand to eradicate "corruption on earth," when taken together with the Qur'ānic
ideology of "enjoining the good and forbidding the evil," represents a basic moral
requirement to protect the well-being of a community. A logical outcome of this
Qur'ānic ideology is the question of *jihād* in Islamic religious law.

In light of the need for eradicating "corruption on earth" and of "commanding
good and forbidding evil," it is plausible to speak of a moral basis for *jihād* in Islam.
Indeed, the ordainment of *jihād,* according to the Muslim exegetes, occurred the
first time in Medina when the Muslims were given permission to fight back against
the "folk who broke their solemn pledges":

> Will ye not fight a folk who broke their solemn pledges, and purposed to drive
> out the Messenger and did attack you first? [9:13]

> If they withdraw not from you, and offer you not peace, and refrain not their
> hand, take them, and slay them wherever you come to them; against them We
> have given you a clear authority [4:91–93).

It is not difficult to adduce a strictly *moral* justification for the permission given
to Muslims to retaliate with force against attacks upon them. The Qur'ān, thus,
justifies defensive *jihād* by allowing Muslims to fight against and subdue hostile
unbelievers as dangerous and faithless, because they are inimical to the success of
God's cause. Furthermore, the Qur'ān requires Muslims to strive to establish just
public order overall. It is at this point that *jihād* becomes an offensive endeavor to
bring about the world order that the Qur'ān seeks.

Offensive *jihād* raises the question about the justification of *jihād* on moral
grounds only, because the Qur'ānic passages revealed in the latter part of the
Prophet's career in Medina require Muslims to wage *jihād* against unbelievers
"until there is no dissension and the religion is entirely God's" (8:39). The question
of offensive *jihād* is not a simple one. It is complicated by the wars of expansion that
were undertaken by Muslim armies up to the end of the Umayyad period (eighth
century A.D.). These wars are regarded as *jihād* by Sunnī Muslim scholars. However, upon careful scrutiny, these wars appear to be political, with the aim of
expansion of Islamic hegemony without the Qur'ānic goal of "religion being entirely
God's." Moreover, offensive *jihād* against "those who do not believe in God and
the Last Day and do not forbid what God and His messenger have forbidden and do
not practice the religion of truth, from among those who have been given the Book,
until they pay *jizya* [poll-tax]" (9:29) points more to the complex relationship and
interdependence of religious-moral considerations in the policy of Islamic public
order vis-à-vis the "People of the Book" than to their conversion to "God's religion," Islam.

There is no doubt that the Muslim jurists conceived *jihād* in the sense of
engaging in a war to increase the "sphere of Islam" (*dār al-islām*) as an integral part
of Islamic faith where Islamic social order is enforced, and where Islamic acts of
devotion are publicly observed.[35] However, the Qur'ānic ordainment of *jihād* in the
sense of fighting "until there is no dissension and the religion is entirely God's"
(8:39) appears to be concerned with the question of eradication of unbelief that
causes a breakdown of Islamic order. Accordingly, the sphere in which this *jihād*

was to be waged was designated the "sphere of war" (*dār al-harb*), with the essential aim of uprooting unbelief and preparing the way for the creation of Islamic order on earth. On the other hand, the *jihād* that the jurists treat in their works is undertaken to subdue the forces of unbelief rather than to convert individuals or even groups to Islam.

In the final analysis, the Sunnī jurists identified *jihād* in the direction of establishing a universal social and political order under a Muslim ruler. But the Qur'ān clearly points toward the establishment of a universal creed based on the affirmation of the oneness of God, the necessity of divinely guided leadership, and the ultimate day of judgment.

It is in the realm of this latter religious justification for *jihād* that the presence of the divinely ordained leader with effective power of constraint has been deemed necessary. It is therefore relevant to understand the problem of unbelief in the Qur'ān and the way it implicitly justifies the use of force against influences inimical to the establishing of divine order on earth.

According to the Qur'ān, humanity has not, by and large, submitted to the divine will, has not followed the guidance with which God has favored it. The Qur'ān indicates that failure to respond to the guidance necessary to attain the purpose for which human beings are created is tied to the narrow-mindedness and stupidity (self-cultivated, not innate) of human beings who do not reflect on the signs that come their way. Preoccupied with worldly affairs, they do not see that their life and all that they have is given by the beneficent and merciful Lord to whom they will have to render an account on the day of reckoning. They live for the pleasure of the moment, not considering that the pleasures of the next world (not to mention the punishments) will exceed their wildest imaginings.

On the other hand, the Qur'ān also upholds the lordship of the almighty, sovereign God. God has the power to make unbelievers into believers, and that is what would happen if God so chose (6:107). In other places the Qur'ān says that those who go astray have been led astray by God, whereas those who follow the "straight path" are able to do so because God helps them (2:6–7). These verses stand alongside others indicating that human beings have been afforded the necessary cognition and volition to further and to realize the purpose for which they are created (91:7–8). There are, moreover, passages that evidently resolve the apparent tension, indicating that in the case of those whom God has misled, God did so because they had already rejected God (59:19).

The relation of these Qur'ānic treatments to the problem of unbelief, or failure to respond to divine guidance, was the object of intense discussion among Muslim theologians following the establishment of Islamic empire. The main issue that developed in this discussion was the question of "misguidance" (*idlāl*). It is important to note that the Qur'ān considers "misguidance" or "leading astray" as God's activity in response to unsatisfactory actions or attitudes on the part of individuals who have chosen to reject faith. As such, they deserve it:

> How shall God guide a people who have disbelieved after they believed. . . .
> God guides not the evildoing people [3:86].

It would not be expedient to take up these discussions at this point. Suffice it to say that the problem of human persistence in unbelief and ingratitude is a major concern, even a preoccupation, for the Qur'ān. Why does a human being not respond to divine guidance? Unbelief became for the Qur'ān and the Prophet not only a denial of truth, to be punished on the day of judgment, but also a threat to the community of the faithful, to be subdued in the present, by the use of force, if necessary. The picture that emerges in the Qur'ān points to a growing need on the part of the community to engage in armed resistance to the threat posed by those who do not share its faith and the socio-political implications of its faith.

As stated above, in my brief remarks about the early ordainment of *jihād* in the Qur'ān, the need to engage in *jihād* became evident when the early Muslims established the first Islamic order in Medina. The ordainment of *jihād,* then, served to resolve the problem of unbelief on the part of humanity in its religious and political dimensions. The opposition of the Meccan tribes was increasingly seen by the Qur'ān as a religious and moral problem that called for political action. As a religious problem, however, unbelief was beyond the jurisdiction of the Prophet and the Muslim community, because belief and its opposite were construed as the work of God. At the same time, unbelief can be malicious—wilful action on the part of human beings who seek to deceive God, or to deprive God of divine rights.

It is at this level that unbelief becomes a problem with moral and religious dimensions. The Qur'ān indicates that various kinds of action are appropriate on the part of the Prophet and the Muslim community to combat this situation. The significant fact to observe here is that the more the moral aspects of the problem are stressed, the more use of force as a form of political action is justified. When unbelief takes on actively hostile aspects in relation to the community of the faithful, the Qur'ān justifies the use of force, even commands it:

> Fight in the way of God against those who fight against you, but begin not hostilities. Lo! God loveth not aggressors. And slay them wherever ye find them, and drive them out of the places whence they drove you out, for persecution is worse than slaughter [2:190–91].

When unbelief among the People of the book takes the form of disregard for the moral standards prescribed by the Islamic public order, the Qur'ān justifies *jihād* until the forces inimical to this order are subdued and are forced to pay the *jizya* (9:29). Reluctance to fight, in 2:190–91, which may be understood in terms of the priority of the rule against killing, is overcome by the security needs of a persecuted and outnumbered community. Consequently, against the aggressive hostility of unbelievers seeking to harm the community, whether by engaging in active hostility or refusing to comply with the norms of Islamic polity, Muslims are commanded to fight.

The command is given in such a way that it overrides certain prohibitions about times and places of fighting as affecting religious duties. Following the justification of warfare in cases of persecution, the Qur'ān commands:

> Fight not with them at the Inviolable Place of Worship until they first attack you there; but if they attack you [there] then slay them. Such is the reward of disbelievers [2:191].

Similarly, in verse 194:

The forbidden month for the forbidden month, and forbidden things in retaliation. And one who attacketh you attack him in like manner as he attacked you. Observe your duty to God, and know that God is with those who ward off evil.

The above passages show that the security needs of the Muslim community, and the demands of justice in 2:194, can override even certain regulations of a religious nature concerning the "Inviolable Place of Worship" or "forbidden months." When unbelief threatens the existence of the faith, such regulations are overridden. One does whatever is necessary to stop aggression—even customary rules of warfare may be suspended at such a time. Thus, the just community, which, according to the Qur'ān, is steadfast in its service to God and receptive of God's guidance, may engage in *jihād* to defend itself and strive to establish justice, or perhaps, to reestablish a violated justice, in the face of aggressive unbelief.

If *jihād* is comprehended within the consistent and coherent notion of the human responsibility to strive to make God's cause succeed (9:41), then sanctioning the use of force against moral and political offenses cannot be regarded as contradicting the Qur'ānic notion of human volition in the matter of faith so explicitly mentioned in the verse: "No compulsion is there in religion" (2:256). The Qur'ān sanctions *jihād* to establish the order that protects the basic welfare of the community, against both "internal" as well as "external" enemies—that is, the "tyrants" (those who take up arms against God and God's messenger [9:33]), and the "unbelievers" (those who break their oaths [9:12]), and those who "do not forbid what God and His messenger have forbidden (9:29) and thereby obstruct the struggle to make "God's cause succeed."

So construed, to advocate the use of force by an Islamic authority in the name of *jihād* would require rigorous demonstration that the purpose of *jihād* is nothing but "enjoining the good and forbidding the evil," the obligation that clearly appeals to the basic moral requirements for the well-being of humanity. In other words, Muslim authority, whether political or juridical, like *al-sultān al-ʿādil*, must shoulder the burden of proof and establish that the *jihād* was not undertaken primarily for territorial expansion, but in order to usher in the ethico-social world order that the Qur'ān requires.

As indicated earlier, Sunnī Muslim jurists regarded *jihād* essentially in the sense of expansion of the Islamic state conceived as the sphere where the Islamic norms prescribed in the Sharīʿa were paramount. This conception of *jihād* was scrutinized by the Shīʿī jurists in the light of their Imams' statements that did not regard the wars of expansion as being motivated by the Qur'ānic injunction. It is this scrutiny of the purpose of *jihād* that has given rise to the question of the authority that can declare *jihād* in Imamite jurisprudence.

That the Shīʿī Imams did not regard the various *jihād*s undertaken by the caliphs as motivated by the Qur'ānic demand to strive to make God's cause succeed is the point of a narrative in which ʿAbbād al-Basrī met ʿAlī b. al-Husayn, the fourth Imamite Imam, who was on his way to perform the pilgrimage to Mecca. ʿAbbād tauntingly reproached the Imam, saying:

O ʿAlī b. al-Husayn, you have abandoned *jihād* and its hardships and have accepted the *hajj* and its comforts; whereas God says [in the Qur'ān], 'God has bought from believers their selves and their possessions against the gift of Paradise; they fight in the way of God; they kill, and are killed; that is a promise binding upon God in the Torah, and the Gospel, and the Qur'ān' [9:111]." [To this 'Alī b. al-Husayn replied:] "If we had seen those who are thus described [in the Qur'ān as engaged in *jihād*], then *jihād* with them is superior to *hajj*."[36]

The original purpose of *jihād*, then, according to the Imamites, was not pre-served under the caliphate. What had caused the *jihād* to drift away from the Qur'ānic purpose was the coming to power of unjust and unrighteous authority claiming to undertake *jihād* in the name of God. Of the two main purposes of *jihād*—namely, to call upon the people to respond to God's guidance, and to protect the basic welfare of the community—the first purpose, according to all the Imamite jurists, required the presence of *al-imām al-'ādil* or the person deputized by such an authority. This was to guarantee that *jihād* against unbelievers was undertaken strictly for the cause of God.

Tūsī in his *Mabsūt* declares that "it is imperative that the Imam should be the one to commence *jihād* against unbelievers (*kuffār*)."[37] This is in line with the Imamite doctrine that at every stage in the struggle against unbelievers the follow-ing condition must be met: the leadership of the infallible Imam; or his permission in the form of his own presence or participation; or his designation of his deputy. The Imam, according to this doctrine, as the divinely appointed authority, is en-dowed with knowledge of the will of God and, consequently, he, unlike the unjust ruler, can guarantee that the *jihād* will not lead to the nullification of the cause for which it is undertaken. Moreover, his infallibility (*'isma*) protects him from destroy-ing or commanding to destroy any life without proper justification.[38]

Imamite jurisprudence consistently underscores the point that the call to *jihād* can be issued only by an individual who is most learned in the purpose and the aim of Islamic revelation, a requirement that is affirmed as a requisite in the exercise of authority (*saltana* and *wilāya*) in the Imamate; and such an individual is the Imam or his deputy who possesses "sound belief" (*imān sahīh*) and "sound knowledge" (*'ilm sahīh*).

This point was made by Imam al-Sādiq at a very crucial point in the history of Shī'ī revolts in the eighth century. Al-Sādiq's cousin, Muhammad al-Nafs al-Zakiyya (the Pure Soul), was expected to rise up as the Mahdī of the community and establish a new order in the last decade of the Umayyad rule. A group of persons from Basra, having regarded al-Nafs al-Zakiyya as being the most qualified candidate for the caliphate, especially following the political turmoil caused by the deposition and assassination of Walīd II, the Umayyad, in 125/743, had initially paid allegiance to him but had subsequently withdrawn their support of him. This group from Basra included the early Mu'tazilite teachers 'Amr b. 'Ubayd and Wāsil b. 'Atā'. Although it is not clear from the sources when this particular group withdrew its allegiance, it was probably at the time when al-Nafs al-Zakiyya failed to convince the people that his revolt in Hijaz was a serious attempt to launch an alternate public order. According to Abū al-Faraj al-Isfahānī,[39] Wāsil b. 'Atā', 'Amr b. 'Ubayd, and other Mu'tazilites of Basra had come to meet with al-Nafs al-Zakiyya in Medina. Having met him, they paid allegiance to him and returned to Basra. This must have been on their earlier visit, before the revolt in 145/762.

Tabarsī, another Imamite scholar, mentions a different visit of the same group of persons from Basra when they met Ja'far al-Sādiq in Mecca. With him they discussed al-Nafs al-Zakiyya's revolt, explaining why they had considered him to be a rightful candidate for the caliphate before the difference of opinion had arisen among al-Nafs al-Zakiyya's followers that led to a rebellion against him. They went to al-Sādiq, seeking to persuade him to assume the leadership.

As part of a long discussion that took place between the Imam and the Mu'tazilites of Basra who had initially supported al-Nafs al-Zakiyya, al-Sādiq cited his father's statement regarding a misguided and false leader who calls upon the people to join him in taking up the sword in spite of the existence of someone more knowledgeable than he regarding the goals of Islam:

> If a persons strikes people with his sword and calls them to himself, and if there is someone among Muslims who is more knowledgeable [about the Will of God] than he, then he is certainly misguided and false.[40]

The implication of al-Sādiq's statement is in conformity with the Imamite view of *jihād,* that only the Imam has the necessary divine grace to avoid any error of judgment in endangering the lives of people and the goals of Islamic revelation. The second purpose of *jihād*—protection of the basic welfare of the community against those who have threatened it—makes *jihād* a defensive measure, and the Imamite jurists have ruled its needfulness unrestrictedly.[41] In the section on *jihād* in *Mabsūt,* Tūsī argues for the necessity of the presence of *al-imām al-'ādil* or his deputy for the *jihād* to become obligatory. But when the Imam is in occultation and his specially designated deputy is absent, the obligatoriness lapses and it is not proper to engage in *jihād* at all. However, when Muslims are unexpectedly attacked by enemies and they fear for the safety of the boundaries and peoples of Islam, at such a time (Tūsī rules) it becomes urgently necessary for them to defend themselves against those who threaten their security. This defense, says Tūsī, should be undertaken with the intention of repelling the enemy, not with the intention of *jihād*—to convert them to Islam—because the latter purpose of *jihād* requires the presence of the Imam or his deputy appointed for that purpose.[42] It is a pernicious error to engage in *jihād* under the leadership of unjust rulers (*a'immat al-jawr*) for the purpose of conversion. Anyone who participates in this type of *jihād* deserves censure and punishment. In fact, Tūsī says, if such a battle is won, the winners do not gain any reward; and if it is lost, the losers are sinners as well. Moreover, in a battle that is fought without the Imam or his deputy, all booty belongs to the Imam alone; no one else has a share in it at all.[43]

It is plausible that the Imamite jurists at all times regarded the *jihād,* in the sense of making God's cause succeed (9:41), possible and obligatory only when *al-imām al-'ādil,* or the one appointed by him for that purpose, summoned the people to take arms. In fact, according to Muhaqqiq al-Hillī, this is the only kind of struggle that deserves to be called *jihād.* This is so because, when a battle becomes necessary out of the need for defense, as happens among enemies when attacked by an external force with Muslims fighting alongside the *ahl al-harb* ("the enemy") to protect themselves, this is not *jihād,* because the Muslim fighters have helped the enemy in order to defend themselves.[44]

In later works of jurisprudence, when the *jihād* was classified into more than the two categories so far seen, the *jihād* against unbelievers was the only one classified as *ibtidā'ī* (initiatory, i.e. offensive), and the only one that became obligatory when summoned by the Imam or his deputy. Thus, ostensible *jihād* against those who attacked Muslims (unbelievers); ostensible *jihād* against those who want to kill and to drive people away from their homes; and ostensible *jihād* against those who rise up against the Imam (the *bughāt*)—all these were known as "defense" rather than strict *jihād* to further the cause of God against unbelief among the human race.[45]

Accordingly, the question of obtaining the permission of the Imam in these cases did not arise, because defense was a moral requirement as much as a religiously justified *jihād,* as the relevant passages of the Qur'ān indicate. It is plausible that in Imamite jurisprudence the term *jihād* was used in two significations with their corresponding theological implications.

 First, *jihād* is used in the sense of furthering the cause of God until it succeeds. This type of *jihād* would include fighting (*qitāl*) against hostile forces, even against dissenting groups within the Muslim community, who are engaged in spreading "discord on earth" and in undermining the creation of the just public order that Islamic revelation demands. In an important *hadīth* report Imam 'Alī explains this meaning of *jihād* and declares that two groups of persons with whom *jihād* is obligatory are the unjust group (*al-fi'at al-bāghiya*), and unbelievers (*al-fi'at al-kāfira*).[46] It is in this signification of *jihād* that Shī'ī jurists have characterized the *jihād* fought by 'Alī during his caliphate (36–40/656–661) against those among the Muslim community who "broke their covenants" (*al-nākithūn*) in disobeying the rightful caliph; those who "deviated from divine guidance" by raising arms against the righteous government (*al-qāsitūn*); and those who became "renegades by seceding" (*al-māriqūn*) from the caliph's camp. Indeed, according to Mufīd, when 'Alī appealed to the Kūfans to join him in his struggle against those who broke their covenants, he reminded them of the significance of this *jihād* by declaring:

> Verily, God has imposed the obligation of waging *jihād* and has glorified it and made [participation in it a sign of] support for Him. I solemnly declare that worldly and religious affairs will not be in order without [participating in] it.[47]

Clearly, the theological implication of this type of *jihād* is that participation in it is obligatory, and to ignore the call of *al-imām al-'ādil* in this regard is tantamount to disbelief in God's ultimate purpose in creation. In a tradition reported on the authority of the ninth Imam, Muhammad al-Jawād, the Imam states: "I do not know of any *jihād* these days, except *hajj, 'umra,* and *jiwār.*"[48] This tradition further corroborates the way the Imamites have interpreted *jihād* in this first signification.

Second, the term *jihād* is applied to all types of defense undertaken to protect the lives and property of Muslims threatened by either external or internal hostile forces. In this type of *jihād* there seems to be a presupposition regarding the morally and rationally derived obligation of self-protection demanded by the Qur'ānic injunction. Accordingly, the Imamite jurists have unrestrictedly allowed Muslims to undertake this form of *jihād.* In other words, when the community is in danger, it has the responsibility to defend Muslim families, children, and property.

The theological implication of this form of *jihād* becomes obvious when one considers the ruling of Tūsī in his *Kitāb al-jihād* section of *Nihāya:*

The presence of *al-imām al-ʿādil* is a precondition [for waging *jihād* against unbelievers]. Without his permission it is not allowed to engage in fighting; nor is it permissible to undertake *jihād* without his presence or the presence of the one appointed by him to conduct the affairs of the Muslims. It is only when these preconditions are fulfilled that it is incumbent upon Muslims to fight when summoned to do so. . . . However, Muslims can fight without *al-imām al-ʿādil*, under the leadership of *al-imām al-jāʾir*, if they fear for their lives and their borders are being attacked by enemies. At such a time, it is obligatory for them to fight. But they should do so with the intention of defending themselves against the enemies and not waging the *jihād;* nor should they fight with the intention of converting others to Islam.[49]

In my discussion above on the ambiguous usage of *sultān ʿādil* and *sultān jāʾir* in the writings of the Imamite jurists, I mentioned Mufīd's classification into *sultān al-zamān,* the authority who manages the affairs of humanity, regardless of his being a just or unjust authority, and the *sultān al-islām,* the ruler of Islam, designated by God, who is necessarily just. It is reasonable to apply Mufīd's categories of *sultān al-islām* and *sultān al-zamān* to the two forms of *jihād* discussed above. The first form of *jihād* must be commenced by the *sultān al-islām,* but the second form can be undertaken by the *sultān al-zamān.* This is corroborated by Mufīd's ruling regarding an Imāmī jurist who is appointed by a *sultān al-zamān* to undertake the responsibility of instituting legal punishment (*hudūd*). Mufīd states that it is necessary for the jurist to assume this responsibility because of the ruler's manifest authority over him; and in addition, he should strive to cause the laws of Islam to be effective, to enjoin the good and forbid the evil, and to wage the *jihād* against those who deserve to be attacked among the wicked. It was in the light of this understanding of the *jihād* and its politico-theological implications that Imamite jurists like al-Shaykh Jaʿfar al-Kabīr Kāshif al-Ghiṭāʾ (d. 1228/1813), the author of *Kashf al-ghiṭāʾ,* during the Qājār period, were able to give their legal decision regarding the necessity of *jihād* against the Russians in the early years of the nineteenth century.

A. K. S. Lambton's article, "A Nineteenth-Century View of *Jihād*,"[50] fails to take into consideration the Imamite interpretation of the *jihād* in the Būyid period, which, she believes, did not have any impact on the juristic decisions made by later Imamites in the area of Imamite political theory. The reason for this situation, according to Lambton, was that the Būyids, having continued to tolerate the existence of the Sunnī caliphate in the face of the largely Sunnī population of the empire, had not forced or required the Shīʿī jurists to rethink the Imamite theory of temporal leadership during the occultation. It is true that the establishment of the Būyid dynasty had not called for a revision of the Imamite theory of political authority in order to legitimize or rationalize the new situation, but it had certainly caused a major rethinking in the matter of *saltana* (exercise of legal and moral authority) by an Imamite, which had no precedents during the lifetime of the Imams. This is reflected, as I have shown earlier in this chapter, in the development of the twin concepts of *sultān ʿādil* and *sultān jāʾir* in Imāmī jurisprudence.

As for the relationship between *jihād* and *saltana,* both the article by Lambton and that by Etan Kohlberg, "The Development of Imāmī Shīʿī Doctrine of *Jihād*"[51] (which in some important ways fills out the historical development of the concept of *jihād*), treat the subject of *jihād* without discussing its theological underpinnings within the Qur'ānic religious-moral framework. It is the Qur'ānic understanding of the question of *jihād* that points to the twofold signification of the term, as it was understood by Imamite jurists and as it was reflected in their opinion about the authority that could summon the people for one or the other form of *jihād*.

In the early part of his article Kohlberg acknowledges the close connection between *jihād* and justice in Shī'ī texts, without expounding on this significant observation, which should have led him to explore the question of the purpose of Islamic revelation (namely, the establishment of a just public order) and the forces that obstruct the realization of the divine plan for humanity. Such an examination is indispensable for the elucidation of the theologico-political concern of the Imamite jurists, who were, on the one hand, engaged in explaining the historical battles fought by the Imams 'Alī, al-Hasan, and al-Husayn in Karbalā' within the Qur'ānic teachings about the religious-moral purport of *jihād;* and on the other, in reformulating, however ambiguously, the question of "just" and "unjust" authority in the wake of the establishment of Twelver Shī'ī temporal authority of the Būyid *sultān*s. Thus, when Mufīd uses the term *al-mutaghallib* for the authority that has come to power by force and appoints an Imamite jurist to administer *hudūd*, one can discern a probable reference to the Būyids who had come to power in this fashion.

It is reasonable to maintain that the Imamite jurists displayed an astute and prudent leadership in understanding the new reality created by the occultation of the twelfth Imam and guiding the Shī'a toward a *taqiyya*-oriented life. This attitude in no way deterred them from formulating theoretical and hypothetical situations in juridical texts, which at times pointed to concrete situations, in providing guidance to the Imamite community. This is clearly demonstrated by Mufīd and Tūsī's use of the titles *sultān al-zamān* and *al-sultān al-'ādil*, and their guidance regarding *jihād* under these authorities. There is no doubt that by distinguishing the two types of *sultān* and the two types of *jihād* the Imamite jurists, from the Būyid period on, set up the possibility for Shī'ī jurists not only to accept office under a *sultān al-zamān*, but also to become a *sultān* in order to wage *jihād* against wicked persons obstructing justice.

There is a tradition, reported on the authority of al-Sādiq, that the Prophet on one occasion declared: "The best form of *jihād* is to utter just words (*kalimat 'adl*) in the presence of a tyrant ruler (*imām jā'ir*)."[52] As pointed out in chapter 1 above, this has been used to bolster the sense of responsibility, among those who are equipped with "sound knowledge" and "sound character," for taking upon themselves to wage *jihād* against tyrants. This understanding of *jihād* must be construed as a further elaboration of the concept in the sense of protection and defense against outside aggressors. In the light of the above tradition about *jihād*, the tradition about *taqiyya* being a form of *jihād* during the occultation must be interpreted in the sense of postponement of the creation of the just order, not its permanent suspension. This interpretation is corroborated by another famous tradition of al-Sādiq in which he mentions the characteristics of a true believer:

> The believer is a *mujāhid*, because he struggles (*yujāhidu*) with the enemies of God, under the false government (*dawlat al-bātil*) with *taqiyya* [precautionary dissimulation], and under the true government with sword (*al-sayf*).[53]

Thus, in *taqiyya* there is a state of alertness and preparedness to launch the final *jihād* under the aegis of *al-sultān al-ʿādil*. This is the connotation of *dār al-taqiyya*, the "abode of precautionary dissimulation," in the following *hadīth* report in which Imam ʿAlī al-Riḍā is reported to have written to the ʿAbbāsid al-Maʾmūn, explaining the *jihād* to him:

> And as for *jihād*, it is obligatory with a just Imam (*imām ʿādil*). . . . It is not permissible to kill any unbeliever in *dār al-taqiyya* except if he kills or behaves unjustly; and this also [is permissible] if you are not afraid of your own safety. *Taqiyya* in *dār al-taqiyya* is obligatory. One should not break his oath in order to defend himself from unjust behavior if he has sworn to employ *taqiyya*.[54]

The use of the phrase *dār al-taqiyya* implies that the Shīʿites must remain alert at all times for their struggle. Moreover, *taqiyya* was ordered by the Imams, as reported on the authority of al-Ṣādiq, to save Shīʿite lives from being lost. But if *taqiyya* might result in the shedding of innocent blood, then the Shīʿites were not required to employ precautionary measures to save themselves, because, as the report says, "There is no *taqiyya* when it comes to blood-spilling."[55] Indeed, there is an underlying moral concern in the *taqiyya* traditions to protect the innocent lives of the followers of the Imams from persecution by the ruling authorities. Accordingly, *taqiyya* was part of *jihād* in the second signification of protecting oneself from the arbitrariness of those in power.

This meaning of *taqiyya* as an obligatory form of defensive *jihād* during the occultation is in conformity with the Imamite juridical decision permitting the *jihād* that must be waged by any *sultān al-zamān* (ruling authority), unrestrictedly, to protect the lives and properties of Muslims. Thus, in his *fatwā* on *jihād*, after explaining four kinds of *jihād* (*jihād* against nonbelievers; against those who attack Muslims; against those who want to destroy lives they are forbidden to destroy; and against those who rise up against the Imam), Shahīd II says:

> *Jihād* becomes obligatory when *al-imām al-ʿādil* or his deputy are present. This deputy is designated particularly by the Imam for the purpose of *jihād*, or sometimes he is appointed as his general [deputy]. In the case of his general deputy, who can be a *faqīh* (jurist), it is not permissible for him to undertake the *jihād* in order to call upon nonbelievers to convert to Islam [i.e., offensive *jihād*] during the occultation. As for other forms of *jihād* [i.e., defensive ones], it is not necessary for the Muslims to engage in these under the aegis of the Imam, or his "special" or "general" deputy.[56]

I have already referred to this exception in the ruling on defensive *jihād* as coming down from the early days of Imamite jurisprudence. That this exception was made on the basis of the Qurʾānic injunctions regarding the ordainment of *jihād* is further corroborated by Sabzawārī (d. 1090/1679), a prominent jurist during the Safavid period. In the following quotation from his *Kifāyat al-ahkām*, it appears that the jurist is disputing the fundamental precondition for any *jihād*—namely, the existence of the Imam—as ostensibly required by the Qurʾān:

In the ruling regarding the incumbency of *jihād*, a precondition has been set that the Imam must be present, or the one designated by him should exist. This is a widespread opinion among the Imamite jurists, and it appears that the *hadīth* reports cited to support this opinion have not attained the level of "soundness," all the more so when they are in contradiction with the general sense derived from the verses of the Qurʾān [on this subject]. Thus, there is a problem of incoherence in the ruling [regarding the precondition] as such.[57]

It is not difficult to discern the moral-religious presuppositions of the Qurʾān in the case of the exception made for defensive *jihād*, and probably Sabzawārī's observation points to the *jihād* undertaken against those who want to destroy innocent lives, for which the Qurʾān ordains without any precondition. Indeed, it is for this form of *jihād* that the community is made responsible to judge the situation and act accordingly.[58] It was in the context of this moral-religious estimation of *jihād* that the Imamite jurists, as the "general" deputies of the Imams, were authorized to act as functional imams and guide the community as the *sultān al-zamān* would have done. The Imamite jurists were fully aware of the implications of their assuming the responsibilities of *sultān al-zamān* in the absence of the *sultān al-islām*, including the implications of the defensive form of *jihād* that could be undertaken by the ruling authority.

On the basis of our sources, it is plausible that there was nothing in Imamite jurisprudence that could prevent jurists from assuming these responsibilities, all the more so when they were acknowledged as the "general" deputy of the *sultān al-islām* because of their "sound belief" and "sound knowledge." And in the theory of the Imamate, as discussed in chapter 1, the emphasis on continuous, uninterrupted, and authoritative divine guidance through the divinely appointed Imam necessitated belief in qualified, Imamlike individuals, designated by the Imams themselves, whether directly or indirectly, who could provide the guidance needed by the Imamite community when the Imam himself was inaccessible.

Defensive *jihād* was regarded as part of the function of this designated authority, which is discussed by Shahīd II in his commentary on Muhaqqiq's ruling on *jihād* in *Sharāʾiʿ*. Muhaqqiq maintains the standard Imamite view regarding the requirement of the existence of the Imam, who should be manifest, invested with power, and able to act without constraint, to declare *jihād*, or to affirm the existence of the deputy appointed for *jihād*. Shahīd II, commenting on the deputy appointed for *jihād*, says that the appointment can be realized through either "special" or "general" designation of the Imam, which would include the *wilāya* (authority) to declare *jihād*. Agreeing with Muhaqqiq that the *faqīh* during the occultation is the "general" deputy of the Imam, Shahīd II adds that although appointed for the general best interests (*al-masālih al-ʿāmma*) of the community, the Imamite jurist is not allowed to undertake *jihād* in its first sense—that is, *jihād* to call non-Muslims to accept Islam.[59]

In the above discussion of *jihād* and its relationship to the *saltana* (exercise of legal and moral authority) there is a tacit confirmation of two points. First, there could be a *sultān al-zamān* (a ruling authority), whether "just" or "unjust," when the *sultān al-islām* (i.e., the twelfth Imam) is in occultation. This means that during the occultation the affairs of the community continue to need management, and this in turn requires investiture of authority (*wilāya*) in a *sultān al-zamān* for that purpose (*li-tadbīr al-anām*). There is no prescribed manner in which this authority should be invested in the *sulṭān al-zamān*. Nevertheless, there is certainly a way in which this authority can *become* just, and that is by upholding the Sharīʿa and implementing the laws of Islam in the administration of the public affairs of the community.

Secondly, there was nothing in the fundamental teachings of the Imamite school to bar Imamite jurists from becoming *sulṭān al-zamān*, all the more so because, as we shall see in the next chapter, in Imamite jurisprudence many functions with theologico-political implications could be executed by jurists in the absence of the Imam. Far more significant in the age when *jihād* came to be limited to the defense of Islamic lands was the authority to administer justice, regarded as the most crucial part of the general obligation of "enjoining the good and forbidding the evil"—that is, the ongoing struggle of Islamic ideology for the creation of just public order. In the sections of jurisprudence that deal with the general obligation of "enjoining the good and forbidding the evil," and the exercise of authority to administer justice and adjudicate among the Imamites, as is discussed in the next chapter, we have the most explicit statements regarding the partial investiture of the Hidden Imam's politico-religious authority in his general deputies.

It is plausible that in Imāmī Shīʿism, in the absence of the political authority of the Imam throughout the period preceding the occultation, and the period following the termination of the manifest Imamate, there always existed the potential for assumption of such authority by a well-qualified jurist—the functional imam of the Shīʿa—when circumstances so demanded. However, there seems to be no theoretical procedure through which the authority of the jurist assuming the role of functional imam could be legitimized except through *niyāba* (deputyship), whether "special" or "general." If the deputyship of the jurist could be established during the Complete Occultation, especially in the absence of the directly designated deputies during the Short Occultation, then there surely was room for growth in the *wilāya* (authority) of such a deputy to include all-comprehensive authority—*al-wilāyat al-ʿāmma*—in the name of the Imam in concealment.

Conclusions

In the treatment of *jihād* in Imamite jurisprudence one can perceive the significance attached to the position of *al-imām al-maʿsūm*, the infallible Imam, in dealing with the problem of unbelief and the way it affects the establishment of Islamic public order. Concurrence among Imamite scholars regarding the mandated existence of *al-imām al-ʿādil* (the just Imam) in the matter of waging offensive *jihād* demonstrates the consistency with which these scholars have upheld the fundamental aspects of the doctrine of the Imamate—namely, that it is only the explicitly designated Imam who is entitled to make binding decisions in matters affecting the welfare of humanity. By virtue of his being an infallible leader and authoritative interpreter of Islamic revelation, the Imam is the sole legitimate authority who could establish the Islamic state.

However, under the impact of historical circumstances, the Imamate became divided into temporal and spiritual spheres. The temporal authority of the Imam was regarded as having been usurped by the ruling dynasty, but the spiritual authority remained intact in the Imam who was regarded as God's (unanswerable) demonstration (of divine omnipotence) (lit. the proof of God), empowered to guide the spiritual lives of his adherents as the true Imam. This spiritual authority was not contingent upon the Imam's being invested as the ruling authority (*sultān*) of the community. Accordingly, the Imamate continued through all political circumstances until the last Imam, the twelfth, Imam al-Mahdī, went into occultation (A.D. 874). It was during this period that questions regarding Imamite political authority during the absence of the Imam began to be treated methodically, especially when, for the first time, following the last manifest Imamate (i.e., of ʿAlī, A.D. 656–660), the *temporal* authority of the Shīʿī Imamite Būyid dynasty was established de facto.

No doubt the jurists who dealt with this new contingency were responding to it individually in accordance with the degree and quality of their relationship and contacts with the de facto powers. There is sufficient evidence to maintain that even among the de facto powers, especially professing Imamite faith, there was a pragmatic recognition by jurists of the existence of a *sultān ʿādil* because of certain qualities or acts that he manifested, not because of any documentation in Shīʿī traditions. Nevertheless, this pragmatic recognition in jurisprudence depended on rational and at times radical interpretation through extrapolation of Imamite sources handed down in the form of precedents from the time of the Imam al-Sādiq onward. The various forms of phrases introduced to designate *sultān ʿādil* (e.g., *sultān al-waqt, sultān al-zamān,* or *imām ʿādil*) are examples of extrapolations introduced into judicial decisions, which were extended to include Imamite jurists—also because of qualities recognized in them. However, there also was some form of documentation suggesting deputization, as delineated in chapter 1.

Accordingly, as will be discussed in the following two chapters, the steps toward the recognition of the jurist as a candidate to occupy the place of *sultān ʿādil* in the juridical writings of the later period (more particularly, Safavid and post-Safavid eras) were adequately taken in the form of open-ended exegesis and methodological expedients applied to the documentation provided during the Būyid period. These rulings must be seen as the product of tensions within the Imamite school created by the occultation of the twelfth Imam and by the interaction between Imamite and Sunnī scholars at this time.

The termination of the manifest Imamate gave rise to the institution of the deputyship of the Imam as the only feasible way to preserve the religious-social structure of the Imamite community. The deputyship, furthermore, became a sort of extension of the Imamate in such a way that not only could deputies assume functions with theologico-political implications; they could actually become functional imams with the potential of becoming *sultān al-zamān* (the ruling authority of the time). As Imamite juridical sources reveal, there is nothing in the rulings of the Imamite jurists to prevent jurists from assuming the authority (*wilāya*) invested in them indirectly by the twelfth Imam. This *wilāya* would lead them to be considered *sultān al-zamān li tadbīr al-anām* (the authority appointed for the management of the affairs of humanity). In fact, as subsequent argumentation will demonstrate further, there was always the potential for the well-qualified Imamite jurist to assume *al-wilāyat al-ʿāmma*—all-comprehensive authority in the name of the last Imam among the Shīʿa—as the deputy of the Imam and as the functional imam of the community. Thus he could be regarded as *al-sultān al-ʿādil,* the Just Ruler.

4

The Deputyship of Jurists
in *wilāyat al-qaḍāʾ*
(Administration of Juridical Authority)

In the previous chapter, on the Imamite theory of political authority, I have alluded
to the relationship of *saltana* (exercise of legal and moral authority through the
demand of obedience) and the concept of justice in Twelver Shīʿīsm when I discussed *jihād*, the struggle to create just order on earth. I have also dealt with the
question of *al-wilāyat al-ilāhiyya* (the divine sovereignty) and the way it was invested in the Prophet and the Imams for the purpose of creating just social order
under the guidance of Islamic revelation.

It is important to bear in mind that, according to the Imamite understanding of
the Qurʾān, the question of divinely guided leadership—possessing the *wilāya* (the
faculty that enables a person to assume authority and exact obedience)—is of
utmost significance in attaining the goal of just social order, because it is only
through divinely guided leadership that the creation of an ideal society could be
actualized. However, during the occultation of the Imam, when the *sāhib al-wilāya*
(the wielder of authority) is in concealment, there arises the question of the exercise
of the *wilāya,* which as I have shown, is concerned with the entirety of the life of the
Muslim community. This is because believers in Islamic revelation have never
relinquished their belief in the identity of religion and government as they saw them
embodied in the Prophet. It was this belief in the goal of Islamic revelation that
necessitated belief in the continuation of the divine *wilāya* in the individuals who
afforded the necessary guidance for the good estate of the community.

The institution of *niyābat al-imām* (the deputyship of the Imam) during the
occultation, from its inception during the Short Occultation (A.D. 874–941), came to
be viewed as metaphysically connected with the source of divine authority (*al-wilāyat
al-ilāhiyya*), the Imam, whom these individuals represented among his followers. As
such, the deputyship of the Imam was always viewed as not only the logical extension

of the religious authority of the Hidden Imam, but also as the authority that possessed the necessary legitimacy through fully accredited documentation.

The Imamite concept of juridical authority, as introduced in chapter 1, demonstrates the necessity of continuity in the Prophetic precedent for religious prescriptions, which was achieved by extending the "apostolic" authority to the loyal associates of the Imams and their subsequent transmitters who became part of that
authoritative precedent admitted as the only valid basis for religious practice in
Shīʿism. What rendered the precedent authentic, then, was the perception of "apostolic" authority through some sort of designation. This designation was sought in
those traditions that spoke about deputyship. Legal opinions were deductively
inferred by the interpretation and extrapolation of the documentation to resolve
practical problems faced by the Shīʿa at a given time in history.

The problem of authority in the absence of the Hidden Imam, more particularly after the stabilization of the Imamite community in the tenth–eleventh century
under the scholarly elite who wrote in the relatively protected environment provided by the Shīʿī Būyids, was not merely an academic issue in the realm of the
ideal. It emerged as a concrete and urgent situation created by the prolonged
absence of the Imam, on the one hand, and the appearance of the Shīʿī power of the
Būyids, on the other, to which the Imamite jurists were responding in their juridical
endeavors to exegetically expound upon meager documentation regarding the deputyship in the early traditions. Their exegetical stratagems in studying these traditions enabled them to produce coherent and adequate judicial decisions that could
establish the legitimacy of the deputyship and its extent during the occultation.

As discussed above, there was nothing in Imamite juridical or theological
writings to inhibit the deputyship from becoming *sultān al-zamān li-tadbīr al-anām*
(the authority to manage the affairs of humanity), pending the Imam's return at the
end of time. It is the documentation regarding the *wilāya of the deputyship, its*
interpretation for relevance to a contemporary situation, and its implications in the
development of *al-niyābat al-ʿāmma* (the general deputyship) of the potential *sultān
al-zamān*, the Imamite jurist, that I propose to examine in the following pages.
Wilāyat al-qaḍāʾ, the authority to administer judgment, as I shall demonstrate,
became a prelude to the much broader *al-wilāyat al-ʿāmma* (the all-comprehensive
authority) that was emerging as a necessary consequence of the prolonged occultation of the twelfth Imam and the necessity for the Shīʿī community to maintain
a concrete sense of continuity in their linear progression in history.

The Qur'ānic Notion of Justice and its Political Implications

In the Qur'ān, there is perhaps no other injunction that matches the emphasis laid
on "establishing justice," basically in the sense of doing justice and giving truthful
evidence. The primary aim of Islamic revelation is to establish a viable social order
on earth that will be just and ethical. However, the creation of this social order, as
shown by the example of the Prophet himself, was dependent on the divinely
guided leadership that could ensure its realization. Immediately after the death of
the Prophet in A.D. 632, the question of the creation of just order became interwoven with the question of Islamic leadership. It was this dual question that formed

the crucial issue that divided Muslims into various factions. According to the Qur'ān, the sole justification for appointing a *khalīfa* (viceroy) on earth was "to judge aright between mankind." Thus, in connection with David's appointment as *khalīfa*, God says:

O David! We have appointed thee as a viceroy (*khalīfa*) on earth; therefore judge aright (*fa'hkum*) between mankind and follow not desire . . . that it beguile thee from the way of God [38:26].

"Judge aright" clearly sets the objective for David's appointment on earth. It is valid to maintain, then, that the Qur'ān presupposes divine guidance through the divinely appointed mediatorship of human agents for the realization of the Islamic public order. Hence, controversy as to whether any *khalīfa* was to be accepted as a divinely designated person, responsible for leading the community toward that goal after the Prophet's death, was inevitable. By the end of the early period of Islamic leadership, when civil wars broke out in 656 A.D., Muslims were confronted with an unfulfilled ideal of a just order, which made the necessity of a divinely guided leadership—to assume the function of guiding the community toward its primary goal—even more imperative. After all, it was obvious that the establishment of a just Islamic order depended on a person who was appointed by divine designation to exercise *wilāyat al-tasarruf* (discretionary authority) and was protected (*ma'sūm*) from becoming a cause for sinful deviation in Islamic polity.

The creation of a just society, which depends on some sort of divine intervention or "grace," has thus become the focal point of the Islamic belief system. In this belief system, as discussed in the previous chapter, the Prophet and his properly designated successors are visualized as representing the transcendental active God on earth—the God who delegated authority (*al-wilāyat al-ilāhiyya*) to them in order for them to rule over humankind aright. Accordingly, it is this basic religious focus on the creation of just order and leadership (*imāma*), which can create and maintain it, that orients the authoritative perspective or worldview of Muslims. It is, therefore, important to discuss this worldview in the light of the conception of justice in Islam, in its political as well as in its religious manifestations. To this end, I will discuss the notion of justice in Islamic revelation, and its implications for political justice where the question of the authority that can administer this justice becomes central.

To understand the notion of justice in the Qur'ān, it is essential to grasp that the Qur'ān was revealed against the background of the tribal society of Arabia and, as such, the moral exhortations to "establish justice" (4:135) or to "judge with justice" (4:58) become comprehensible within the context provided by Arab usage before Islamic revelation defined the scale of divine justice. Undoubtedly, the exhortation to "establish justice" in 4:58 (or in 4:135) refers to the notion of "justice" as an objective and universal moral truth, on the basis of which one can affirm it to be a universal and natural mode of guidance to which humankind in general can be called upon to respond. In other words, "justice" is a moral prescription that follows from a common human nature and is regarded as independent of particular spiritual beliefs, even though all practical guidance regulating interpersonal human relationships springs from the same source—namely, from God. This observation regarding the objective nature of "justice" is important to bear in mind because the Qur'ānic notion could not become intelligible without some reference to an objective state of affairs.

The objective nature of the concept of justice in the Qur'ān is summarized by Ibn Manzūr, the author of a classical Arabic lexicon, in his entry on the meaning of ʿadl, where he cites a letter written by an early jurist-theologian of Medina, Saʿīd b. Jubayr, in reply to an inquiry about the meaning of the term ʿadl by the Umayyad caliph ʿAbd al-Malik (d. A.D. 705):

> ʿAdl [justice] may have four significations. [First,] al-ʿadl in the administration of justice (al-hukm), in accordance with [the sense implied in] God's command [in the Qur'ān]: ". . . and when you judge between the people, judge with justice" [4:61]. [Secondly,] al-ʿadl in speech, as construed in this command: "and when you speak, be just" [6:153]. [Thirdly,] al-ʿadl [in the meaning of] a ransom (al-fidya)¹ [as understood in what] God said: ". . . and beware a day when no soul will in aught avail another; and no counterpoise (ʿadl) shall be accepted from it [the soul], nor any intercession shall be profitable to it" [2:113]. [Fourthly,] al-ʿadl in the sense of attributing an equal to God (al-ishrāk) [as implied in] this saying: ". . . the unbelievers ascribe equals (yaʿdilūn) to their Lord" [6:1]—that is, they are committing an act of "association." As for the passage [where God says]: "You will not be able to be equitable (taʿdilū) between your wives, be you ever so eager" [4:128], ʿUbayda al-Salmānī and Dahhāk are of the opinion that [the human inability to be equitable] meant [the inability to be so] in respect to love and intercourse. [Moreover, when someone says] "so-and-so has done justice to so-and-so," it means that one is equitable to the other. [Also one says: "To us nothing is equal to you," meaning: "Nothing stands with us in your stead."²

This explanation of the notion of "justice," with reference to the Qur'ānic passages where the notion occurs, demonstrates an important point that "justice" in revelation, which denotes moral virtues like "fairness," "balance," "temperance," and "straightforwardness," should be understood as universally objective values "ingrained in the human soul" (91:8). The term ʿadl has a real, concrete, physical signification first, and an extended, abstract signification secondly. As such, it becomes comprehensible by logically appealing to the universally self-subsistent value of "justice." The idea of justice/balance as a self-subsistent principle running through all creation is sometimes called "cosmic harmony."

In another important passage, the Qur'ān recognizes the universality and objective nature of moral virtue ("goodness"), which transcends different religions and religious communities, admonishes human beings "to be forward in good work," and holds them accountable for their deeds regardless of their religious differences:

> To everyone of you [religious communities] We have appointed a law and a way [of conduct]. If God had willed, He would have made you all one nation [on the basis of that law and that way]; but [He did not do so] that He may try you in what has come to you; therefore, be you forward [i.e., compete with one another] in good works. Unto God shall you return all together, and He will tell you [the truth] about what you have been disputing [5:48].

There is a clear assumption in this verse that certain basic moral requirements, like "being just" or being "forward in good work," are self-subsistent and apply to all human beings, regardless of differences in religious beliefs. Interestingly enough, the ideal human being is conceived of as combining moral virtue with complete religious surrender:

> Nay, but whosoever submits his will to God, while being a good-doer, his wage is with the Lord, and no fear shall be on them, neither shall they sorrow [2:12].

Undoubtedly, we have here a clear basis for a distinction between religion and morality in the Qur'ān, where moral virtues are further strengthened by the religious act of "submission" to sacred authority.

It is in the realm of universal moral truth that human beings are treated equally and held equally responsible for responding to the call "to be forward in good work." Furthermore, it is this fundamental equality of all humanity at the level of moral responsibility that directs humankind to create just public order on earth, and makes it plausible that the Qur'ān manifests something akin to the Western notion of natural law:

> . . . by the soul, and That which shaped it and inspired it to [know the difference between] lewdness and god-fearing, [Whereby it can guard itself against moral peril]. Prosperous is he who purifies it [through obedience to God and sincerity in faith,] and failed has he who seduces it [9:7–10].

God, according to this passage, endowed human beings with the necessary cognition and volition to further their comprehension of the purpose for which they were created and to realize it by using their naturally acquired knowledge. Moreover, the verse also makes it plain that the distinction between "lewd" (evil) and "god-fearing" (good) is ingrained in the soul in the form of inspiration, a form of guidance with which God has favored human beings. This is the divine trust given to humanity, which I discussed in the context of *al-wilāyat al-ilāhiyya* in chapter 3. It is through this guidance that human beings are expected to develop their ability to judge their actions and to choose what would lead them to "prosperity" (i.e., the creation of just public order). But this is not an easy task to fulfil. It involves spiritual and moral development, something that is most challenging in light of the basic human weaknesses indicated by the Qur'ān: "Surely man was created fretful, when evil visits him, impatient, when good visits him, grudging . . ." (70:19–20).

This weakness reveals a basic tension that must be resolved by further acts of guidance by God. It is at this point that God sends the prophets and the "books" (revealed messages) to show human beings how to change their character and bring it into conformity with the divine plan for human conduct (2:2, 5). Guidance from God signifies the "direction" God provides to procure the desirable human society, first by creating in the soul a disposition that can guard against spiritual and moral peril, if a person hearkens to its warnings, and then by further strengthening this natural guidance through the Book and the Prophet. "Guidance" in the sense of "showing the path" is a fundamental feature of the Qur'ān and is reiterated throughout to emphasize the fact that this form of guidance is not only part of normative human nature, but is also universal and available to all who aspire to become "god-fearing" and "prosperous."

However, insomuch as human beings are free agents, they can reject this guidance, although, because of their "innate disposition" (*fitra*)[3] prompting or even urging them subtly to believe in God, they cannot find any valid excuse for this rejection. Even then, their rejection pertains to the "procuring" or "appropriating" of what is desirable, and not to the act of apprehending in the first place what is desirable. However, when human beings choose to reject this guidance, God denies further guidance to them: "Those that believe not in the signs of God, God will not guide them" (16:104). This denial of guidance clearly pertains to the guidance that would lead to the "procurement" of the desirable end, *not* to the initial guidance that is originally engraved in the hearts of all human beings, in the form of an innate disposition, to guide them toward "prosperity."

Because the question of guidance is related to the source of knowledge of ethical values, such as justice, in the classical as well as in modern works on Qur'ānic exegesis, I have taken some care to explicate the various forms of guidance in the Qur'ān. Significantly, it is at this point that theological differences among Muslim scholars become conspicuous. These differences are rooted in two conflicting conceptions of human responsibility in the procurement of divine justice. The discussion of Qur'ānic material in this connection was dominated by the proponents of the two major schools of Muslim dialectical theology: the Mu'tazilite and the Ash'arite.

The basic Mu'tazilite thesis is that human beings, as free agents, are responsible before a just God. Furthermore, good and evil are rational categories that can be known by reason, independent of revelation. God created the human intellect in such a way that it is capable of perceiving good and evil objectively. This is a corollary of the main Mu'tazilite thesis, that God's justice depends on the objective knowledge of good and evil, as determined by reason, whether the Lawgiver pronounces it so or not. In other words, the Mu'tazilites asserted the efficacy of natural reason as a source of spiritual and ethical knowledge, maintaining a form of rationalist objectivism.[4] Thus, the Mu'tazilites emphasized the complete responsibility of human beings, in responding to the call of universal guidance through natural reason as well as guidance through revelation.

The Mu'tazilite standpoint was bound to be challenged. The Ash'arites rejected the idea of natural reason as an autonomous source of ethical knowledge. They maintained that good and evil are as God decided them to be, and it is presumptuous to judge God on the basis of categories that God has provided for directing human life. For the Mu'tazilites, the Ash'arites argue, there is no way, within the bounds of ordinary logic, to explain the relationship of God's power to human actions. It is more realistic to maintain that everything that happens is the result of God's will, without explanation or justification.

However, argued the Ash'arites, it is important to distinguish between the actions of a responsible human being and actions attributed to natural laws. Human responsibility is not the result of free choice, a function that, according to the Mu'tazilites, determines the way an action is produced; rather, God alone creates all actions directly, but in some actions a special quality of "voluntary acquisition" (*kasb*) is superadded by God's will, and this makes the individual a voluntary agent and responsible. Consequently, human responsibility is the result of the divine will nown through revealed guidance. Otherwise, values have no foundation but the will of God that imposes them. This attitude of the Ash'arites to ethical knowledge, s, according to Hourani, theistic subjectivism. This means that all ethical values are lependent upon the determinations of the will of God expressed in the form of evelation, which is both eternal and immutable.[5]

Both of these theological standpoints were based on the interpretation of Qur'ānic passages, which undoubtedly contain a complex view of human responsibility in procuring divine justice on earth, as referred to above. On the one hand, it ontains passages that would support the Mu'tazilite position, which emphasized he complete responsibility of human beings in responding to the call of both natural guidance and guidance through revelation. On the other hand, it has passages that could support the Ash'arite viewpoint, which upheld the omnipotence of God, and hence denied humans any role in responding to divine guidance. Nevertheless, it allows for both human volition and divine will in the matter of accepting or rejecting the faith that entailed the responsibility for procuring justice on earth.[6]

The Concept of Justice as Embodied in the Sharī'a

In Islam, the Sharī'a is closely intertwined with religion and morality, which are regarded as expressions of God's will and justice. But, whereas the aim of religion and morality is to define and determine goals and provide practical guidance to achieve salvation, the function of the Sharī'a is to teach humans *the way* to achieve their salvation by virtue of which God's justice and other goals are realized. Thus, the Qur'ān commands Muslims to worship God, to arbitrate with justice, to give truthful evidence, to fulfil one's contracts, and to return a trust or deposit to its owner. All these commands have one purpose in mind—namely, to create a new society on a religious and moral basis to replace Arabian tribal society. Again the Qur'ānic prohibitions of gambling, of drinking wine, and of charging interest are directed against ancient Arabian standards of behavior, with the purpose of creating what is to be an ethically based society. An underlying concern of Qur'ānic legislation is that of enunciating ethical principles in the form of moral exhortations that ought to be followed in the administration of justice in the Islamic public order.

Islamic revelation, which included both the revealed text and the Prophetic precedent, was, thus, to provide an impetus to structure the Islamic public order on the basis of the administrative and legislative functions of the Prophet. Accordingly, the Qur'ān, in addition to the Prophet's personal elaboration and practice based on it, determines the ultimate course of direction the community ought to follow in organizing and regulating its public order. The Prophet's elaborations in communicating revealed guidance and his own practice formed the "model pattern of behavior"—the Sunna—for his community and was in the subsequent period promulgated as an authoritative precedent for religious-moral prescriptions as treated by the Sharī'a. Through the promulgation of the Sunna as a normative Prophetic precedent, Islamic revelation as the source of juridical prescription came to be composed of two authoritative texts, the Qur'ān and the Sunna, both of which were regarded as being the embodiment of God's will and justice.

These two sources provided the raw material on the basis of which later Muslim jurists, together with the use of a third derivative source based on human reasoning called *ijtihād,* laid down the definitive law of the Sharīʿa. The foundation of ideal Islamic social order was conceived as firmly based on the fundamental principles of the Sharīʿa and the elucidations of the succeeding generations of Muslim jurists, which came to be regarded as a system sanctioned by God. The ultimate adherence to the Islamic system, which provided the only valid basis for religious life, was achieved through the consensus (*ijmāʿ*) of the Muslim jurists who had concerned themselves with the transformation of local administrative law into a unified religious law of Islam. As a result, their consensus, which became part of the paradigmatic precedent, came to be regarded as another source for deriving decisions on points of Sharīʿa. Moreover, through their concern with the preservation of the Islamic way of life, the jurists came to be acknowledged as a group of pious persons, specializing in points of law and morality, who could advise the community about correct Islamic behavior. Furthermore, it was this group of pious specialists who became the defenders of the Islamic ideal of a socio-political order against corrupt behavior of the ruling class, whether the caliph or his governors, which was regarded by them as detrimental to the creation of justice and equity. As such, the jurists became the guardians of the Sharīʿa, and made and enforced the laws of God, sometimes as the delegates of the caliph and his governors and at other times, when they were dismissed by them at will, they acted as the "heirs of the Prophets," as implied in the famous *hadīth:* "the *ʿulamāʾ* are the heirs of the Prophets." Indeed, the Muslim jurists were required, like the Shīʿī Imams, to possess moral probity (*ʿadāla*) in order for them to defend good persons from injury, to aid the oppressed, vindicate the innocent, and justly mete out legal punishments (*hudūd*) to those deserving them.

With the weakening of the caliph's authority and the accentuation of the already imposed limitations on his person as merely the protector of the law of the Sharīʿa, not its legislator or executor, as had been the case with the early caliphs who made laws for the community and also executed them, the jurists became in fact the executors of the justice of God. Consequently, *qadāʾ,* administration of justice, which was headed by these Muslim jurists, the *ʿulamāʾ,* became one of the most important institutions in preserving the popular sense of divine justice. At times of political turmoil, the administrators of *qadāʾ* were regarded by the populace as the protectors of the people against the license of those in power.[7]

It is plausible, then, that the persistent Qurʾānic challenge of creating a just society and the repeated failure of those in power in the Islamic empire to realize divine justice on earth, especially when the central power of the caliphs had disintegrated, immeasurably increased the prestige of *al-qādī* or *al-hākim,* the administrator of justice. This was the only religious administrative institution committed to preserve the Sharīʿa as the embodiment of divine justice. Upholding the Sharīʿa as the embodiment of divine justice meant that any government, in order to be regarded as a legitimate Islamic government, had to be based on constitutional principles sanctioned by divine revelation, whose interpretation depended upon the Muslim jurists.[8] The interpretations of the Muslim jurists, necessarily based on exegetical materials in the Qurʾān and the Sunna, set the tone of Islamic political theory in medieval times. These interpretations, always involving distribution of power through delegation, are most evident with respect to certain administrative functions of jurists in their capacity as specialists in Islamic law. These administra-

tive functions included *al-qaḍā'* ("administration of justice"), by far the most important duty of the caliph, according to the Qur'ān.

Delegation of office for the administration of justice was particularly important
for the Shī'ī jurists whose Imam was in occultation. This gave rise to the problem of
determining the process by which the Imam's authority to administer justice was
delegated to the Imamite jurists. The problem of determining the process of
delegation—itself a fait accompli—was crucial for the survival of the Imamite community: without such a legitimation there would always remain a question about the
lawfulness of the Imamite jurists to assume the administration of justice in the
absence of the Imam.

It is to this legitimation of the authority of the jurists in the administration of
justice that I now turn. In the following pages it will become evident that the
Imamite jurists could not have assumed the *wilāya* (the authority) of the Imam
without authoritative precedent furnished by accredited documentation to that
effect in the traditions attributed to the Imams.

Shī'ī Jurists and the Administration of *al-qaḍā'* in Imāmī Shī'ism

In Imāmī Shī'ism the administration of *al-qaḍā'* has been regarded as the constitutional right of the Imam because of *al-wilāyat al-ilāhiyya* (the divine authority)
invested in him through Prophetic designation in the form of *al-naṣṣ*. Accordingly,
al-qaḍā' is considered as *al-wilāyat al-shar'iyya* (the authority based on revelation)
possessed by the Imam in the matter of *al-ḥukm*—that is, possession of the authority to judge—on the basis of divine revelational sources—on matters pertaining to
public welfare.[9] Because *al-qaḍā'* is among the most important functions of human
society on which the well-being of the people depends, it is also regarded as part of
the Qur'ānic obligation imposed on the Imam to control "discord on earth" (9:33–
34). "Discord" here means breakdown of social and moral order because of general
lawlessness. The Imam is responsible for the protection of the people against tyranny and hence his concern for *al-qaḍā'* is based on a much wider requirement of
the Qur'ān—namely, "commanding the good and forbidding the evil" (3:104, 110;
9:71).[10] This latter requirement emerges as the central aim of the Islamic ideology,
and forms the basis for the Imam's *wilāya* in general. As a leader of the community
invested with the authority to uphold this central aim, the Imam is responsible for
creating conditions for the implementation of the Qur'ānic injunction. Thus, the
Imam's general *wilāya* in matters pertaining to the spiritual and temporal affairs of
the Muslim community serves as a starting point for *wilāyat al-qaḍā'*. The main
purpose of *wilāyat al-qaḍā'* is to defend good persons from harm and to settle
disputes (*taḥkīm*). As such, it presupposes detailed, accurate information about the
rulings of the Sharī'a.[11] It is for this reason that in Shī'ism *wilāyat al-qaḍā'*, the
authority to mete out just punishment, was closely linked with *wilāyat al-ḥukm*, the
authority to judge by forming an opinion on the basis of knowledge of the rulings of
the Sharī'a. In Imamite jurisprudence *al-qaḍā'* has been taken as synonymous with
al-ḥukm, "judgment on a matter, [in the form of] verbally [expressed opinion] and
[the subsequent sentence] in action."[12]

It is important to understand *al-ḥukm* in the context of *al-qaḍā'* in Imamite jurisprudence, because the noun *al-hākim* ("the judge") has been interpreted both politically and juridically in the writings of the Imamite jurists, depending on the period in which their works were composed. As we shall see below, in the classical age of Imamite jurisprudence (tenth–eleventh centuries), *al-hākim* was taken strictly in its juridical signification, whereas later on when Shīʿite jurists became the sole representatives of the religious institution, *al-hākim* in a Shīʿite state was sometimes given a much broader political signification of "guardianship" of the Imamite community in the absence of the rightful Imam.

The Qurʾān sets the tone of Islamic revelation for a judicial decision in a single verse, where the Prophet is called upon to act as an arbitrator and settle disputes.

> But no, by thy Lord! They will not believe till they make thee the judge (*yuhakkimūka*) regarding the disagreement between them, then they shall find in themselves no impediment touching that which thou decidest (*qadayta*), but shall surrender in full submission [4:65].

In the above passage the first verb refers to the arbitrating aspect of the Prophet's function, and the second one stresses the binding character of the decision made by him. The term *qādī* is derived from the latter verb. The primary acceptation of *hukm* and its other derivations in the sense of judicial decision are also preserved in the historical event, in the early Medina period, when Saʿd b. Muʿādh is mentioned as *al-hākim* in the case of the Jewish tribe of Banū Qurayza (A.D. 624).[13]

The position of *al-hākim* under Islam was certainly more authoritative than what had been the case in Arabian tribal society, as is evident in the Prophet's archetypal position as the divinely ordained lawgiver of his community. It was probably this divine nature of the appointment of an arbitrator in the Qurʾān that made *al-hākim*'s responsibility in Imamite jurisprudence a God-ordained office requiring a person to be in possession of perfect reasoning, sound opinion, knowledge, piety, and the ability to undertake the position of *al-hākim*.[14] There is no doubt that *al-hākim*, in the sense of an "authoritative arbitrator" or "judge," always appears in the signification of a juridical authority, in Qurʾānic usage, as preserved in the early works of jurisprudence. In some medieval texts *al-hākim*, is also assigned political functions, such as administering state affairs and conducting peace negotiations. It is accurate to say that gradually *al-hākim* of early Islam evolved to become *hākim al-sharʿ*—that is, a jurist who is well-qualified to decide on legal matters and supervise the affairs of Muslims in the area of Sharīʿa.[15] In the latter function he was also *nāzir*, a "guardian" who could protect Muslims' rights against rulers or other individuals.[16]

In the later works of Imamite jurisprudence these functions of *al-hākim* were classified under *wilāyat al-faqīh*, the authority of the Imamite jurist, who was believed to be the deputy of the last Hidden Imam. In some of the works that were written during the Safavid period *hākim al-sharʿ* is equated with *al-imām al-ʿādil* (the just Imam—i.e., the Imam properly speaking),[17] or the one who has been deputized by him, with the implication that *al-hākim* is the "functional" imam in the absence of the actual Imam.

That the position of *al-hākim* was inherently a constitutional one belonging to the head of the Islamic state, who alone could and did delegate it, at least in theory,

:an be discerned from the following tradition of Imam al-Sādiq. He warned his associates against assuming the position of *al-hākim,* saying:

> Beware of the *hukūma* [administration of justice]. Indeed, *al-hukūma* belongs to the Imam who is knowledgeable in matters of judicial decisions (*qaḍā'*) and who is the just one (*al-'ādil*) among the Muslims, like the Prophet or his legatee.[18]

This statement of Imam al-Sādiq makes it evident that the Imams saw administration of justice (*hukūma*) as their constitutional right in their capacity as the legatee of the Prophet in his function as the head of the Islamic polity. It was this doctrinal concern about the Imam's constitutional propriety that underscored discussions, and at times differences of opinion, among Imamite jurists on the question of exercising the Imam's authority, as his deputy, in all its aspects.

Hākim al-sharʿ and the Question about his Authority in *hukūma*

In Imamite jurisprudence, from its early days, there has been much debate as to the power accruing to the holder of the office of *hākim al-sharʿ,* who, during the absence of the twelfth Imam, is invested with *wilāyat al-hukm* and *al-qaḍā'*. In a sense, it is a question of including the Imamite jurists, in their position as the deputy of the Imam, in the "apostolic" authority of the Imams. Such an inclusion in the period of occultation would not only afford legitimacy to the claims of *wilāya* with theologico-political implications; it would also lead to the recognition of the *fuqahā'* as the necessary paradigmatic link with the actual source of validity for juridical prescription in Imamite jurisprudence. This paradigmatic link, as I have shown in chapter 1, was the most important aspect in maintaining the normative character of juridical prescriptions, which had come to be acknowledged as being revelational and hence immutable.

In other words, the debate among Imamite scholars concerns the fundamental question: Can *hākim al-sharʿ* assume the juridical position of the Imam, with its theologico-political implications, during the period of the Complete Occultation? That *wilāyat al-qaḍā'* was authority with much greater theologico-political implication relating to the "apostolic" authority of the Imam than merely that of giving judgment or executing a decision vis-à-vis litigants is to be construed from the opening statement of Ibn Bābūya's section dealing with the administration of justice in his *fiqh* work *al-Muqniʿ*. The purpose of Ibn Bābūya is certainly to demonstrate that *wilāyat al-qaḍā'* belongs exclusively to those who are invested with *al-wilāyat al-ilāhiyya*. He says:

> Beware of *al-qaḍā'* [sitting in judgment]; abstain from assuming it, because *al-qaḍā'* is the most difficult position in religion.[19] No one assumes it except the Prophet or the legatee (*wasī*) of the Prophet. [This is confirmed by] Amīr al-muʾminīn's [ʿAlī b. Abī Tālib's] statement [of warning] to Shurayh [the *qāḍī* of Kūfa]: "O Shurayh! you are sitting in a place where no one but the Prophet or the legatee of the Prophet or the *shaqī* [the one who has wrongly assumed the position of a judge] sits."[20]

In another place Ibn Bābūya refers to the tradition cited in the preceding section, in which Jaʿfar al-Sādiq warns his associates against presiding over affairs involving judgment (*al-hukūma*) because, declares the Imam:

> Indeed *al-hukūma* belongs to the Imam who is knowledgeable in matters of *qaḍāʾ*, and who is the just one among Muslims, such as the Prophet or his legatee.[21]

Clearly, the authority to administer justice in both these traditions is confined to persons invested with *al-wilāyat al-ilāhiyya* (the divine sovereignty)—namely, the Prophet and the Imams—by virtue of their being in possession of "sound knowledge" and "sound character" (*ʿadāla*). Moreover, anyone who assumes the position of *qāḍī* besides the Prophet and the Imams is declared unjust. The emphasis laid on "sound knowledge" in *al-qaḍāʾ* assumes considerable significance because the Imams' authority in Shīʿism, more particularly in the absence of their political investiture, was grounded on their infallible knowledge, which commanded recognition among their followers. It was this source of authority, in the form of "sound knowledge," that was transmitted to their close associates and made it possible for such qualified Imamites to not only hope for the implementation of juridical prescriptions based on Imamite sources, but also to undertake the administration of justice as the knowledgeable deputies of the Imams.

The possibility of administering Imamite law, however restrictedly, came into effect under the Būyid rule. Major works on Imamite jurisprudence were written and judicial decisions were issued to legitimize the acceptance of the office of *qāḍī* under the de facto government as long as the principle of justice was honored. Favorable circumstances under the Būyids prompted traditionists like Ibn Bābūya to consider the practical aspects of accepting the office of magistrate, on the one hand, and to remind qualified individuals of judicial authority in Imāmī Shīʿism, on the other. In this way, the right of the Imam to exercise juridical authority was reasserted and the qualifications for assuming the office in his absence were reaffirmed.

It is in this light that detailed sections on the administration of justice in the absence of the supreme judicial authority of the Imam, in both the traditions and the applied jurisprudence of the Imamites, make sense. The necessary justification for assuming the *wilāyat al-qaḍāʾ* was sought in the "sound knowledge" and "character" of the Imamite jurist.

In his classification of the *quḍāt*, the judges, Ibn Bābūya stresses the importance of judging rightly (*bi al-haqq*)—that is, deriving a judgment (*al-hukūmāt*) from "sound knowledge":

> As for judges, they are of four kinds:
> [1] A judge who executes a decision that is incorrect while knowing that he is wrong. Such a person is doomed to be in Hell.
> [2] A judge who carries out a judgment that is unsound while not knowing that he is incorrect. He too will be in Hell.
> [3] A judge who judges correctly, while not knowing that he is sound [in his judgment]. He too will be in Hell.
> [4] A judge who carries out a judgment correctly, being fully aware that he is right. He will be in Paradise.[22]

Thus, according to Ibn Bābūya, only a person who is thoroughly grounded in knowledge of religious sources and who is confident about the soundness of his judgment can take up the position of a judge. In fact, according to some Imamite jurists, it is imperative that *al-ḥākim* be knowledgeable not only in his judgments, but also in all that is referred to him, because he is responsible to administer justice and protect the rights of the people.[23] As was seen in chapter 1, "sound knowledge" can be derived only from Imams who have inherited Prophetic knowledge and who alone are regarded as authoritative interpreters of Islamic revelation. But in the absence of the explicit designation (*nass*) of a *qāḍī*, how can a *qāḍī* ascertain that his knowledge is protected from error, as the Imam's knowledge is through the very act of explicit designation, as maintained in the Imamite theory of Islamic leadership?[24]

This question about the source of knowledge of a *qāḍī* has received serious treatment in Imamite jurisprudence, because if it is established that the Imamite *qudāt* are the successors of the Imams in their juridical authority, then it is established that these *qudāt* can execute judgment on the basis of their knowledge—just as the Prophet and the Imams did, unrestrictedly—without their depending on evidence that might be based on supposition (*zann*). Moreover, as pointed out earlier, possession of "sound knowledge" gave authority to the jurists. As such, in the matter of executing judgment, the nature of the knowledge of judicial authority generated dispute about the level of "certainty" that could be attributed to the judgment of any person besides the Imam. The question of the nature of the knowledge of a jurist underlay the entire development of Imamite jurisprudence where it was important to ascertain that fulfillment of religious prescription was based on certainty (*'ilm*, *qat'*) and that judicial decisions inferred deductively led to the certainty of obligation, not merely a supposition (*zann*).

The function of the *usūl* (the theoretical basis of Islamic law) was to provide a method to ascertain the soundness of judicial decisions as much as to generate the necessary confidence in the authority that applied principles to derive rulings. As shall be discussed in this chapter, the *usūl* methodology provided a means of ascertaining that opinions inferred deductively from Islamic revelation through *ijtihād* were "sound" and that the authority engaged in this intellectual movement was fit to assume the Imam's deputyship as the functional imam of the community. In some important ways *wilāyat al-qaḍā'* was going to be a test case to establish tangible authority within the Shī'ī community in the absence of the Imam.

It is important to note that a judgment based on the statement of two witnesses was considered by jurists to be based on supposition, whereas a judgment based on knowledge as attained by observation and perception was regarded as being at the level of certainty (*'ilm*). The underlying principle appears to be that of forestalling judgments based on a judge's personal knowledge. However, this was not a universally accepted principle among jurists who treated the problem from the perspective of searching for truth. According to them, justice might not be achieved if a judge were prohibited from giving a judgment on the basis of prior knowledge. There is no dispute among the jurists that the Imams can execute a decision on the basis of their knowledge. But some jurists have denied that ability to the Imamite jurists in all cases pertaining to the "rights of God" and "rights of humankind," because, according to these scholars, a jurist does not possess the Imam's prerogative of sitting in judgment and delivering a judgment that is infallible and irrevocable.

Among the earliest Imamite scholars, Abū 'Alī Ibn Junayd has been associated with the explicit opinion regarding the restricted nature of the knowledge of *al-hākim* to rule on any legal issue or to administer legal punishment (*al-hudūd*) on the basis of his knowledge. Sharīf al-Murtadā refuted Ibn Junayd's opinion, saying that he had depended on his own personal judgment, especially in light of an overriding consensus among Imamite scholars expressing the opposite view. It is interesting to read Sharīf al-Murtadā's argument here, referring as it does to the incident of the Fadak estate. Fātima, the daughter of the Prophet, claimed it as her right. Abū Bakr denied the claim despite his certainty that the Prophet had in fact given the estate to her. Abū Bakr's decision to withhold Fadak from Fātima, says Sharīf al-Murtadā, has been disliked by all the Imamites, and he has been criticized for having asked Fātima to produce a proof for her claim on the estate when he was fully aware of her *'isma* (infallibility) and purity, and could have given judgment in her favor on the basis of his knowledge regarding Fātima's right to inherit Fadak.[25]

The problem with Ibn al-Junayd, according to Sharīf al-Murtadā, was that he differentiated between the Prophet's knowledge and that of his successors and *hukkām* (administrators of justice). Such a differentiation is unwarranted, says Sharīf al-Murtadā, because the knowledge is in relation to a case under consideration and as such it does not matter who the person is as long as the knowledge is authentic. It is similar to the knowledge of the Prophet and the Imams when they have witnessed a man committing an act of adultery or theft. Just as they know about the incident with certainty, so their *hukkām* could also have similar knowledge if they had been witnesses of what had happened.

The gravity of the question of the knowledge of a person sitting in judgment is further corroborated by Ibn Junayd's theological argument as cited by Sharīf al-Murtadā. Ibn Junayd says that he has found in the Qur'ān that God considered it necessary to declare null and void certain laws regulating relationships between believers, on the one hand, and nonbelievers and those who had become apostates, on the other. These laws dealt with inheritance and marriage between believers and nonbelievers, and the lawfulness of eating animals slaughtered by them. God informed the Prophet about those who were hiding disbelief and were outwardly professing Islam, thus making it possible for the Prophet to be informed about their inner state. On the basis of this kind of Prophetic knowledge the Prophet gave judgment forbidding believers to marry them or to eat animals slaughtered by them. Sharīf al-Murtadā refutes this argument. He maintains that Prophetic knowledge had no bearing on the decision made by the Prophet. Sharīf al-Murtadā says that the Imāmiyya do not agree that God informed the Prophet about the inner state of the hypocrites, or the display of faith and concealment of disbelief in the community. Even if, for the sake of argument, says Sharīf al-Murtadā, we concede Ibn al-Junayd's proposition that the Prophet had knowledge about the inner state of the people, it does not necessitate Ibn al-Junayd's conclusion. Why? Because the prohibitions regarding marriage or inheritance or the eating of slaughtered animals—as between believers, on the one side, and nonbelievers or apostates, on the other, becomes applicable only when persons make their disbelief or apostasy manifest. As such, these laws do not apply to concealed disbelief or apostasy. Indeed, public welfare demands that the application of these laws be conditional upon outward manifestation of disbelief and apostasy, otherwise they would be disruptive of the well-being of other individuals in society. Thus, concludes Sharīf al-Murtadā, the Prophet is not obligated to expose the hidden state of those who apostatize or disbelieve, because giving judgment is not dependent upon the concealed state; rather, it depends on outward manifestation. On the other hand, execution of decision in cases involving adultery, theft, and drinking of wine, are clearly cases in which legal punishments (*ḥudūd*) are to be instituted on the basis of knowledge about their occurrence, whether manifest or secret.[26]

Muslim jurists have distinguished between the "rights of God" and the "rights of humankind" in Islamic penal law. The "rights of God" are regarded in Qur'ānic usage as a prelude to the "rights of humankind," whose welfare is the prime concern of divine sanctions. Crimes against the "rights of God" are regarded as deserving *ḥadd* ("fixed") punishments because they are serious religious crimes against God's purposes for humanity. By committing these crimes an offender has disrupted the ethical public order that God's purposes require. Consequently, the Imam, as the head of the Islamic order, was required to institute these punishments because, according to the doctrine of the Imamate, it was the Imam endowed with divinely protected knowledge who could implement the general purposes of God for the Muslim community.

Some Imamite jurists have regarded it permissible for the *ḥukkām* to sit in judgment only in cases involving the "rights of humankind"—that is, legal judgments to prevent conduct prejudicial to the good order of Muslim society; whereas the Imam can rule on all cases. This is so because the jurist does not have the overriding, personal discretion to determine, according to time and circumstances, how the purposes of God for the Islamic community might best be implemented. This is the opinion held by Ibn Idrīs al-Ḥillī, whose *Sarā'ir* consistently limits the juridical authority of jurists in all areas of the *fiqh* where, implicitly or explicitly, it is possible to argue for the comprehensive *wilāya* of the jurists through further interpretation of the textual evidence used in support of the limited *wilāya*. In his section dealing with *al-ḥukūma* Ibn Idrīs affirms the Imamite position that execution of justice belongs to the Imam alone or his deputy (*nā'ib*) among the Shī'ites, when the Imam cannot assume this responsibility. This deputy possesses certain qualifications, besides being the most knowledgeable (*al-a'lam*) about the truth and the most able in executing a decision. These qualifications, as I shall discuss presently, indicate Ibn Idrīs's hesitation in extending the Imam's "apostolic" authority to jurists who lack proper legitimation through specific designation by the Hidden Imam.

First, and most importantly, according to Ibn Idrīs, *al-ḥākim*, acting as the deputy of the Imam, must belong to *ahl al-ḥaqq* (the people of the truth—i.e., the Imamites). In other words, he must acknowledge the Imamate of the twelve Imams and must adhere to their teachings to qualify as *al-ḥākim* and hold office on their behalf. This acknowledgment of the Imams is what I have identified earlier in this study as "sound belief," possession of which admitted a person to membership in the *ahl al-ḥaqq*. The absence of such belief would lead *al-ḥākim* to judge falsely because he would be depending on knowledge derived from non-Imamite sources, sources designated by Ibn Idrīs as "misguided." Moreover, by not accepting the rightful *walī al-ḥukm*—that is, the Imam—*al-ḥākim* would be deriving his authority from a corrupt source, which would render his *wilāya* unjust, because designation to the position of *al-ḥākim* must be derived from the Imam. It is for this reason that some Imamites, according to Ibn Idrīs, have declared the undertaking of *al-ḥukm* prohibited, even by the Shī'ites, for *al-ḥākim* does not fulfill the fundamental prerequisite in assuming *wilāyat al-hukm*—namely, designation in the position of deputy by the Imam.[27]

There is no doubt, however, that the discussion about the extent to which a *qādī* can depend on his knowledge to make judgments is indirectly related to the question of the extent of the prerogative of taking up the Imam's position during his absence. The question, in the final analysis, turns on the problem of the designation of jurists as deputies, and the theologico-political implications of such a designation in the context of the Imamite theory of religious-political leadership. The problem of the deputyship of the jurists assumed great significance in the later period of Imamite history when temporal authority was invested in the professing Imamite rulers. But in the classical period (ninth–tenth centuries), as our sources indicate, the deputyship of the jurists was limited to areas where Imamites living under Sunnite authorities could, independently of the Sunnī government, administer their own affairs.

Thus, for instance, during the later period of the ʿAbbāsid caliphate, especially when leading Imamite figures like Sharīf al-Murtaḍā and Shaykh al-Ṭāʾifa Tūsī were acknowledged, by Sunnite authority, head of the Imamite community, they exercised some forms of *wilāya*, and there was sufficient documentation of this in the *fiqh*. It included the *wilāya* to collect *khums* and *zakāt* and distribute them among the followers of the Imams as specified in Imamite law. Moreover, a jurist could act as guardian (*walī*) for Shīʿites who did not have a legal guardian. Hence, even in the administration of justice, *wilāyat al-qaḍāʾ*, the question of sitting in judgment on behalf of the Hidden Imam was superseded by a more urgent, practical consideration—that is, whether an Imamite *faqīh* should agree to administer justice on behalf of a de facto political authority, whether Imamite or non-Imamite, considered to be an unjust authority.

Before I treat the solution to this question offered by the jurists, it is important to examine the sources to determine the textual references that offer, on the one hand, the earliest and most explicit statements establishing the juridical authority of the jurists; and, on the other, the extent to which these statements have been interpreted or reinterpreted to prove the extent of that authority in the Imamite community.

There is no doubt that it is in this area of Imamite jurisprudence that one can observe the exegetical extrapolation applied to the crucial documentation by the jurists to respond to the needs and aspirations of the Imamite community and to preserve the sense of continuity of the Imam's guidance for the contingencies of time. Indeed, in these juridical works dealing with the future of the Imamite community, the jurists emerge as the authoritative paradigm for the "sound" religious response to the situation created by the perpetual concealment of the actual paradigmatic authority of the Imam. Moreover, these prescriptions in the form of judicial decisions formed the basis for the legitimation of not only the *wilāyat al-hukm*, but also the "general" deputyship, which lacked specific documentation for the assumption of wider theologico-political implications of the "special" deputyship during the Short Occultation (A.D. 874–941). Most of the Imam's functions could be assumed by his "special" deputy. In order for the jurist to assume those functions, his "general" deputyship had to be interpreted, at least functionally, as being "special." This was done by their *ijtihād* based on the precedent provided in the traditions to which I now turn.

Traditions Regarding the *wilāya* of *al-hākim*

There are basically three traditions most commonly cited as documentation in support of *wilāyat al-hukm* of the Imamite jurists. It is important to note that the same three traditions have been cited by later jurists to establish the "general deputyship" (*al-niyābat al-ʿāmma*) of the *fuqahā'* in matters going beyond *al-hukm*—namely, al-siyāsa (political administration) and *tadbīr al-nizām* (management of public order) on behalf of *al-sultān al-ʿādil* (the Just Ruler).[28] In fact, in the study of these opinions it is not difficult to discern the development in the authority of the jurists from *wilāyat al-hukm* to *al-wilāyat al-ʿāmma*, because *al-hukm* has served as the stepping stone for other kinds of *wilāya*. In other words, if *wilāyat al-hukm*, juridical authority, could be established on the basis of the documentation found in the traditions, then, by further interpretation of these same traditions, other forms of *wilāya* could be derived. This follows because, in the final analysis, the purpose of all forms of *wilāya* is the well-being of an individual in society, a rational argument derived through *ijtihād*.

In Imamite jurisprudence, *ijtihād* (personal, independent judgment of a jurist to infer precepts from authoritative sources like the Qur'ān and the Sunna) has arisen more frequently in the carrying out of *sharʿī* injunctions (*tanfīdh al-ahkām al-sharʿiyya*)—dependent on *wilāya* (authority)—than in the ordinances themselves. Consequently, the interpretation of the traditions on the deputyship (*niyāba*) of the jurists, whether in the area of *al-hukm* or *al-iftā'* (the giving of the legal decisions), has depended directly on the development of *al-usūl al-ʿaqliyya* (the principles laying down procedures and rules for rational analysis), on which *ijtihād*, as a rational process of deriving a decision, is founded. It is, therefore, possible to trace the early development of *usūlī-akhbārī* (*akhbarī*, derived from *akhbār*, meaning "traditions") positions in discussion about the *wilāyat al-hukm* of the jurists.[29]

The role of *akhbār* (traditions) to support the jurist's claim to the *wilāyat al-hukm* can be understood by studying the rational procedures devised to extrapolate the general or limited sense of some key terms in the actual wording of communications attributed to the Imams. Without these rational procedures expounded in the *usūl* (theoretical foundation of Islamic law), it would have been impossible for the developing authority of the jurist to attain maturity in subsequent assertions of the all-comprehensive *wilāya*.

The problem with the authentication of the traditions that dealt with the authority of the jurists was recognized early on in the well-known disputation among the Imamites on the validity of religious prescriptions based on a "single" tradition (*khabar al-wāhid*). Consequently, rational evidence had to be furnished to elevate the supposition (*zann*) based on such "single" traditions for their admittance as a valid proof necessary for authoritative judicial decisions in the matter of the juridical authority of the jurist. In the light of this intellectual movement in the direction of determining the extent of juridical authority through the *usūl*, one can consult the *ijtihād* to ascertain the validity of communications dealing with the delegation of the Imam's authority to Shīʿī jurists.

The *akhbārī-usūlī* debate at this formative period of Imamite jurisprudence was rooted in the extrapolation of the literal sense of these traditions designated as "investiture traditions" (*riwāyāt al-nasb*), rather than the methodological devices developed to ascertain their authenticity. Both *akhbārī* and *usūlī* jurists had a keen interest in the latter intellectual exercise, because their ability to issue a valid judicial decision depended on the authentication of the documentation set forth in a tradition.

In the matter of the tradition dealing with investiture, the *usūlī* jurists, depending on *dalīl ʿaqlī* (rational evidence reached through a process of reasoning) to interpret the *dalīl sharʿī* (textual evidence provided by reports [*akhbār*] on the authority of the Imams), were willing to concede far wider authority to the *fuqahāʾ* than merely *wilāyat al-hukm*. But the *akhbārī* jurists were not inclined to interpret the traditions beyond their explicit, outward sense; and hence they did not believe that the *wilāya* of *al-hukm,* authorized in some traditions, could be interpreted as a prelude to the all-comprehensive *al-wilāyat al-ʿāmma.*

Without doubt, the *usūlī* approach, based on rational analysis, was an inevitable development in the Imamite school of jurisprudence, especially in the absence of the juridical, absolute authority of the Imam, who had entered the Complete Occultation. But even so, the *ijtihād* based on *al-usūl al-ʿaqliyya* was never independent of the *akhbārī* procedures of deriving a decision based on strict textual evidence without inferential interpretations. As a matter of hard fact, reason (*al-ʿaql*), which has been regarded as an authoritative source of law in Shīʿī *usūl al-fiqh,* functioned as the "discoverer" of opinions and judgments in the Qurʾān and the Sunna through the creation of fresh terminology derived by interpreting revelation to make it relevant. *Al-ʿaql* supplements the essential preliminaries that underlie Islamic revelation through the application of *ijtihād,* which establishes relationships between rational propositions and revelational proofs (*nusūs*) in such a way that judicial decision based on them would demonstrate certainty in matters that otherwise would have remained at the level of mere supposition. Hence, the *akhbār* (the traditions), being part of revelational proof, play an equally decisive role in both *usūlī* and *akhbārī* jurisprudence.

In the light of this fact, the conclusions of both *usūlī* and *akhbārī* jurists on the issue of *wilāyat al-hukm,* in its limited sense of administration of juridical authority, are strikingly similar, in that both the supporters of *dalīl ʿaqlī* and those of *dalīl sharʿī* endorse the Imam's appointment of the *fuqahāʾ* or the *hukkām,* those who can sit in judgment and execute decisions between litigants. However, it is plausible that, without insisting on strict categorization between the two methodological trends in Imamite jurisprudence, the majority of the early jurists interpreted the authority of administering justice as a prelude to the much wider authority implicitly endorsed in the Imam's appointment of the *fuqahāʾ* or the *hukkām,* as found in the traditions to be discussed presently. A minority tended to limit the authority of the jurists to *al-hukm* in its specific signification, apprehensive of the possibility that any wider interpretation of this *wilāya* might become an encroachment upon the authority of *al-imām al-maʿsūm,* the infallible one, who alone will establish on earth the divine just order at the end of time. Moreover, as Sallār al-Daylamī and Ibn Idrīs, the two prominent *fuqahāʾ* of the classical age (who represent the early usūlī methodology, though cautiously demonstrating concerns of the *akhbārī* trend), argued that the main reason justifying *fuqahāʾ* to assume *wilāyat al-hukm,* which strictly belongs to the Imam proper, was the need for the management of the public order of the Shīʿites during the occultation (*ghayba*).

Inasmuch as traditions provide the required revelational proof for further interpretation to extrapolate the relevant sense needed to expound the *wilāya* of the
jurists, it is important to reiterate some of the essential points made about the *isnād*
(chain of transmission) and its criticism in chapter 2. *Isnād* was one of the important
features of a tradition, which, if found to be authentic, cleared the way for further
determination of its reliability through textual and contextual study. The procedure
for determining the authenticity of *hadīth* reports was governed by the principles
laid down in the *usūl al-fiqh* where the textual evaluation of these reports became a
prominent feature of *al-ijtihād al-shar'ī* (the reasoning based on revelational evidence). This *ijtihād* gave rise to the terminology for the evaluation and classification
of traditions as "sound," "weak," and so on, that has remained an important aspect
of all inquiry into *hadīth* literature. More importantly, the *ijtihād* provided authentication of *khabar al-wāhid* ("single" tradition), which formed a major portion of the
Imamite traditions. Thus, *ijtihād* became a methodological tool for ascertaining
continuity between the Prophetic paradigm in the form of *hadīth* report as such and
the source of that paradigm, the Imam, to whom ultimately the *isnād* was traced
back. In the absence of such an undertaking in the area of *isnād*, no Imamite jurist
would have been able to establish the authenticity of the substance of his teaching,
or the position he himself was occupying in the chain of "apostolic succession." As
discussed in chapter 1,[30] it was the knowledge of *hadīth*, combined with knowledge
of its source, that gave a position of weight and influence to some of the prominent
jurists and theologians of the Imamite school for the exercise of their authority in
that community.

It is worth noting that the *hadīth* reports that deal with the question of *wilāyat
al-hukm* did not receive much attention in the early works of demonstrative *fiqh*.
The reason for this lack of attention would seem to be twofold. First, careful
examination of *fiqh* works in the early period shows that there was no dispute
among the Imamite jurists on the question of *wilāyat al-hukm*, which came to be
regarded as a requisite for the fulfillment of the duty of *al-'amr bi al-ma'rūf wa al
nahy 'an al-munkar* (enjoining the good and forbidding the evil), and hence, rationally necessary. Secondly, the *wilāya* at this stage was still regarded as limited to *al-
hukm* and *al-qaḍā'* and did not include *al-nizām* or *al-siyāsa*, so as to raise questions
about concrete documentation for such a broad interpretation. Consequently, it is
reasonable to maintain that the *hadīth* reports dealing with *al-hukm* became crucial
at the time when the Imamite jurists had to provide legitimacy for their claim to *al-
wilāyat al-'āmma* through properly authenticated textual evidence. The search for
textual evidence to corroborate *al-wilāyat al-'āmma* was found in the exegetical
treatment of the institution of the Hidden Imam's deputyship (*niyāba*) in the traditions that validated *wilāyat al-hukm*. However, because most of these traditions
were "single" (*wāhid*), the *isnād* in most of these reports, although not entirely
weak, suffered from "wounded" ('*jarh*') reliability; but this could be "legitimized"
(*ta'dīl*) through *ijtihād*, because the transmitter's authenticity in such cases had
never been fully disproved. As we shall see in chapter 5, on *al-wilāyat al-'āmma*,
critical examination of the *isnād* enabled jurists to declare the inner consistency of
our traditions under consideration and to regard them as rationally proven documentation for *al-wilāyat al-'āmma*.

It is relevant to discuss how a "weak" or "wounded" documentation could be
used to issue a legal decision through the process of *ijtihād*. The fundamental

consideration in *usūl al-fiqh* (theoretical basis of Islamic law) is to ascertain, as far as possible, that the proof being advanced to support a legal decision has been thoroughly accredited. In the process of investigating the sources of the Sharīʿa, a jurist might come across two types of evidentiary material: first, that which becomes authoritatively binding because it generates absolute certainty in the mind of the jurist who determines this certainty through rational inquiry; secondly, that which cannot become binding and can be regarded presumptive because it lacks absolute certainty. The first kind of material, known as *dalil qatʿī* (absolute proof), can be used for making legal decisions; the second kind, known as *dalīl zannī* (presumptive proof), in itself is invalid for use as a proof to deduce a *sharʿī* decision. "Presumptive proof," however, can attain the level of "absolute proof" if the presumption in regard to the document is elevated to the level of certainty by means of another "absolute proof." Thus, for instance, a *hadīth* report related through a single "way," irrespective of the number of transmitters (*khabar al-wāhid*, "single" tradition) is regarded as "presumptive proof," even when that report includes a reliable authority in its *isnād*, because there is always a possibility that what is being reported through a single "way" might not be true—might not be capable of verification. Nevertheless, the Qurʾān provides an "absolute proof" by declaring that a report by a single reliable, trustworthy person must be accepted as a valid report.[31] As a result, a *hadīth* report related by a single, thoroughly reliable person is elevated to the level of "absolute proof" by the Qurʾān and used as documentation for a legal decision. In the following discussion we will see the way a "presumptive proof" is elevated to become an "absolute proof" in order for it to be used in making an important legal decision through *ijtihād* in the matter of the *wilāya* of the Imamite jurist.

The most important *hadīth* report with regard to *wilāyat al-hukm* is the one known as *maqbūla* (in the later works of *fiqh*) of ʿUmar b. Hanzala of Kūfa, a disciple of Imam al-Sādiq. *Maqbūla* in *hadīth* classification is a tradition that has been "approbated" authoritatively by approving its text and putting the terms of its text into practice, without probing into the authenticity or weakness of all the transmitters who appear in the *sanad*.

The process of "approbation" of a *hadīth* report involves three stages: first, examination of the chain of transmission (*sanad*); secondly, determination of the congruity of the text (*matn*); finally, consideration of the external factors that corroborate the authenticity of the report. The examination of *sanad* ascertains the number of transmitters who appear in a tradition or traditions that will be used as evidence in a legal decision. A tradition transmitted with a long chain of narrators appended has preponderance over one with a short chain. The reason is that the longer *sanad*, it is contended, would probably lead to the certainty necessary for the inclusion of this tradition as evidence for a legal ordinance. The long *sanad*, in addition, would reduce the possibility of error on the part of numerous transmitters of the same text. On the other hand, the *isnād* that is traced back to the Imam with fewer intermediaries is regarded as superior to one that has passed through too many hands, because, again, the possibility of an error is greatly diminished when the report is narrated by a few contemporaries of the Imam. Furthermore, a *sanad* is examined to see if it includes a transmitter who has been praised for his piety, knowledge, and other such qualities. The inclusion of such a transmitter in an *isnād* renders the tradition preferable to one that does not include such a narrator.

Following the examination of the *isnād,* a hadīth report in need of approbation is examined to determine the congruity of its text. The main point in this regard is to establish whether the transmitted text contains the actual wording of what is attributed to the Imam, or at least that it contains the gist of it in the transmitter's own words. Of course, the former has preponderance over the latter. Next, the eloquence of the text is studied, for it is maintained that the Imams would use the most eloquent form of expression in conveying their teachings. Then the text is studied for inner consistency and is rationally analyzed to determine the strength of its content.

Textual criticism is followed by consideration of the external factors that corroborate the authenticity of the tradition. In this regard, the main issue is to determine whether the early jurists, especially those who lived close to the period of the Imams, agreed upon the content of the tradition as being worthy of having practice based on it. Once that is established, the tradition attains "approbation" (*maqbūliyya*).[32] The approbation of a particular *hadīth* report, then, necessarily characterizes the tradition as "sound," because, by definition a "weak" tradition is one whose transmission and content have not been approved authoritatively.

On the other hand, a *maqbūla* tradition has been authenticated as coming from the Imam himself, at least by demonstration of the validity of its text in comparison with other similar sound traditions. However, the problem of a weak *isnād* of a *maqbūla* is not resolved by the approbation of its text. Accordingly, not all the Imamite jurists have resorted to *maqbūla* as textual evidence to derive a legal decision on the broader implication of such traditions. The fact is that even ʿUmar b. Hanzala, who has been regarded by some later jurists as reliable,[33] was not authenticated by the early authors of *rijāl* works and hence the authoritativeness of *maqbūla* in the area of the *wilāya* has been questioned by the scholars of *usūl al-fiqh.*

In the *usūl al-fiqh,* the *maqbūla* of ʿUmar b. Hanzala has been cited as evidence for resolving the difficulty of authentication of a "single, individual" (*wāhid*) tradition that has preponderance but is "weak" in transmission. It is in the area of resolving this type of contrariety (*taʿārud*) in the *sanad* and *matn* of a tradition, and then accrediting it, that a form of independent reasoning (*ijtihād*) has been skillfully employed by the jurists.[34] As pointed out earlier, without this rational procedure much of the *hadīth* literature in the area of *wilāya* would have remained unapplicable in the derivation of legal decisions.

The use of the *maqbūla* of ʿUmar b. Hanzala as an example of authentication in the *usūl* cannot be regarded as fortuitous. The main purpose of developing the *usūl* methodology was to furnish jurists with a decisive exegetical terminology that could extrapolate revelational proof in juridical prescriptions to respond to the sociopolitical experience of the Shīʿa during the occultation. The *hadīth* part of revelational proof was the most problematic as far as authentication was concerned. Consequently, in the absence of the Imam, *hadīth* required the utmost exercise of juridical authority to settle questions of contrariety for its ultimate use in a legal decision. *Usūl* provided the methodology and the tradition of ʿUmar b. Hanzala provided the revelational justification to corroborate the prerogative of jurists to undertake the resolution of such matters in their jurisprudence on the basis of their "sound" knowledge.

In order to accredit an important tradition like the *maqbūla* of ʿUmar b.

Hanzala in *usūl al-fiqh,* jurists distinguished two aspects of a tradition that would give it preponderance: the qualities of the *rāwī* (the transmitter), and the reputation enjoyed by the *riwāya* (transmission) itself. After a meticulous investigation of the qualities of the transmitter and whether he merits the confidence put in his transmission, the jurist moves on to study the tradition itself in order to establish whether the tradition deserves the reputation it has and whether it is in conformity with the Qur'ān and the teachings of the Imams. It is only after this highly scholastic process, as outlined above in some detail, that the jurists make their decisions with regard to the authoritativeness of the *maqbūla,* giving preponderance to the transmission or the transmitter, for the purpose of authenticating it for use in legal decision-making.

The fact that the *maqbūla* of 'Umar b. Hanzala has been cited as an example of *al-tarjīh* (the preponderance) in all the *usūl* works shows that the text of the tradition was decisive in the matter of *wilāyat al-hukm,* on which depended the superstructure of the administrative authority of the jurists. *Maqbūliyya* (approbation) does not apply to conflicting *riwāya* (here, a generic term for the transmitted teachings of the Imam); rather, and more importantly, it deals with the question of juridical authority and its qualifications to administer justice among the Imamites.

The technical discussion above was necessary to fully comprehend the theologico-political connotation of the *maqbūla* that I will now proceed to quote in full. The earliest and most detailed version of this *maqbūla* has been preserved in the compilation of Kulaynī. The report appears in the early section of the *usūl* where he discusses the merit of acquiring and diffusing knowledge of religious sciences. More specifically, the tradition appears in a subsection of that discussion, where Kulaynī cites several traditions explaining the method of resolving contradictory traditions attributed to the Imams. In other words, this subsection deals with what I have discussed above in relation to *al-tarjīh* (the preponderance) in *usūl al-fiqh,* with the difference that Kulaynī is not citing the tradition to resolve the question of who or what should be given preponderance in the tradition. He is not even classifying the tradition on the basis of *isnād* or *matn* so as to declare it *maqbūla.* His main interest in this tradiiton, as the title of the subsection reveals, is to demonstrate the considerations that govern the decision to validate contradictory traditions. I cite the full *isnād* to show where, according to the scholars of *rijāl,* the problem lies:

> [It is related on the authority of] Muhammad b. Yahyā, [who heard from] Muhammad b. al-Husayn, from Safwān b. Yahyā, from Dāwūd b. al-Husayn, from 'Umar b. Hanzala, who said: "I asked Abū 'Abd Allāh [Ja'far al-Sādiq] about two persons belonging to our community, disputing about a debt of inheritance and referring to the ruling authority (*al-sultān*) and the judges (*al-qudāt*) [appointed by him] for a judgment. Is that lawful?" [The Imam] said: "Whosoever refers to them for a judgment in a matter, whether in the right or in the wrong, has actually referred to *tāghūt* [a tyrannical and insolent leader][35] to give judgment. And whatever judgment is given to him, it will be considered as unlawful [for him to take possession of a property] even if he was entitled to it. The reason is that he has obtained it through the judgment of the *tāghūt,* whereas God has commanded that they should reject him (*tāghūt*). God, the Exalted, has said: 'They desire to take their disputes to *tāghūt,* yet they have been commanded to disbelieve in them' (4:60)."

I said: "What should they do, then?" The Imam replied: "They should refer to one among you who relates our traditions and who has examined what is lawful and what is unlawful according to us and has a deep insight in our decrees. They should take him as their arbitrator since I have appointed him a *ḥākim* [judge] over you. If such a person judges according to our ruling and the person concerned does not accept it, then he has shown contempt for the ruling of God and rejects us; and he who rejects us, actually rejects God and such a person is close to [committing the sin of] associating (*shirk*) [a partner] with God." I inquired: '[What happens] if each of the two men [litigants] agreed on two arbiters of our community and they both would agree to look into their case but they would differ in their judgment and would differ over [the transmission or interpretation] of your tradition?" The Imam said: "The ruling of the one who is more 'righteous' (*'ādil*), more learned, more truthful in relating tradition, and more pious shall prevail. And no attention shall be paid to the decision of the other." I asked: '[What] if both are equally righteous and equally accepted by our associates and neither of the two is regarded more excellent?" The Imam replied: "[In that case] the tradition related on our authority by each of them and according to which they would render their judgment should be investigated. That which is agreed upon by your associates should be accepted as our ruling and that which is rare and unknown among your associates should be abandoned, for that which is agreed upon is not subject to doubt. Traditions (lit. *umūr*, 'articles,' 'matters of faith') are of three types:
(1) That whose integrity is self-evident; such should be followed.
(2) That whose error is plain; such should be avoided; and
(3) That which is difficult to comprehend. Such a tradition should be returned for elucidation to God and His Prophet. [This is the point of the tradition in which] the Prophet said: 'There are ordinances that are evidently lawful and evidently unlawful, and the ambiguous ones between the two.'

"The one who abstains from the latter will save himself from the unlawful; whereas the one who acts upon them will be guilty of having performed the unlawful and as a result destroy himself because of not knowing [that he should have abstained in the first place]."

I asked: '[What] if both traditions related on your authority are well known and are related by trustworthy narrators from you?" The Imam said: "They should be examined. The ruling of the one which accords with the ruling of the Book and the Sunna and [even if it] contradicts [the ruling of] the Sunnīs (*al-'āmma*) must prevail. On the other hand, the ruling which contradicts the ruling of the Book and the Sunna and [even if it] agrees with that of the Sunnīs must be abandoned."

I said: "May I be your sacrifice! What happens if both the jurists (*faqīh*) established their judgment and the ground of an evidence from the Book and the Sunna, but we find that one of the two traditions accords with the [traditions of the] Sunnīs and the other contradicts them? Which one should be followed?"

The Imam replied: "The one that contradicts the Sunnīs [should be followed] for therein lies guidance."

I [further] inquired: "May I be your sacrifice! [What] if both traditions accord with the Sunnīs?" The Imam replied: "It should be seen which one of the two is preferred by their rulers and judges. [The one] preferred by them should be abandoned and the other must be followed." I said: "[What] if their rulers [and judges] accept both traditions?"

The Imam said: "In that case put off [judgment] till you meet your Imam. It is better to stop over a doubtful thing than to rush blindly into destruction."[36]

The discussion between Imam al-Sādiq and his close associate, ʿUmar b. Hanzala, lays down principles for the acceptance of a ruling rendered by qualified Imamite judges, as well as the method of resolving the *taʿārud* (contradiction) that might occur in the transmitted teachings of the Imams. Furthermore, it directs the Imamites to ignore rulings derived by the Sunnī regime and their appointed administrators of justice, who are declared *tāghūt* (tyrannical, unjust). The most important statement of the *hadīth,* which endorses the general appointment of qualified Imamite jurists as the deputy of the Imam to execute justice, is given in the form of a requirement for all the Imamites, who should refer their cases to such a qualified transmitter of the teachings and rulings of the Imam, because, says the Imam: "I have appointed him (*jaʿaltuhu*) a *hākim* over you." This is the statement that establishes *wilāyat al-hukm* and its implications in other areas of the administration of justice (*al-qadāʾ*), including the derivation of rulings inferentially from the two sources of the Sharīʿa: the Qurʾān and the Sunna. It is also this statement that has caused differences of opinion among Imamite jurists from the early days, especially in regard to the identification of the *fuqahāʾ* (jurists) as the *hukkām* (administrators of justice). The question has been centered on the extent of *al-hukm,* whether it is a matter of giving judgment only or whether it includes carrying it out. It is the latter aspect of *al-hukm*—namely, "carrying it out"—that has been found lacking in proper investiture from the Imams.

The apparent sense derived from the statement "I have appointed him a *hākim* over you" points to the particular *wilāya* (authority) of sitting in judgment as the deputy of the Imam, rather than the general authority that would include the exercise of other politically related functions of *al-hukm.* One of these politically related functions, which requires some measure of delegation of executive power to the *faqīh,* is the implementation of legal penalties, the *hudūd.*

Jurists and the Execution of Legal Penalties

In Imamite jurisprudence there seems to be a consensus among jurists that any exercise of authority that may involve "harm" or "injury" to someone else, whether in the process of instituting legal punishments or, more generally, in "enjoining the good and forbidding the evil" (*al-ʾamr bi al-maʿrūf* and *al-nahy ʿan al-munkar*), requires delegation of authority in the form of "permission" (*idhn*) from the Imam himself. In all the classical works of *fiqh* the authority of jurists to execute *hudūd* has been relegated to the general discussion about the extent of the obligation of "enjoining the good and forbidding the evil" and the deputization of jurists to undertake the implementation of this duty.

The obligation of "enjoining the good and forbidding the evil" constitutes the all-embracing socio-political dimension of Muslim communal life. Accordingly, all those obligations that go toward the creation of ideal Muslim community—such as the administration of justice, the execution of legal penalties, and so on—have always been regarded as stemming from this general obligation of *al-ʾamr* and *al-nahy.* In various subtle ways this obligation is connected with the question of the leadership of the community, the leader being directly responsible for providing adequate means of fulfilling this obligation. It is significant that Tūsī discusses the question of *al-ʾamr* and *al-nahy* in the doctrinal section of his *Iqtisād* and makes a strong case for its relationship to the question of leadership, as we shall see below. The leader (Imam), in the Imamite theory of leadership, is the *luṭf* ("grace") of God, whose main function is to lead the community to establish a just public order. This function of the Imam is established by reason, which requires that God provide the best possible means for humankind to attain the purpose for which it has been created. Thus, *luṭf* as exemplified in the existence of the Imam, signifies the divine guidance that God provides in order for humanity "to draw closer to obedience and away from disobedience."[37]

The corollary of this notion of *luṭf* is that the Imam should be available to human beings as God wills them to fulfil their obligation toward the specific goal of creating a just society. Moreover, all the obligations that are ordered by God for this purpose are equally *luṭf* and their obligatoriness is established by "reason" as much as by "revelation." In other words, the duty of *al-'amr* and *al-nahy* is established both by reason (*'aql*) and by revelation (*sam'*). In the former case, there is *luṭf* of God in it for humankind to further the establishment of a just public order; and in the latter case, the performer of good or evil is promised reward and punishment, respectively.

Among the early Imamite jurists, Sharīf al-Murtaḍā, Abū Salāḥ al-Halabī, and those who followed them maintain that *al-'amr* and *al-nahy* became incumbent through revelation only, because there is nothing in reason that could establish its obligatoriness. According to 'Allāma, Sharīf al-Murtaḍā had argued that had *al-'amr* been incumbent through reasoning, then reasoning would have further required that good should never be eliminated at all and evil should never occur, and that it is incumbent on God to act to this end, otherwise God would have failed to do something obligatory. But this conclusion, says Sharīf al-Murtaḍā, is invalid because it would necessitate coercion on the part of God to require human beings to act in a particular way.[38]

Ṭūsī, on the other hand, although strengthening this position of Sharīf al-Murtaḍā in his *Iqtiṣād*, proceeded to maintain that the duty of "enjoining the good and forbidding the evil" is established on the basis of reasoning, only because there is *luṭf* in doing so. He contended that it is not sufficient to say that this obligation is derived only from revelation, whether in the form of consensus, verses of the Qur'ān, or well-accredited traditions, about which a believer is informed with certainty that reward or punishment accrues to the person who has done good or evil, respectively. Otherwise one would be constrained to hold that the Imamate is not necessary on the basis of reason too, and that all that is required is to maintain certainty with regard to reward and punishment for our moral behavior. Beyond that, it is proper to maintain that it is recommended, but not incumbent, that there should be an Imam at all times.[39]

'Allāma has preferred Ṭūsī's opinion in this matter, maintaining that the duty of enjoining good and forbidding evil demands *luṭf*, and therefore, reason says, it is incumbent on God to bestow *luṭf*. Hence, to obligate doing *al-'amr* and avoiding *al-nahy* is rationally incumbent. According to 'Allāma, Sharīf al-Murtaḍā's opinion requires further examination because surely what is incumbent on humankind, as far as "enjoining the good and forbidding the evil" is concerned, is not the same as what is incumbent on God. Thus, when this obligatoriness refers to believers, it is clear that those who can fulfil this obligation should do so with heart, tongue, and hand; whereas those who cannot do so with heart, tongue, and hand, must do with heart, if not more. On the other hand, obligatoriness in the case of God means that God can promise punishment and threaten those who oppose the divine will and God can promise reward and strengthen those who follow the divine commands. In this way God can make the divine purpose clear for humans and enable them to carry out what is in their best interests, as the obligation of bestowing *lutf* requires God to do.[40] In the *fiqh* works of Shahīd I and Shahīd II, the "rule of correlation" (*qā'idat al-mulāzama*) between "that which is required by reason and that which is required by revelation" was developed in greater detail in the *usūl al-fiqh*. There it is maintained that the obligatoriness vis-à-vis *'amr* and *nahy* is established on the basis of reason, for revelation does not contradict the judgment of reason, which is capable of deriving a self-evident verdict without any need for further corroboration from a textual source.[41]

This does not mean that the textual source was not regarded as all-comprehensive. On the contrary, in the area of moral obligations that were religiously ordained—as was the case with the duty of "enjoining the good and forbidding the evil"—there was recognition of the fact that it was reason upholding the general welfare of humanity that acknowledged the comprehensiveness of the religious ordainment by engaging in its interpretation and discovering all the principles necessary to make it relevant at a given time in history.

In addition, in general accord with their rational theology, Imamite jurists recognized a fundamental need of interpreting a moral prescription in revelation by reason in the absence of the authority who was empowered to undertake the fulfillment of the moral obligation of "enjoining the good and forbidding the evil." However, it was not for just anyone to undertake the responsibility for guiding the community by interpreting the extent of the obligation. It assuredly needed authorization from a divine source, a sort of designation that could generate confidence in the community to obey and carry out the moral-legal prescriptions connected with the "enjoining" and "forbidding." It was only such an authorized person who could, during the concealment of the Imam, assume the authority that accrued to the Imam as the rightful ruler (*al-sultān al-'ādil*) in Islam to promote human welfare. But, in view of the prolonged occultation and the absence of a special designation from the Imam during this period, any realization of the ethical public order based on the obligation of "enjoining the good and forbidding the evil," according to some Imamite scholars, was not possible. This opinion can be attributed to those jurists who denied the rational incumbency of this obligation; whereas, jurists like 'Allāma and Shahīd I and Shahīd II, who were responding to the experience of the community in their respective socio-political contexts, maintained the intrinsicality of a ruling derived from the authority of revelation in that derived from the authority of reasoning. Thus, according to them, the rational obligation of "enjoining the good and forbidding the evil" is intrinsic in revelation.

However, whether the obligation of *al-'amr* and *al-nahy* was established by reason or revelation, there was no difference of opinion that the "use of force" in fulfilling this obligation could be employed only by the Imams or the one permitted by them to undertake it. The reason for such a requirement becomes clear when one takes into account the authority of the Imam based on his designation from the Prophetic source, which renders inerrant his judgment involving force. In fact, according to 'Allāma, some Imamite scholars have contended that *'isma* (the divine protection that renders the knowledge of the Imam free from error) is necessary in the matter of *'amr* and *nahy* by force, because use of force may lead to the spilling of blood, and blood may not be shed unjustly.[42] Some other Imamite scholars have required the Imam himself to undertake this obligation when the use of force reaches the level of bloodshed (*al-qatl*). Thus, Shahīd I in his *Durūs* has delegated this level of "enjoining the good and forbidding the evil" to the Imam alone, saying, "It is more correct to hand this obligation over to the Imam."[43] Shahīd II concurred with him in this opinion, adding that they both were of the opinion that *inkār* (prohibition) by use of force, especially *al-qatl*, must guarantee that the desirable effect will be realized in society, and such a guarantee is afforded only by the judgment of the Imams. There is consensus on this ruling among the jurists of all the periods of Imamite jurisprudence, which certainly indicates the theoretical recognition of the Imam's overriding, personal discretion to determine, according to time and circumstances, how the purposes of God for the Islamic community might best be implemented through the use of force.

Tūsī, commenting on the requirement about the delegation of the Imam's authority in the form of "permission" to his Shī'a to undertake the kind of interdiction involving harm or even loss of life, says that Sharīf al-Murtadā had disputed the matter of requiring the Imam's "permission", maintaining that it is permissible to inflict harm or injury in order to fulfil the obligation of *al-'amr* and *al-nahy* as required by the law, without the Imam's permission. Sharīf al-Murtadā argues that the requirement for the Imam's permission has no other purpose behind it than to ensure that what is done is in accordance with the Imam's intention, for by not carrying out the obligation of doing *al-'amr* and avoiding *al-nahy* one is failing to fulfil what is intended for the Muslim community's well-being. When the purpose of interdiction by force is to defend someone's right or to obstruct someone from doing some harm to others, then what occurs is in accordance with the Imam's intention.[44]

Reference to the Imam's intention in fulfilling a public obligation has often been invoked in jurisprudence in areas where there has been doubt about the "permission" of the Imam for someone other than himself or his special deputy to assume a politico-theological responsibility. Sharīf al-Murtadā, who himself held the official position as an administrator of justice classified under the general rubric of "enjoining the good and forbidding the evil," regarded his acceptance of the position in the ʿAbbāsid administration as fulfilling the Imam's intention to uphold the public welfare of humanity. Sharīf al-Murtadā provided much needed practical guidance in dealing with the question of working for an unjust government. His rational interpretation of the Imam's intention in such matters was compatible with his rational theology. In that theology God requires humanity to undertake the responsibility of creating a just public order by ascertaining that desirable effects result in society by fulfilling the duty of "enjoining the good and forbidding the evil," even by use of force.

Tūsī concurred with this opinion of Sharīf al-Murtadā in *Tibyān*, his commentary on the Qur'ān. But in *Nihāya*, his work on *fiqh*, he maintained with other Imamite scholars that one has to limit the interdiction to the heart, unless one has the "permission" of *sultān al-waqt* (the reigning authority—i.e., the Imam) appointed to administer the affairs of the community. Tūsī maintains that when the duty of doing *al-'amr* reaches the level of forcing the people to do something through chastisement and prevention, or through killing and corporal punishment, this cannot be done by anyone without the permission of the ruling authority.[45] In *Iqtiṣād* his opinion is even sharper. There he states explicitly that the *sharī'a* does not require a person to adhere to an obligation if the prerequisites attached to such adherence are not fulfilled.[46]

Sallār al-Daylamī, Ṭūsī's contemporary, maintains that killing and injuring in *al-inkār* (interdiction by force) can be exercised by *al-sultān* (as defined above), or the one appointed by him. Abū Salāh al-Halabī does not make reference to *al-sultān* a precondition in this obligation. According to 'Allāma, Sharīf al-Murtadā's opinion mentioned in *al-Iqtiṣād* has preponderance, because there is a generalness in the obligation of "enjoining the good and forbidding the evil" as ordained in the Qur'ān,[47] and hence limiting it to the Imam or his appointed deputy is unnecessary.

However, fully aware of the implications that the implementation of the obligation of commanding and interdicting by force had for the authority of the Imam (where he was not available to take charge of the matter himself, or where it was not possible for any jurist to secure his explicit appointment during the Complete Occultation), the majority of jurists maintained that the *fuqahā'* in general were delegated to execute this obligation. The reason was that they were permitted to execute legal punishments provided there was no impediment from the direction of the Sunnī authorities. As long as there was one *faqīh* available to take upon himself the enjoining and forbidding, it was not necessary for the others among the Shī'a to undertake it, because the Qur'ān says: "Let there be one nation of you, who should undertake to enjoin the good and forbid the evil" (3:104).[48] The Shī'a are required by the Imams to assist the *fuqahā'* in executing the obligation and ensuring that the *fuqahā'* are executing their task responsibly.[49]

'Allāma's endorsement of Sharīf al-Murtadā's ruling alludes to the recognition of the Imamite jurist's general deputyship in matters in which a *sultān* (invested with political power) could exercise his personal discretion to determine how the welfare of the community might best be maintained. In other words, the jurist, like a *sultān,* exercises his personal discretion to carry out the interdiction by force "representatively" (*kifāyatan*), acting on behalf of the community to effect God's purposes. The implications of such a judicial formulation were that the Shī'a were required to support jurists in their effort to maintain the Muslim public order; and conversely, as Sallār points out, if they did not carry out the task, then the Shī'a should refuse to support them in their wielding of authority unjustly.

The occultation of the Imam was accepted as a long-term, divine arrangement. In the meantime, when opportunity arose, as happened under the Būyids or the Ilkhānids, jurists ruled on the basis of the Imamite doctrine of divine justice, according to which it was impossible for God to abandon the welfare of the Muslim community during the occultation and allow human society to degenerate under the corrupt authority of political rulers and princes. It was, therefore, both a religious obligation and a moral necessity to work toward creating an ethical society.

As stated earlier, the importance of the discussion about *al-'amr* and *al-nahy* in the Imamite *fiqh* is due to its implications for the socio-political dimension of Muslim communal life. Part of this obligation relates to the administration of justice, more specifically, to the execution of legal punishments (*al-hudūd*).

In the Islamic penal code, Muslim jurists discuss the performance of certain acts, which have been forbidden or made subject to sanction by punishments in the Qur'ān, as crimes against religion in the particular sense of violating the "right" or "claim" of God (*haqq allāh*). These acts constitute *hudūd* offenses inasmuch as the Qurʾān takes appropriate cognizance of them. Jurists differ as to the number of *hudūd* offenses. Some consider the following seven as *hudūd* crimes to be punished by mandatory penalties: unlawful sexual intercourse; false accusation of such unlawful intercourse; theft; highway robbery; drinking wine; apostasy; and rebellion. Some jurists omit rebellion; others restrict the list to the first four crimes, for which the Qurʾān prescribes specific penalties,[50] and classify apostasy and drinking wine as *taʿzīr* offenses—crimes for which there are no specified penalties in the Qurʾān, so that it is left to the ruler or *al-hākim* to determine *taʿzīr* (chastisement) as deemed necessary in the public interest.[51]

Apostasy is mentioned in the Qurʾān, where it is stated that for those who have disbelieved after they believed, "there shall rest on them the curse of God and of angels and of man . . ." (3:86–90 and also 16:106). But the death penalty for the crime stems from traditions attributed to the Prophet—"Whoever changes his religion, kill him"—and is sanctioned by precedents set by the early successors of the Prophet.[52] Hence, it is correct to maintain that although this punishment is not, properly speaking, *hadd* (capital punishment), because of the absence of any specific text in the Qurʾān sanctioning the death penalty for apostasy, it is regarded as such by some jurists because it involves the claim of God by virtue of which the prescription becomes necessary for the protection of a fundamental public interest.[53]

It is significant that, according to all the schools of Islamic law, in order to apply these punishments the presence of a legitimate ruler (*al-sultān al-ʿādil*) is necessary. Legal doctrine explained the divine plan in terms of rights and duties of individuals, and established certain inviolable standards of conduct, but the all-comprehensive and supreme obligation of the just ruler was protection of the public interest (*al-masālih al-ʿāmma*), which he did by "commanding the good and interdicting the evil." Administration of justice, including the institution of penalties for violations affecting the "right of God" or the "right of humankind" came under this obligation. Consequently, *hudūd* (sanctioned in revelation) and *taʿzīr* ("discretionary") punishments are regarded as means to prevent conduct prejudicial to the good order of Muslim society.

It is important to note that even when a violation affecting the "right of God" is regarded, in Qurʾānic usage, as a prelude to the "rights of humankind," whose welfare is the prime concern of divine sanctions, the Muslim jurists have regarded crimes deserving *hadd* ("fixed") punishments as serious religious crimes against God's purposes for humanity. Construed in this sense, then, by committing these crimes an offender has disrupted the ethical public order.

This is an underlying concern in regarding apostasy as a *hadd* crime in the majority of jurisprudential works because the act of apostasy, like other *hadd* crimes, infringes on private or community interests of the public order. Furthermore, jurists have upheld the duty of public authorities, such as the Imam or his deputy, to lay down rules penalizing conduct that seems contrary to the interests of the community or to the public order. So construed, apostasy becomes a *hadd* offense punishable by the death penalty in order to protect society against dangerous and incorrigible individuals. This points toward moral, rather than religious, grounds for the death penalty although the religious character of apostasy is undeniable.

The connection between the administration of *hudūd* and the "enjoining of the good and forbidding the evil" in the interest of Islamic public order is very obvious.

Hence, the theologico-political issues that were raised in regard to that obligation when it reached the level of enjoining or forbidding by use of force were most critically raised in the question of *hadd* offenses. In crimes deserving *hadd* the issue of pain, harm, injury, or even loss of life to someone else through the implementation of legal punishment was even more precise. Anyone designated to execute it had to be made responsible to the legitimate authority acknowledged by the public.

As pointed out earlier, the legitimate authority (*al-sultān al-'ādil*) was provided with an overriding, personal discretion to determine, according to time and circumstances, how the divine plan for Islamic society might best be effected. It was for this reason that the *wilāyat al-hukm* (the juridical authority to administer justice) was dependent upon designation by the official for that purpose. Sunnī jurists writing during the absence of the legitimate power of the caliph in the eleventh century had to rationalize the existence of the de facto rulers who were not qualified to exercise such a discretionary authority to implement and supplement the principles established by legal doctrine for the exercise of supreme judicial power. On the other hand, the Imamite jurists, for whom the situation during the occultation was similar to the period when the Imams were living and were not invested with political power, had to deal with a persistent problem of justifying the holding of office under de facto governments while preserving the doctrinal underpinnings of the Imamate. The concept of "general deputyship" and its relationship to the comprehensive authority of the jurist must be viewed from this politico-theological perspective created by the existence of the *wilāya* (the right to demand obedience) and the absence of the *saltana* (the power that enforced and exacted obedience) of the Imam. The perpetuity of the *wilāya* allowed it to be delegated in the deputyship of the Imam; but the *saltana* was contingent on the actual wielding of authority by the Imam himself. If that *saltana* would come into existence without the Imam at the head of the government, what should be the religious prescription for that exigency was the subject matter of the political jurisprudence of the Imamites.

In a tradition reported on the authority of Imam al-Sādiq, Hafs b. Ghiyāth questioned the Imam as to who could administer *hudūd*, the ruling authority or the *qādī?* The Imam replied: "Administration of *hudūd* belongs to the one who possesses [*wilāyat*] *al-hukm*."[54] Consequently, most Imamite jurists from the very beginning have concurred that because the *fuqahā'* possess the *wilāya* of *al-hukm*, they can also administer *hudūd*.

Tūsī in his *Nihāya* has summarized the Imamite position on the matter of executing legal punishments during the absence of the twelfth Imam. Tūsī's opinion is in the form of a concise statement of the Imamite position on the issue, starting from the time of the Imams who had not exercised their temporal authority when the caliphs were in power. He says:

> It is not permissible for anyone to execute the *hudūd* except the authority of the time [i.e., the Imam] designated by God, or the one appointed by the Imam to carry this out. It is not permissible for anyone except those two to apply the *hudūd* in any case. However, permission had been given, during a period when the true Imams do not have discretionary control over the affairs of the community and when unjust persons gained the upper hand, for the people to institute the *hudūd* on their children, their families, and slaves, if they do not fear any harm from the direction of the unjust authorities and if they are safe from harm

from them. Whenever they are not safe from the retribution of unjust authority, they should not try to undertake the obligation, under any circumstances.[55]

Ibn Idrīs, in his section on "enjoining the good and forbidding the evil," including *hudūd*, in *Kitāb al-jihād*, maintains that Tūsī's statement above is not meant to be a *fatwā* in the sense of binding jurists to the legal decision derived by him. The reason is that, according to Ibn Idrīs, there exists a validated *ijmāʿ* among the Imamites and among other Muslims to the effect that it is not permissible for anyone except the Imam and his appointed *hukkām* to administer the *hudūd*. This opinion points to my earlier observation that, in the realm of exercising the supreme judicial authority, proper designation establishing official delegation for assuming *wilāyat al-hukm* was a precondition. Tūsī's opinion could not be regarded as a "legal decision" (*fatwā*), because, according to Ibn Idrīs, it lacked authenticated documentation to validate his independent judgment in the matter of delegation. Ibn Idrīs, as I noted in the Introduction, was critical of Tūsī's use of "single" traditions which had been considered by Sharīf al-Murtadā as generating doubt and conjecture.

As for others besides the Imam and his specifically appointed *hukkām*, Ibn Idrīs rules that it is not permissible at all for them to undertake this duty, nor can they circumvent this validated *ijmāʿ* by resorting to "single" traditions. The only way for them to justify their decision to act contrary to this *ijmāʿ* is to find an equally validated *ijmāʿ*, or a proof from the Qurʾān, or a well-documented *sunna* in support of their *fatwā*. Thus, Ibn Idrīs believes that one can undertake to apply *hudūd* only to his slaves in the absence of the Imam and his *hukkām*, when it is possible to do so in this way, but not otherwise.[56]

However, in his section on the execution of judicial decisions in *Sarāʾir*, Ibn Idris maintains the necessity for this on the part of qualified Shīʿa jurists who, having been appointed by the Imams, should undertake to ensure that the ordinances of the Sharīʿa (*al-ahkām*) are carried out; because, he explains, the purpose of the *ahkām* is to generate, by observance, of the ordinances devotion to the service of God. However, the validity of the performance of the ordinances will depend on acknowledgment by the community in general of the jurist whose ruling in these matters is sound and who is able to render his legal decision effective. If this requirement is established, then it can be advanced that the performance or realization of the *ahkām* pertaining to the Sharīʿa (which are necessary for generating devotion to God), is among the duties of the Imams, being restricted to them alone, and that no one else can undertake it without being properly qualified to do so. When it is possible for the Imams to carry out the ordinances of applying *hudūd* themselves, and they are afraid to undertake its performance for some reason, then it is not permissible for other than their followers, the Shīʿa (among whom the jurists have been appointed for this purpose by the Imams), to undertake this responsibility. For a legal decision the Shīʿa must appeal to these qualified jurists and arrive at truth through their ruling, voluntarily following them in their exposition of the ordinances. The duty of applying the *hudūd* belongs to the jurist who meets the qualifications which are necessary in order to become the deputy of the Imam for assuming the office of *al-hukm*.[57]

This opinion would seem to contradict Ibn Idrīs's earlier opinion in *Kitāb al-*

jihād where he limits the undertaking of instituting *hudūd* only on the slave and then also by the master who owns the slave, whereas administration of *hudūd* in principle belongs only to the Imam or his specifically appointed deputy.

Ibn Idrīs faces a clear conflict between theoretical and practical considerations. The Imam and his properly designated deputy, theoretically, are the only recognized authorities in the Imamite theory of public order; but there is no other practical solution to the growing need among the Imamite community for a leader who can be legitimately regarded as the general deputy of the Imam without proper designation. Ibn Idrīs, it is relevant to recall, had opposed the opinions that implied the greater degree of discretionary authority of an Imamite scholar and might be construed as authorizing the assumption of wider politico-religious functions of the Imam during the occultation. But practical considerations required the jurist to assume comprehensive authority to undertake *wilāyat al-hukm* and related functions, which could be delegated through proper appointment. Faced with that the absence of such a designation *and* the rule of need that dicated the preservance of the public interest, a solution had to be extrapolated by restricting the exercise of such a *wilāya* to qualified individuals in the community. It is for this reason that, to ensure the necessary legitimacy, Ibn Idrīs includes a list of qualifications that must be fulfilled by any Shīʿa member who intends to exercise this *wilāya* in the absence of explicit designation by the Imam.

In other words, as noted in chapter 1, the possession of "sound" belief, knowledge, and character made up for the lack of specific designation and confer the required legitimacy to assume the *wilāya* in the absence of the Imam. Not only that, even when a Shīʿite, because of his preparation in jurisprudence, was asked by an unjust authority (*sultān jāʾir*) to assume such a responsibility, he had to agree to hold the office and apply the *hudūd*, bearing in mind that:

> What he does, he does at the command of the rightful *sultān* [the Imam], and it is incumbent upon the Shīʿa to cooperate with him and to consolidate him as long as he keeps within the bounds of what is right in that over which he is appointed, and does not go beyond what is legal according to the Sharīʿa of Islam.[58]

In his *Qawāʿid*, ʿAllāma has reiterated the opinion of the early jurists and has declared more specifically that during the occultation of the Imam the right to administer *hudūd* belongs to the jurists of the Shīʿa. But as to whether a *faqīh* should undertake to administer *hudūd* willingly when appointed by an unjust ruler, ʿAllāma believes that it is permissible only when the ruler forces him to do so. Even then, the *faqīh* is obliged to undertake the application of *hudūd* only as long as it does not lead to killing and oppressing the innocent, for one cannot resort to *taqiyya* when *hudūd* would do harm to innocent life. Moreover, to the jurists belong *wilāyat al-hukm* among the people, if they have nothing to fear from tyrannical rulers. They can also undertake to distribute *zakāt* and *khums,* and issue legal opinions, provided they have the qualifications necessary to become the *muftī*.[59]

Muhaqqiq has also conceived a relationship between *al-hukm* and the *hudūd:* just as *al-hukm* can be administered by a well-qualified person who has thorough grounding in the teachings of the Imam and can inferentially derive new rulings through the exercise of his rational faculty, so there is a comparable case for a person who is qualified to apply the *hudūd*.

It must be borne in mind that, according to the *hadīth* cited by Tūsī, "the execution of the *hudūd* belongs to the one who can give judicial decisions (*al-ahkām*)."[60] Shahīd II, commenting on Muhaqqiq's opinion regarding the permission for the knowledgeable *fuqahā'* to execute *hudūd*, says:

> This is the opinion of the two Shaykhs [Mufīd and Tūsī] and a group of Imamite jurists who have based their opinion on the *hadīth* related on the authority of al-Sādiq.
>   The tradition suffers from weakness in the various ways (*turuq*) by which it has been transmitted. But the report related by 'Umar b. Hanzala supports the opinion [held by Muhaqqiq and others]. After all, administration of *hudūd* is a sort of *al-hukm*. In this undertaking there is general welfare [of the public], *lutf* to refrain from prohibited acts, and termination of the spread of corruption. This is sound opinion. Nevertheless, it is quite obvious that the *faqīh* can undertake it only when he is safe from harm to himself and other believers.[61]

It was important to establish the authority of the jurists in this regard through reference to *wilāyat al-hukm* as based on the traditions, including the one discussed above on the authority of 'Umar b. Hanzala. The permission granted in this tradition formed the basis for further development in the authority of the jurists, because, in the Imamite jurisprudence almost any Shar'ī obligation dependent on the position of the Imam requires an Imam's "permission." And if the Imam's permission could be established in one general administrative sphere such as *al-hukm*, its extension to other areas of similar consequences would not be problematic. Even in the case of *hudūd* when "injury" and "killing" were inevitable, its implementation could be regarded as essentially forming part of the obligation of "enjoining the good and forbidding the evil" rather than the execution of *hudūd* itself, in which case the permission of the Imam would not be a precondition—as, for example, in the obligation of engaging in defensive *jihād*, as discussed in chapter 3. (In the Islamic laws of *jihād*, offensive *jihād* cannot be undertaken in the absence of the Imam or his designated deputy, but such a precondition is waived in the case of defensive *jihād*.)[62]

Qualifications of a Jurist for Exercising *wilāyat al-hukm*

Ibn Idrīs, in his section on *tanfīdh al-ahkām* (the execution of judicial decisions), enumerates several qualifications necessary in the deputy of the Imam for carrying out the duties of *al-hukm* among the Shī'ites. He must be:

> . . . most knowledgeable about the truth in the administration of justice on matters that are referred to him. He should be capable of executing his decisions. He should be in possession of intelligence, sound opinion, and resolution. He should be a man of learning and should possess breadth of knowledge about the rulings [of the Imams], and should have insightful grasp of legal cases. He should be persistent in his legal opinions and should uphold them. He should be conspicuous in righteousness and piety in giving judgment, and should demonstrate capacity to stand by his judgment and to put the thing in its proper place (that is, deal justly).[63]

It is possible to classify the above qualifications under three subheadings, as Tūsī and others have done in their discussion on the qualifications of the *walī* of *al-qadāʾ: al-ʿilm, al-ʿadāla,* and *al-kamāl.* (*Al-ʿadāla* and *al-kamāl* will be taken up in the next section; the remainder of this section treats *al-ʿilm.*)

In order for a judgment to be sound, it has to be based on truth and justice as explicated in the teachings of the Imams. In fact, in the absence of the discretionary control of the Imam due to his occultation, what justifies the undertaking of *wilāyat al-hukm* by a *faqīh* is his "sound knowledge," in addition to his well-attested piety (*waraʿ*).[64] In a tradition reported by Dāwūd b. Husayn, Imam al-Sādiq instructs his followers regarding the person whose judgment becomes effective among the Shīʿa. He says that if two righteous jurists are in disagreement on a ruling, the one who is more knowledgeable about the teachings of the Imams and more pious has preponderance over the other.[65] It is, moreover, according to the Imamite jurists, the possession of "sound knowledge" that justified the waiving of the precondition about the Imam's explicit appointment of the *qādī,* because, as discussed above, in the famous tradition of ʿUmar b. Hanzala the general deputyship of a well-qualified *faqīh* to assume the role of a judge is well established.

In his introduction to the section in *Mabsūt* on the administration of justice, Tūsī regards *al-qadāʾ* as representatively (*biʾl-kifāya*)[66] incumbent upon knowledge-able persons. They should assume the execution of judicial decisions because they are responsible for protecting the right of the weak in society. However, he issues a warning to those knowledgeable persons who do not judge in accordance with truth, or those ignorant ones who undertake to execute decisions in spite of their lack of knowledge. According to the tradition of the Prophet, "Out of three judges one will enter paradise, whereas two will go to hell." As for the one who will be rewarded with entry to paradise, says Tūsī, it will be because of his knowledge of truth and his concern for justice. On the other hand, a person possessing knowledge but ruling carelessly without any concern for truth commits an act of injustice, which will take him to hell. Similarly, a person who judges in spite of being ignorant will also go to hell. Tūsī then proceeds to enumerate in detail what the *qādī* should know in connection with his authority to administer justice. The Book of God comes at the head of the list of the texts the *qādī* must know. There are five pairings of declarations in the Qurʾān that the *qādī* must learn to distinguish:

1. *al-ʿāmm wa al-khāss:* passages that are intended to be understood in their general sense, and those that are intended to be taken in their particular application.

2. *al-muhkam wa al-mutashābih:* communications that convey "explicit" signi-fications, not needing to be interpreted by special reference to the context or through cross-references in the Qurʾān, and verses that are "ambiguous," with an implicit meaning; as such they must be interpreted and clarified in order to discover their meaning.

3. *al-mujmal wa al-mufassar:* declarations of the Qurʾān that are compre-hended immediately, and those that can be comprehended in detailed fashion only through exegesis.

4. *al-mutlaq wa al-muqayyad:* passages that are intended to be unrestricted in their application, and those in which restrictions are specifically applicable as indi-cated by the wording of the ordinance.

5. *al-nāsikh wa al-mansūkh:* verses that abrogate, and verses that are abro-gated by them.

Tūsī, who wrote one of the earliest and most complete works of Qur'ānic exegesis, provides the necessary method for theoretical discussion of legal verses. He underscores the importance of the systematic interpretation of legal pronouncements relevant for deducing religious ordinances.

After the Qur'ān, the *qāḍī* must learn to recognize the authentic Sunna under five categories of pairings. But, whereas the text of the Qur'ān is well-established, the Sunna, made up of *ḥadīth* reports, is problematic. It could have problems of textual ambiguity or problems of authenticity. In both cases the *qāḍī*'s task was difficult.

I have discussed the compilations of the major works of Shī'a *ḥadīth* in chapter 2. Here I concentrate on the way the exercising of *wilāyat al-hukm* necessitated meticulous classification of the Sunna along particular lines.

1.1. *al-mutawātir wa al-āhād*

Ḥadīth reports are generally divided into two categories: *al-mutawātir* and *al-āhād*. *Al-mutawātir* is a tradition that has been reported by numerous transmitters in each degree of its transmission to such an extent that it would be absurd to suppose that all these transmitters concurred to report a falsehood. In other words, a *mutawātir* tradition is one that has been attested to in each generation without any break.

There are very few *ḥadīth* reports that can be categorized as *mutawātir*. A number of these traditions fulfil the condition of being related by numerous authorities, but only in the first degree of their transmission; the subsequent transmission lacks uninterrupted linkage. However, for the few traditions that can safely be categorized as *mutawātir,* the fact of their being reported by thoroughly reliable transmitters, however few, has outweighed the consideration of their not being reported by numerous narrators uninterruptedly.

There are two types of *mutawātir ḥadīth: lafzī* and *ma'nawī*. The *mutawātir* traditions that have been preserved verbatim (*lafzī*) are extremely few indeed;[67] those that have been transmitted in the rational import (*ma'nawī*) of the communication are far more numerous. In fact, most of the communications dealing with well-established daily worship and other forms of religious obligations, such as fasting, almsgiving, pilgrimage, and so on, are not *mutawātir,* and in some cases they are even "single" traditions (*al-āhād*).

This brings us to the consideration of *khabar al-wāhid,* to which I referred in the context of *usūl al-fiqh* in chapter 2, above. *Khabar al-wāhid* is basically a tradition that is not *mutawātir*—that is to say, it has not been transmitted by so many traditionists as to preclude the likelihood that they have all agreed to a falsehood. There have been problems in accepting the authority of this type of tradition as a proof for a binding legal decision. The authoritativeness of *khabar al-wāhid* as a legal proof is dependent upon the reliability of the transmitter, just as is the case with a *mutawātir* tradition, but with the difference that it is sufficient for the *wāhid* narrative to be related by one reliable person on the authority of the Prophet himself. This requirement for the *wāhid* narrative is derived from the Qur'ānic warning against accepting any information from an "ungodly man" (*fāsiq*) (49:6), which implicitly approves the opposite—that is, accepting information from a "godly man" after ascertaining that the person was godly. On the basis of the reliability of its transmitters, and the frequency with which it was cited, a *wāhid* tradition was classified under several subcategories:

1. *Mustafīd* is a "thorough" (*mustafīd*) *wāhid* tradition transmitted by more than two or three narrators in each generation of its transmission.

2. *Mashhūr* ("famous") is synonymous with *mustafīd*, according to some scholars, with the slight difference that whereas the latter has to be "thorough" in transmission in each generation, a *mashhūr wāhid* tradition need not be so in each generation. Thus, the tradition that "deeds are to be judged in accordance with their intentions" is a "famous" but not *mustafīd* tradition. Sometimes *mashhūr* is applied to something that has become merely well known among the people, even if the chain of transmission cannot be traced back to more than one individual or any at all.

3. *Maqbūl*, "satisfactory" or "acceptable," is another type of *wāhid* tradition. It has been accredited through practice as based on its contents, and as such it resembles a tradition that is "surrounded" by evidence supporting its soundness. Some scholars have regarded a *maqbūl wāhid* tradition as being almost a *sahīh* (thoroughly sound); others have regarded it as a *hasan* (fair, good) tradition—that is, a tradition in which the "righteousness" of its transmitters has not been established in all instances, although they are all regarded to be "in good standing."

It is important to note the fundamental difference between the Imamite and non-Imamite conception of the *mutawātir* traditions. In the non-Imamite sources *mutawātir* is applied to those traditions that have been related directly on the authority of the Prophet and the Imams, uninterruptedly, by persons whose well-attested trustworthiness absolutely excludes doubt of the truth in the texts. These directly cited traditions from the Imams were the basis for *usūl* works among the Imamites, discussed above in chapter 2. Accordingly, although the arduous requirement regarding the "righteousness" (*ʿadāla*) of the transmitter is insisted upon in the *mutawātir*, the number of this type of tradition is greater in Imamite collections, even though they relate the import of the communication (*maʿnawī*) rather than the actual wording (*lafzī*). Jurists are required to attest the reliability of narrators before they can authenticate the text through a detailed examination of the language and various contextual points. For the former purpose they investigate the *rijāl* works where the persons who appear in the *sanad* are discussed in terms of their reliability or weakness. On the basis of *sanad* criticism, *hadīth* reports are classified into several categories (sections 2a–2d, below), important for a *qādī* to learn.

2a. *al-mursal wa al-muttasil*

Al-mursal is sometimes a tradition whose proper *sanad* has been omitted; at other times it is a tradition that has been transmitted by someone who has not heard it directly from the Prophet or the Imams, and further has related it without naming the associate who had heard it from them.[68] There is, however, a difference of opinion among Imamite scholars as to whether *mursal* traditions could be used as proofs to support a legal decision. The majority of the early jurists maintained that such traditions were usually transmitted by thoroughly reliable authorities like Abū ʿUmayr, al-Bizantī, Safwān b. Yahyā, and other associates of the Imams al-Bāqir and al-Sādiq, who, according to Kashshī, even when they narrated in the form of *nursal* were regarded by the concurrence of the Imamites as reporting with full authenticity.

In fact, Kashshī gives us detailed information about three generations of the Imamite transmitters of *hadīth* who reported in the form of *mursal* (i.e., a tradition traced back without interruption to one of the associates of the Imams). The first generation included the most prominent associates of the Imams al-Bāqir and al-Ṣādiq, among whom six were regarded as the leading *fuqahā': Zurāra*, Maʿrūf b. Kharrabūdh, Yazīd b. Muʿāwiya al-ʿIjlī, Abū Basīr al-Asadī, al-Fuḍayl b. Yasār, and Muḥammad b. Muslim al-Thaqafī.[69]

As for the second generation, these were the *fuqahā'*, other than the six mentioned in the first generation, who were the associates of Imam al-Ṣādiq. Again, in their case also, what they authenticated was regarded as authentic even if it was related in the form of *mursal*. These included Jumayl b. Darrāj, ʿAbd Allāh b. Muskān, ʿAbd Allāh b. Bukayr, Ḥammād b. ʿĪsā, Abān b. ʿUthmān, and Ḥammād b. ʿUthmān.

The authentication of *mursal* traditions was crucial for the development of applied jurisprudence, all the more so if such traditions were to be admitted as valid precedents for juridical prescription. More importantly, in the light of the discussion in chapter 1 about the continuity of the "apostolic succession," which was extended to Shīʿī transmitters, acknowledgment of transmitters' juridico-religious authority in the absence of the Imam was contingent upon their being rightfully linked to the Prophetic paradigm through the Imams.

As for the third generation, these were the *fuqahā'* from among the associates of the seventh and eighth Imams, al-Kāzim and al-Riḍā. They too authenticated *mursal* traditions through their own prestige and trustworthiness, and also because of their eminence in Imamite learning. There were also six *fuqahā'* in the third generation: Yūnus b. ʿAbd al-Raḥmān, Safwān b. Yaḥyā Bayyāʿ al-Sābirī, Muḥammad b. Abū ʿUmayr, ʿAbd Allāh b. al-Mughayra, al-Ḥasan b. Maḥbūb, and Aḥmad b. Muḥammad b. Abū Naṣr.[70]

These *fuqahā'* in three generations have formed the group known as *aṣḥāb al-jmāʿ*. Their consensus has been regarded as authoritatively binding, and their *nursal* traditions have been regarded as attaining the level of *sahīh* ("sound"), and hence have been efficacious in making legal decisions.

Al-muttasil or *al-mawsūl,* on the other hand, is an "uninterrupted" *hadīth* report in which the chain of transmission is linked, in ascending order, either to the close associates of the Imam (from whom the first transmitter had heard the tradition directly and had received permission to spread it), or to the Imam himself. Sometimes a *muttasil* tradition is restricted in designation only to those reports that go back to the Imams and their close associates, and to no one else besides them. In this case the *hadīth* is designated "unrestrictedly" *muttasil*. Where the link is restricted to a particular transmitter it is known as *muttasil al-isnād*—that is, a tradition that can be traced back to a certain transmitter and no further.

The classification of *muttasil* traditions into "restricted" and "unrestricted" types indicates the importance attached to the authenticity of the precedent which, even in *ijtihād* where there existed possibility of extrapolation, constituted the only valid basis for a judicial decision as much as the right of that person to engage in deductive inference of a ruling. As noted above, some of these transmitters formed the group known as *aṣḥāb al-ijmāʿ* who had participated with the Imam in arriving at a judicial decision. Accordingly, a 'restricted' *muttasil* tradition more than often was restricted to one of these eminent associates who actually became the source of authenticity for the *hadīth*.

2b. al-musnad wa al-munqati'

Musnad is a tradition that has been provided with a "sound" and uninterrupted *isnād* going back to the Prophet and the Imams, which affords it "authenticity" and "support." When there is a missing link in the *sanad* the tradition is rendered *munqati'* ("interrupted"). Sometimes a *munqati'* tradition ends with an "associate" in the second generation and contains his opinion as well as the communication of the Imam. The "weakness" of such traditions is probably because of the missing link in the *sanad*. On the other hand, a tradition that has fewer links between the Imams and the transmitter is regarded favorably by jurists because the probability of error in transmission is greatly reduced by the reduction of links. Such a tradition is known as *'āli al-sanad,* a tradition provided with eminent *sanad*. In this classification there was recognition of the inevitable insertion of one's own exegesis in the wording of the *hadīth* or even transmission of the purport rather than the actual wording used by the Imams as it passed through numerous hands.

2c. al-'āmm wa al-khāss ("general" and "specific")

This classification of *hadīth* reports is used in determining, for instance, whether a declaration of unlawfulness has a "general" application or is a "specific" prohibition bearing on specific circumstances. Moreover, some traditions, although indicating a general order of prohibition, are restricted by other traditions that state specific circumstances under which a general ordinance must be applied. The *qādī* must carefully study the traditions to determine the general meaning provided in some, and the specifications provided in still others, so as to enable him to issue a sound legal decision.

2d. al-nāsikh wa al-mansūkh (the "abrogating" and the "abrogated")

This classification of traditions is important in the textual study of the ordinances derived from them. It is also regarded as one of the most difficult areas of study in the religious sciences because of the difficulty of determining the "abrogating" traditions chronologically.

An "abrogating" tradition indicates the elimination of a certain ordinance in the Sharī'a with the proviso that the lawgiver (the Prophet) himself eliminated it. An "abrogated" tradition contains the eliminated ordinance and stands as textual evidence in support of the "abrogating" tradition. The only way to ascertain "abrogating" and "abrogated" traditions in ostensible conflict is to find a text from the Prophet to prove the elimination of an ordinance. In the course of history, however, the *ijmā'* (consensus) of the early associates also came to be regarded as a source for ascertaining an abrogation, with the stipulation that the *ijmā'* was discoverer of the Prophet's opinion in a given matter, and not in itself the abrogating agent.

I have summarized the main kinds of *hadīth* reports that qāḍīs must learn to distinguish when they undertake to investigate the *sunna* to formulate a judicial decision. This is crucial for an adequate comprehension of the methods by which jurists apply "general" and "specific" terms derived from Islamic revelation—the Qur'ān and the Sunna—to formulate their opinions. Furthermore, investigation leads to the discovery of "consensus" and "disagreement" among jurists in various matters of application. This is the third area of learning that a *qāḍī* needs in order to exercise *wilāyat al-qaḍā': knowledge of *al-ijmā' wa al-ikhtilāf.*

This is an important area of Islamic jurisprudence. It reveals the flexibility demonstrated in the early stages of the systematic formulation of Islamic law by recognizing the individual nature of legal inquiry in the sources of law. In addition, differences among scholars underscored an important acknowledgment—namely, that judicial decisions made by scholars were subject to variation if the documentary evidence could be exegetically extrapolated in support of a different conclusion. The only way to put an end to further speculation on the interpretation of the sources and provide a uniform system was agreement on the validity of a particular decision. Accordingly, scholarly concurrence provided the required uniformity in the Islamic legal system for its implementation by the sovereign, who nevertheless reserved his right to override any such decision if he deemed it contrary to the public interest. Dissension and concurrence among scholars, furthermore, demonstrates the level of *ijtihād* of each scholar in interpreting the sources in order to elaborate on their legal content and formulate decisions to provide ascertainable prescription for the religious life of the community.

3. *al-ijmā' wa al-ikhtilāf*

The search for *al-ijmā'* (consensus) and *al-ikhtilāf* (differences of opinion) in matters of the Sharī'a entails a meticulous examination of the systematic works of individual jurists, from the earliest generation on, to discover the authoritative and binding opinion of the Prophet and the Imams on a particular question in dispute. Research is, in fact, geared to discovering the *sunna* that must illuminate *al-ijmā'* or its antithesis, *al-khilāf,* with certainty, to render the discovered opinion authoritative. It is only when the opinion of the infallible Imam is revealed with certainty that "consensus" or "disagreement" can be used as evidence for the execution of a decision. In this sense, the process of discovering *al-ijmā'* or *al-khilāf* on a particular question is no different from a search for a *mutawātir* tradition, handed down by numerous chains of narrators on the authority of the infallible Imam, because the function of such a tradition is to direct a jurist to the opinion of the infallible Imam.

It is for this reason that in Imamite law there is a category of reliable narrators who collectively form a group known as *ashāb al-ijmā',* which I referred to above in connection with my discussion on the *sunna.* In practice, they are the prominent *fuqahā'* who have related the Imams' opinions with certainty.[71] *Ijmā'* pronouncements are usually mentioned in the systemic works of *fiqh,* but there are works dealing exclusively with *khilāf,* among which the most significant are Tūsī's *al-Khilāf fī al-fiqh,* Sharīf al-Murtadā's *al-Intisār,* and 'Allāma Hillī's *al-Mukhtalaf.* There were practical considerations that prompted such comparative research by Sharīf al-Murtadā, Tūsī, and 'Allāma into the divergences of legal schools. Primarily it was the interest that the ruling powers had shown in consulting the Imamite doctors on the

religious aspect of public law that called for informative works on the exact differences between the Imamite school and other schools of law. Accordingly, these
works include differences of opinion among all schools of Islamic law, both Shīʿī and
Sunnī, and as such are extremely important sources for the study of different interpretations of legal ordinances among jurists of different schools. Recent scholarship has
shown that there are gaps in Tūsī's *Khilāf* that can be filled and information supple-
mented by ʿAllāma's *Mukhtalaf* and his *Tadhkirat al-fuqahāʾ*.[72]

4. Arabic

Tūsī sums up the qualifications of *al-qāḍī* in the area of *ʿilm* by requiring that he
have a thorough knowledge of Arabic for a profound comprehension and interpretation of all the sources of Islamic law.[73] It is with this type of comprehensive knowledge that the *qāḍī* can employ his "independent reasoning" (*ijtihād*) by personally
investigating legal sources to deduce his own decision.

The exercise of *ijtihād* has been regarded as a requirement in the assuming of
wilāyat al-qaḍāʾ because, according to ʿAllāma Hillī, *al-qaḍāʾ* administration of
justice constitutes *wilāya sharʿiyya*—that is, religiously sanctioned legal authority
instituted for the welfare of the general public. The purpose of *al-qaḍāʾ* is to settle
disputes, and as such, by the very nature of this function, judgment given in it
should remain incontestable by means of *ijtihād,* through the exercise of which the
original judgment was made. The *qāḍī,* says ʿAllāma, differs from the *muftī,*
mujtahid, and *faqīh* in the exercise of his authority, although the qualifications they
need are also required in him. A *qāḍī* is known as *qāḍī* because of his judgment (the
fundamental sense of the Arabic root *qdy*) and his ability to execute that judgment.[74] It is not sufficient for him to know the rulings (*fatāwā*) of other *ʿulamāʾ;*
rather, it is obligatory that he should be knowledgeable regarding all that he has
been invested with, remembering it and memorizing all that is incumbent upon him.
In fact, ʿAllāma declares, if forgetfulness overcomes him or if he needs to be
reminded, then it is not permissible for him to undertake the *qaḍāʾ*. Then ʿAllāma
adds: "The prerequisite that he should know how to write (*al-kitāba*) is dubious,
and so is the prerequisite concerning his ability to see."[75]

Commenting on ʿAllāma's ruling on writing, Muhammad b. al-Hasan al-Hillī.
ʿAllāma's son and the author of *Īdāh,* says that God, who perfected the message of
the Prophet and made the ordinances and the teaching and recording of them
ramifications of divine prophecy, did not make writing a requirement for him.
Indeed the Prophet, says Hillī, was *ummī* (illiterate). Accordingly, inasmuch as
kitāba was not made a requirement in principle, it is even more appropriate that it
should not be made obligatory in its ramifications. However, writing is a means of
recording, and the *qāḍī* needs to record and to remember the details of a case.
Writing is fundamental for both these purposes. Hence, rules Hillī, it is more sound
to make writing a precondition in the *qāḍī* as a measure of precaution. This, he
says, is also Tūsī's opinion in *Mabsūt.* The reason for such a precondition in a *qāḍī,*
according to Hillī, is that the Prophet, being infallible, was endowed with infallible
memory, and was not in need of the art of writing for recording and remembering.
Moreover, forgetfulness and inadvertence were prevented in him through continuous revelation. The *qāḍī* is not protected from these shortcomings and hence it is
appropriate for him to learn the art of writing.[76]

I have already noted 'Allāma's opinion about the independence of *al-qādī* in the matter of formulating his own opinion through *ijtihād*. It is possible to discern, from a virtual consensus among Imamite jurists on the matter of the necessity of employing *ijtihād* in the matter of *al-qadā'*, that administration of justice was regarded as symbolizing *wilāya 'āmma* (general deputyship) derived through reference to the deputyship (*niyāba*) from the Prophet and the Imams. They designated well-qualified persons to this position both for the specifically stated purpose of exercising *wilāyat al-hukm* and for the administration of the affairs of the community, especially in view of the prolonged absence of the Imams' discretionary control. The most frequently cited statement that *al-qadā'* cannot be effected without the *'ālim (faqīh)* who can issue a *fatwā* independently, without depending on the judicial decisions of other jurists, implies that it is only a well-qualified *mujtahid* who can administer justice with truth and equity—that is, the function attributed to the Imam himself. Moreover, it is in *ijtihād,* based on the teachings of the Imams, the ultimate infallible authorities of Shī'ī law, that one can locate the "permission" from the Imams to their Shī'a to refer to those among them who are the repositories of their authentic teachings. In other words, *ijtihād,* which depended on rigorous training in religious sciences, justified the right of the jurist to assume the *wilāyat al-qadā'* of the Imam.

It would be correct to surmise from these rulings that in the absence of the Imam, *wilāyat al-qadā'* could be exercised by the most qualified among their Shī'a. And the most qualified person was obviously the jurist who, through his meticulous and lifelong learning, had become the repository of "sound knowledge."

However, in the later writings on Imamite jurisprudence the requirement of *ijtihād* in *al-qādī* was challenged. Najafi, writing during the Qājār period, argued in favor of some form of positive *taqlīd,* in the sense of accepting the authoritative works of past generations as equally important for the *qādī* who, thus far, seems to have been required to remain independent of the judicial decisions of other *fuqahā'.* In practice, a *mujtahid,* because of his education as discussed above, especially in the matter of *al-ijmā'* and *al-ikhtilāf,* based his *ijtihād* on the past judicial decisions of other *fuqahā';* but in theory, the process was regarded as based on his own deductive inferences. According to Najafī, *al-qadā'* cannot be effected without an *'ālim* who is qualified to issue the *fatwā.* Moreover, it is not sufficient for him to know the rulings of other *'ulamā';* rather, he should be able to give independent judgment based on his thorough knowledge of the sources. However, this does not mean that he cannot adopt the sound judgment of past *'ulamā',* because both conscientious *ijtihād* and carefully researched *taqlīd* would lead to sound judgment. Communications from the Imams indicate the permission from them for their Shī'a to follow the rulings of those who have adhered to their teachings, which have been preserved authentically by these learned individuals by means of sound *ijtihād* or *taqlīd.*

Najafī's opinion reflects the situation of the Imamite jurisprudence during the Qājār period, when jurists for the most part were engaged in employing their *ijtihād* in writing commentaries on the works of past eminent scholars like Muhaqqiq and 'Allāma. Moreover, by this time the concept of *taqlīd* (accepting the authority of those who could exercise *ijtihād*), for those who were themselves incapable of deducing laws from the sources, had become a logical necessity when only few could or chose to become the practitioners of *ijtihād*. In light of this development,

Najafī ruled against requiring *ijtihād* in a *qādī*, who could accept the authority of the *ijtihād* of the eminent scholars by *taqlīd* in executing his judgments.

According to Najafī, the assertion that one who is not a *mujtahid*, and who rules in accordance with the rulings of another *mujtahid*, cannot exercise *wilayat al-qadā'* is baseless. On the contrary, argues Najafī, the evidence is against such an opinion. It is even possible to establish a definite disproof of such a disqualification. That a *mujtahid* should be appointed for *al-qadā'* during the absence of the Imam, says Najafī, is based on the apparent sense derived from the texts of some traditions, which, however, does not necessarily exclude someone other than a *mujtahid* from holding that position.

It is possible to surmise from these traditions that the purport of general designation to administer the affairs of the community is based on the understanding that the designation covers administration of all that on which the Imam himself has authority. Accordingly, the statement of Imam al-Sādiq in regard to his disciple, "I have appointed him *hākim* over you"—that is, *walī*—includes discretionary power in the matter of *al-qadā'*, as well as other executive offices that form other types of *wilāyāt*. In addition, this is also the purport of the statement of the twelfth Imam regarding "future contingencies" (*al-hawādith al-wāqi'a*), in which he advised his Shī'a:

> As for events that may occur [in the future when you may need guidance in religious matters] refer to the transmitters (*ruwāt*) of our teachings who are my *hujja* (proof) to you and I am the Proof of God to you all.[77]

This statement makes clear that the "transmitters" are the proof of the Imam in whatever matter he acts in as the Proof of God. Moreover, the statement does not exclude permission for other than the *mujtahid* to exercise *wilayat al-hukm*, especially one who has knowledge of particular rulings. Nevertheless, admits Najafī, such a person does not possess general leadership nor is he the kind of *qādī* appointed for arbitration (*al-tahkīm*).[78]

There is nonetheless advantage in having the *mujtahid* exercise *wilāyat al-hukm*, for he can designate a *muqallid* (follower) for the administration of *qadā'* in accordance with his *fatwā*, which forms the rulings for the Shī'a in matters lawful and unlawful in the Sharī'a. In such a situation, says Najafī, the ruling of this *qādī* will be the ruling of the *mujtahid* whom he represents; and the ruling of the *mujtahid* will be the ruling of the Imams; and the ruling of the Imams, the ruling of God; and "its rejection will constitute rejection of God."[79]

It is plausible that by the time Najafī wrote his multivolume *Jawāhir*, which is a very detailed and thoroughly documented commentary on Muhaqqiq al-Hillī's *Sharā'i' al-islām*, the position of a *mujtahid* in the matter of exercising various types of *wilāyāt*, as specifically located in Shī'ī law, had been well consolidated. This is corroborated by the fact that Najafī's elaborate rulings regarding *wilāyat al-hukm* have surpassed the brief rulings of the classical works like those of Abū al-Makārim, Muhaqqiq al-Hillī, and others. For the classical jurists the position of *al-qadā'* during the occultation belonged to someone who could demonstrate "sound knowledge" and "sound character," in addition to the ability to put judgments into effect. But for jurists during Najafī's time, when the temporal authority in Qājār Iran had been of Imamite allegiance for over three centuries, there was no question

)f the efficaciousness of the judgment given by the Imamite *mujtahid*, whose "sound knowledge" and "character" had come to be acknowledged as virtually mpeccable.[80] The problem for these later jurists, as appears from their detailed works (both in the form of independent works on jurisprudence and of commentaries on ancient works during this period), was to effect a proper *isnād* between "contemporary" and "ancient" jurists. This meant that even when *ijtihād* was a prerequisite in *wilāyat al-qadā'*, *taqlīd* would also be an inseparable part of the *ijtihād* based on the *nusūs* (texts and rulings) of the "ancients" in the form of further elaborations and elucidations, with the result that *ijtihād* within *taqlīd* would provide a continuity in the teachings of the *ahl al-bayt* generation after generation.

Ijtihād in Imamite law, accordingly, has become the effort to comprehend the sense of a relevant text appearing in the works of the jurists of the classical age onward, and to modify a ruling by reinterpreting the sources on which it was based originally, and by extending or restricting its implications in such a way that a new situation could be subsumed under it by a new ruling. This is known as *al-ijtihād al-shar'ī*,[81] the *ijtihād* based on reinterpretation of a past text or precedent by examining all the proofs that were used as evidence to infer the initial ruling, such as those based on *al-ijmā'*, *al-qiyās*, *al-istislāh*, *al-istishāb*, and also the proofs derived from practical application used to deduce laws when textual proofs were not available in certain cases.[82] *Al-fiqh al-istidlālī*, the "demonstrative" *fiqh* of the Imāmiyya up to the modern age, is based on *al-ijtihād al-shar'ī*.

This fact can be discerned from the published lectures on *usūl al-fiqh* by one of the most eminent present-day jurists, Āyat Allāh Abū al-Qāsim al-Khū'ī. He maintains that there is no difference of opinion on the legality of employing *ijtihād* in deducing *shar'ī* decisions between the supporters of Usūlī and Akhbārī methodologies, because the *ijtihād* that the Usūlī jurists employ is none other than *al-ijtihād al-shar'ī*.[83] This *ijtihād* aims at obtaining absolute proof (*al-hujja*) by exerting the faculties of the mind to the utmost through the study of the Qur'ān and the Sunna for the purpose of forming a legal injunction (*al-hukm al-shar'ī*). Such an intellectual procedure is applied even by the Akhbārī jurists who correctly rejected the *ijtihād* that merely aimed at obtaining presumptive proof (*al-zann*) to deduce a legal decision. With this explanation of *al-ijtihād*, Khū'ī requires that jurists who wish to employ *ijtihād* must equip themselves with *'ilm al-rijāl* (the religious science that determines the authenticity of transmitted communications of the Imams by examining the authenticity of the transmitters), besides thorough knowledge of the Arabic language and the *usūl al-fiqh* (principles of Islamic law).

Furthermore, this *ijtihād*, based as it is on rigorous knowledge of revelation—whether the Qur'ān, the *hadīth*, or *ijmā'* (consensus)—does not require the study of logic (*'ilm al-mantiq*) at all. For what is important in logic, says Khū'ī, is to train a person to formulate a syllogism and deduce a conclusion based on the major and minor propositions. Such a logical formulation can be performed by anyone who possesses a sound rational faculty.

Most legal decisions, says Khū'ī, are derived from the *akhbār* (traditions) because the legal injunctions in the Qur'ān are few and concise. Application of *ijtihād*, then leads to a close examination of the traditions to establish their authoritativeness for use as evidence in a legal decision. Consequently, both Usūlī and Akhbārī jurists depend on the same traditional sources in their search for absolute proof, with a substantial difference: Akhbārī jurists regard all the traditions in the four major compilations of the Imamites as definitely originating from the Imams, but Usūlī jurists do not share this uncritical evaluation of the traditions. They insist on careful research through '*ilm al-rijāl*. In fact, concludes Khū'ī, no Imamite scholars in their *ijtihād* have ever put blind faith in all the traditions attributed to the Imams. They investigate thoroughly textual and contextual factors determining the evidentiary use of them for deducing a legal injunction. It is for this reason that Imamite demonstrative jurisprudence has always upheld *al-ijtihād al-shar'ī*.[84]

Al-'adāla and al-kamāl
in wilāyat al-qadā'

In Imāmī Shī'ism "sound knowledge" without "sound character" has always been regarded as lacking in rectitude. Human perfection (*kamāl*), in Islamic works on ethics (*akhlāq*), has always been located in the perfection of the two faculties of the rational (*nātiqa*) soul: the theoretical faculty (*quwwa 'ilmī*) and the practical faculty (*quwwa 'amalī*). Perfection of the theoretical faculty lies in its perception of all sorts of knowledge and the acquisition of sciences, according to the measure of one's ability, so that it may attain perfect knowledge of the true goal and purpose for which humankind has been created. Perfection of the practical faculty lies in organizing and ordering one's behavior to attain the true goal, which would regulate human life at all levels of its existence:

> So the first perfection is connected with speculation (*nazar*) and is [as it were] the form, while the second perfection can be regarded as matter (*mādda*). Just as form without matter, or matter without form, can possess no stability or permanence, so theory without practice is abortive and practice without theory absurd. Theory is the starting point and practice the conclusion. The perfection composed of both is that which we have called the "purpose of human existence," for "perfection" and "purpose" are approximate in sense, the difference between them being that established by relationship: "purpose" is what is still in the state of potency (*quwwat*); when it reaches the state of act it becomes "perfection."[85]

From the above passage cited from the *Akhlāq-i Nāsirī* of the Shī'ī savant Khawja Nasīr al-Dīn Tūsī, it can be seen that human perfection is dependent on the perfection of the two faculties of the rational soul. This interdependence between the two faculties is consistent with the Shī'ī religious belief system, which is dominated by the idea of the Imam of the Age (*Imām al-zamān*). According to this system, it is this Imam who can guide human society toward that perfection. As such, the Imam of the Age provided the criteria of perfect human behavior, at the individual as well as the collective level, which should be emulated by those who represented him in the exercise of authority (*wilāyāt*) in the areas constitutionally assigned to him alone or his explicit legatee. As a matter of fact, the legitimacy of any claim to exercise authority requiring the Imam's explicit designation was made dependent on possession of an Imamlike character—that is, possessing '*adāla* and *kamāl* (perfection of the practical faculty), besides having "sound belief" and "sound knowledge" (perfection of the theoretical faculty). It must be remembered that *al-qadā'* was regarded as part of *al-wilāyat al-'āmma* (all-comprehensive authority) on the basis of deputyship (*niyāba*) from the infallible Imams, both in particular and in general.[86] Consequently, *al-qadā'* exercised by a jurist who was not *'ādil* was not effective. In other words, "sound knowledge" became identified with "sound character."

'Adāla has been defined with much care in *fiqh* texts in the discussion about *al-wilāyāt*. Definition of *'adāla* also occurs in the sections dealing with the *imāma* of daily worship where, consistent with the Shī'ī conception of *al-wilāyat al-shar'iyya*, it is maintained that the Imamate of the *salāt* should be assumed by a righteous (*'ādil*) believer. The question of *'adāla* in the *salāt* will be dealt with in chapter 5 in the section on the Imamate of the Friday *salāt* and its politico-theological implications. Here it is relevant to discuss the prerequisite of *'adāla* in connection with *wilāyat al-qadā'*.

The requirement of *'adāla* in a jurist occurs early in the history of Imāmī Shī'īsm, as can be construed from the tradition of 'Umar b. Hanzala. In this long tradition, cited and discussed above, Imam al-Sādiq lays down the principles for selecting a jurist whose legal decisions would be binding. This includes his being *'ādil*.[87] The tradition acknowledges the difficult task of determining the *'ādil* jurist among many equally qualified to judge among the Shī'a.

In another tradition this difficulty was alleviated by Imam al-Sādiq when asked by Ibn Abī Ya'fūr how one could recognize *'adāla* in a person. He replied that *'adāla* could be known in a person who dressed himself properly, being chaste, and who guarded his physical and mental activities, keeping them within the boundaries set by God. It could also be recognized in a person who avoided committing the grave sins for which God has promised severe punishments, such as drinking wine, adultery, usury, disobedience to parents, deserting the battlefield, and so on. This abstention from grave sins functioned as a cover for his other shortcomings, and Muslims were prohibited to search beyond these grave sins to discover his other faults and weaknesses and to investigate his character in extreme detail. On the contrary, if a person shuns the grave sins, it is incumbent upon believers to attest his integrity and credibility, and make his *'adāla* known to all. Furthermore, this person can be recognized because he also keeps the five daily acts of worship at their specific times and leads Muslims in their congregational prayers, and, with the exception of those who are sick, everyone attends prayer behind him. If such a leader was available, then it was positively incumbent on the people to attend the mosque services at the prescribed five times a day. The *'adāla* of the person could further be attested by asking those who know him, such as members of his tribe or his locality, who could declare without any hesitation that his behavior was beyond reproach and that he was very assiduous in the performance of daily worship at the proper times in the mosque. When this is recognized in a person, then his testimony cannot be questioned, nor his *'adāla* among Muslims be challenged, because his acts of worship serve as his shield and as expiation for any misdeeds. Al-Sādiq assures Ibn Abī Ya'fūr that had it not been so, it would have been impossible to ascertain a person's *'adāla* in order for that person to act as a witness, or take on any other responsibility requiring such an aptitude. Consequently, a person who prayed at home and whose piety could not be recognized in public could not undertake the position that required *'adāla,* because *'adāla* must be recognizable through the avoidance of the grave sins and through the confidence others demonstrate in that person by worshiping behind him in the mosque.[88]

The fact that 'adāla, although consistently defined as an "innate" or "natural" disposition (*malaka nafsāniyya*) that made a person follow piety and *muruwwa* (manly virtues), had to be objectified is further attested by the insistence that such a person enjoy among his contemporaries a reputation for "good outward behavior" (*husn al-zāhir*).[89] In fact, there is consensus on this among Imamite jurists, based upon the following authenticated tradition reported on the authority of the Imam al-Sādiq, which specified that 'adāla must be required of the Imams who lead congregational prayers:

> Do not pray behind an extremist (*al-ghālī*), even if he believes in what you believe and who, both in his private as well as public life, acts wickedly, while claiming that he is a believer.[90]

Another tradition reported on the authority of Imam al-Ridā advises the Shī'a not to pray behind an Imam who lets himself be tempted by a sin when he is fully aware of the sinfulness of the act in question, because such a person is a perpetrator of a sin.[91] 'Adāla, thus, is the opposite of *zulm* ("wrongdoing"), an objective category of evil action. 'Adāla has the precise signification of "sensual probity and freedom from sensual deviation" necessary to develop the spiritual probity required in the comprehension of religious and legal matters in the Sharī'a. Accordingly, 'adāla has levels, with 'isma (infallibility) being at the top, whereas *fisq* (wickedness) is at the bottom of the scale, when 'adāla is completely absent. As persons move upward on the ladder of 'adāla, their mental aptitude produces firmness in religion and diligence in piety, thereby perfecting their intellect and their character through pursuance of dauntless virtues. 'Adāla is, thus, like all the virtues, that by which one shuns all acts of disobedience.[92]

In studying 'adāla and *kamāl* it is important to remember that *wilāyat al-qadā'*, as treated above in my introductory remarks, was seen as the function of the Imam himself or his directly appointed deputy. In the absence of the authority of the Imam or his deputy, the superhuman qualities demanded of their indirect representatives in the form of 'adāla or *kamāl*, in all the authoritative sources of Imamite jurisprudence subsequent to the occultation, demonstrate an important concern in the minds of the "ancient" jurists to limit the number of jurists who could exercise this *wilāya* authoritatively. This was a necessary precaution in the light of the struggle for the leadership of the Imamite community that followed the occultation in A.D. 874. Moreover, although jurists were willing to devise practical solutions to ensure the survival of the nascent Imamite community by allowing the *wilāyāt* to be exercised by well-qualified Imamite jurists, they did not and could not allow any compromise of the basic Shī'ī doctrine of the Imamate of the Hidden Imam. Even in occultation, the twelfth Imam remained the Imam of the community.

In the meantime, Imamite jurists played the role of functional imam, carrying on an impressive and ingenious work of guiding the Shī'a community through the vicissitudes of history. What legitimized their role as "functional imam" and ensured public approval of their leadership was the triad of qualifications recognized by the community in theory and in practice—"sound belief," "sound knowledge," and "sound character." Hence, the *walī* for the administration of justice—a position that became a stepping stone for much wider politico-theological functions—called for singular piety, rectitude, and virtue. Only someone so qualified could occupy a place that, according to ʿAlī b. Abī Ṭālib, was reserved for prophets or their successors. However, according to this tradition, there always remained a question regarding the legitimacy of assuming the authority that lacked a designation for undertaking an all-comprehensive *wilāya* of the Prophet or his executors. The requirement of *ʿadāla* and *kamāl* could not resolve the theoretical problem of not having specific designation (*nass*) from the Imams to assume *al-wilāyat al-ʿāmma,* as will be discussed in the next chapter. The absence of a theoretical designation for anyone other than the Imams to occupy the place of the Prophet's "executor" had to be compensated for by investigating other pieces of documentary evidence that could be interpreted and extrapolated to prepare the way for the recognition of all-comprehensive, tangible Imamite authority during the occultation. A number of these traditions whose texts have been exegetically extrapolated to demonstrate the requisite theoretical designation of an Imamite jurist to assume *al-wilāyat al-ʿāmma* will be treated in the next chapter.

In the area of *kamāl*, jurists were required to possess perfection in two things: perfection in "(physical) constitution" (*khilqa*) and perfection in "(judicial) judgments" (*ahkām*). As for the perfection in physical constitution, they should possess all the sensory organs, especially eyes, because the blind cannot adjudicate independently (i.e., without visual information about the plaintiff and the prosecutor), nor can they see what is written in their presence, and so on. As for perfection in judgments, jurists should be mature, intelligent, free, and male. A woman cannot administer *qaḍā',* although some jurists, according to Tūsī, have permitted women to become judges. But this decision, he says, is not soundly based. As for those who have allowed women to become judges, they have allowed them in those areas in which their testimony would be acceptable—namely, all areas except *hudūd* and *qisās* (i.e., in areas requiring the exaction of corporal punishments).[93]

Ibn Hamza, in his *Wasīla,* adds one more thing to the attainment of perfection in physical constitution—*al-idtilāʿ bi al-'amr* (to possess power sufficient for the affair).[94] Although excellence in virtues and judiciousness could ensue from *ʿadāla* as well, it is the power or "strength" to bear the burden of *wilāyat al-qaḍā'* that clearly determines the level of *kamāl.* Indeed, without this "power" the demand for perfection remains a mere potential in a position of leadership; it is not considered applicable to the socio-political situation of the Shīʿī community.

The reference to *al-idtilāʿ* in the perfection of a jurist is closely related to the question of *tanfīdh,* "carrying out" the decision made by a qualified judge. It should be remembered that the sections on *adab al-qaḍā'* in the works on *fiqh,* where the question of qualifications of the *qāḍī* is discussed, have an important purpose—namely, to establish the efficacy of the judgment made by such an individual in areas where general permission of the Imam is available for *al-faqīh* to assume the juridical function of the Imam in the absence of a specific designation from the Imam. The oft-quoted communication of the Imam al-Sadiq, "Beware of judging (*al-hukūma*), because it is for the Imam, knowledgeable in the matter of *al-qaḍā',* to judge with fairness among Muslims," cautions a jurist not to undertake *hukūma* unless he is well-qualified to put Sharīʿa ordinances into effect.[95] According to Shahīd II, if a jurist is well qualified, he does not even need an explicit designation to undertake *al-qaḍā',* because his being *mujtahid* makes it possible for him to undertake any case and rule on it effectively. It is therefore necessary that a *qāḍī* be a *mujtahid* at all times.[96] In other words, his qualifications determine his *wilāya* in the administration of justice.

This insistence on the qualifications of a jurist is what the Imamite jurists have maintained throughout their works on *al-qadāʾ*, essentially in the form of "permission" (*idhn*) from the Imam. In *Sharāʾiʿ*, Muhaqqiq holds that in order to prove *wilāya* it is necessary that it derive from the permission of the Imam, or of the one in whom the Imam has invested such authority. Thus, if the inhabitants of a city seek to appoint a *qādī* without the Imam's permission, then the *wilāya* of such a *qādī* in that area is not proven. However, in the absence of the Imam, says Muhaqqiq, the administration of justice undertaken by a jurist from among the Shīʿa of the *ahl al-bayt* (i.e., the followers of the Imams) becomes effective. Such a jurist is the one who possesses all the necessary qualifications in the issuance of legal opinion (*fatwā*), just as the Imam al-Sādiq has said: "Make him the judge, for surely I have made him a judge. Thus, take your disputes for judgment to him."[97] The equation of "permission" with qualifications is further connoted in *Qawāʿid* of ʿAllāma Hillī, where he clearly says that the *tawliya* (appointment) of *al-qadāʾ* is established by the permission of the Imam or his deputy, and cannot be established by the designation of the populace of a city or region. And then he adds:

> During the *ghayba* [occultation], the *qadāʾ* administered by a well-qualified jurist, who can issue legal decisions (*fatwā*), is effective, and one who turns away from him, [appealing for justice] to the judges appointed by unjust rulers, is a sinner.[98]

Here ʿAllāma seems to be implying that "permission" of the Imam is available to jurists who are well qualified to issue legal decisions, although he does not explicitly say so. This equating of "permission" with "qualification" has a precedent in the appointment of the well-qualified associates of the Imams to the position of *al-hākim*, as seen in the tradition of ʿUmar b. Hanzala. It is in this tradition that the idea becomes well established that a more learned, more pious, and more righteous jurist has "permission" to exercise *wilāyat al-qadāʾ* among the Shīʿa. As a result, the tradition of ʿUmar b. Hanzala has been most frequently used as one of the main pieces of documentation for supporting the qualified juridical authority of Imamite jurists as having the sanction of the twelfth Imam. Significantly, this tradition is known as *riwāyat al-nasb*—that is, the "tradition [that establishes] investiture."

A well-qualified jurist has not only the "permission" of the Imam to exercise *wilāyat al-qadāʾ*, but also the obligation to make himself known to assume this reponsibility. In a section on those who can or cannot undertake *al-qadāʾ*, Tūsī says that there are three kinds of persons to be considered in the matter of undertaking the administration of *al-qadāʾ*: (1) those on whom it is incumbent to undertake *wilāyat al-qadāʾ*; (2) those for whom it is prohibited; and (3) those who have permission to assume it. As for those on whom it is incumbent to administer it, they are the ones on whom the duty of assuming this obligation has become individual and personal (*ʿaynī*) by virtue of their being thoroughly reliable, and belonging to the group of the leanred (*ahl al-ʿilm*), especially when the Imam cannot find any other well-qualified person for the position. In that situation, it is for the Imam to appoint such a person and it is for such a person to administer justice, considering it to be a personally obligatory act. If the Imam does not know this person, then it is his duty to introduce himself to the Imam in order for him to appoint him an administrator of justice.

The reason for this obligation on the part of the well-qualified person is that *al-qadā'* is classified as a "representatively" obligatory act, the fulfillment of which by a qualified person excuses other individuals from fulfilling it; a typical example would be funeral prayers.[99] Thus, when a person dies and there is no one else to perform the funeral rites except one person who knows how to do it, then it is incumbent upon that person to perform the final rites connected with shrouding and burial. Moreover, Tūsī argues, *al-qadā'* is part of the general obligation regarding "enjoining the good and forbidding the evil." If there is no one who can uphold this duty except one person, then it is incumbent upon that person to do so.

The one prohibited from undertaking *al-qadā'* is either someone who administers justice when he is ignorant, regardless of whether he is reliable or not, or someone who is corrupt (*fāsiq*), irrespective of whether he is among the *ahl al-'ilm* or not. Some jurists, according to Tūsī, are of the opinion that if a person is reliable, it is permissible for him to administer justice, even if he does not belong to the group of the learned (*ahl al-'ilm*). However, according to the Imamites, says Tūsī, this opinion is not valid, because there is a Prophetic tradition to the effect that an ignorant person judging among the people will end up in hell.

As for the one who has permission and is not prohibited from assuming *al-qadā'*, this is an individual who happens to be among the group of jurists who are all equally qualified to administer justice. If the Imam calls upon one of them to undertake this responsibility, then, according to some scholars, it is incumbent upon that person to assume the responsibility. But others have ruled that it is not incumbent upon him, because of the presence of many equally qualified persons to respond to the call of the Imam. This latter opinion, according to Tūsī, is more sound and it is the Imamite position on the question.

As a corollary to this position, Tūsī raises the issue of whether taking up the *qadā'* under those circumstances, if not incumbent, is still a recommended (*mustahabb*) duty for that person. Tūsī's response is that the situation can be analyzed thus: the obligation of *al-qadā'* can be either representatively or nonrepresentatively incumbent. If it is a nonrepresentative obligation, then *al-qadā'* becomes a recommended duty for him to assume, because by doing so he is obeying God by attending to the needs of the poeple. In return for this service there is maintenance for him, whereas if he does not undertake *al-qadā'* he might have to seek his livelihood from some other permissible means, such as trade, and so on. Indeed, concludes Tūsī, seeking livelihood through obedience to God is better for him than seeking it from other permissible means.

On the other hand, if *al-qadā'* becomes a "representative" obligation for him, then the situation is twofold: either he is famous or he is unknown. If he is famous for his learning, and well known for the fact that others refer to him seeking his legal opinion and they learn from him, then it is recommended for him not to undertake the administration of justice, because teaching and instructing are also forms of obedience and worship, with peace and security from hazard. *Al-qadā'*, even when it is a form of obedience to God, carries with it a risk, as the Prophet has declared: "One who undertakes the administration of justice, will be butchered without a knife." Accordingly, safety from such risks may be seen as better for his religion and his good faith.

However, if a person is unknown, and his knowledge and excellence of character are unrecognized, and others have not benefited from his knowledge, then it is recommended that he should undertake al-qadā' in order to guide the people toward himself, to manifest his excellence, and to benefit the people with his knowledge. Some, according to Tūsī, have gone as far as to suggest that it is recommended for him to spend some wealth for getting the necessary recognition, so that he would be known and the people would be aware of his excellence, and benefit from him. But the former opinion is sounder, because spending wealth for this purpose is not permissible; nor can the Imam take anything in return for appointing such an individual in the position of al-qadā'.

Tūsī then raises another related question regarding the permissibility of receiving recompense for undertaking al-qadā'. In this connection, there are two kinds of qadā': when it is personally incumbent to undertake the qadā', and when it is not personally incumbent. If the person concerned is one who has permission to exercise the authority of al-qadā', but is not obligated personally to undertake the authority, and if he is one who is either "representatively" or "nonrepresentatively obligated to undertake al-qadā', then in the situation where he is nonrepresentatively obligated, he can receive recompense for his service; whereas if he is representatively obligated to assume the authority, then it is *recommended* that he should not accept payment; but it is permissible for him to accept payment. To award payment for the service performed in administering justice is permissible, as established through consensus (ijmā'), because, ultimately, the public treasury exists for the well-being of society; and al-qadā' is not only part of its well-being, it is something that the community has great need of in settling disputes, in seeking legal opinions in matters of law, and in assisting the oppressed and restraining oppressors.

This is the situation when the qadā' is not personally incumbent on a particular person. But when it becomes personally incumbent, that person could be undertaking it because it has become incumbent either "representatively" or "nonrepresentatively." If he undertakes it as a representatively incumbent duty (that is, substituting for a larger body of persons), then it is prohibited for him to receive payment for assuming the responsibility, because he is performing an act that has become personally obligatory for him. And for undertaking this obligation, it is not lawful for him to receive reward when he is able to do without it. But if he undertakes it when it is not representatively obligatory on him, then it is lawful for him to receive payment, because it is his duty to provide for his family, and undertaking the qadā' is another duty beyond that. As such, when he receives payment he is actually combining the two obligations, because receiving recompense relieves him of earning a livelihood. Indeed, combining the two obligations is better than failing to fulfil either of them.

If the Imam knows that a region needs a qādī, then it is necessary for him to dispatch one. This is supported by the Prophetic practice when Muhammad sent 'Alī to Yemen and 'Alī sent 'Abd Allāh b. 'Abbās to Basra as a qādī. There is concurrence on this point among jurists. Moreover, if the Imam knows the most suited person for this position who will benefit the people, he must appoint him. If he does not know such a person, he should summon learned persons (ahl al-'ilm) and have them confront each other, and he should interrogate them to find out the most qualified person for the position. After having selected him from among the learned group, he should send someone to the neighborhood where he lives, to his house, to his mosque, to persons around his marketplace, and to those who know him. Then he should investigate his *'adāla* (righteousness), just as the judge investigates the righteousness of witnesses. When he has done all that is necessary to ascertain that he is among those who can administer justice, he should appoint him and write instructions for him, to the effect that he should fear God and obey God in his own person and in his opinions. The Imam should also order him to pay close attention to the character of witnesses; attend to the welfare of minors; take care of pious endowments, and other such matters that are usually assigned to administrators of justice.[100]

The way in which Tūsī discusses the Imam's duties in appointing the judge for a region leaves an ambiguity: whether the Imam he refers to in his *fiqh* rulings is the theological Imam or the political Imam, who need not be divinely designated to perform the functions of the real (*al-asl*) Imam. This ambiguity, as discussed in chapter 3 in greater detail in connection with the use of the titles *sultān al-zamān* and *sultān al-islām,* is never resolved in Tusi's works or in the works of other Imamite jurists who were writing under *taqiyya,* and at a time when precautionary ambiguity in the use of the word "Imam" was deemed expedient. Thus, for instance, in dealing with the question of accepting an office (*wilāya*) to implement *hudūd* on behalf of an unjust ruler, Tūsī rules that if a person is appointed by an unjust ruler to implement legal punishments on a people, it is permissible for him to do so to the best of his ability; and in his heart he should consider the performance of this duty as under the order of *sultān al-haqq* (just authority). Furthermore, it is necessary for believers to assist and support such a person in the performance of his duties.[101] The reference to *sultān al-haqq*—just authority—is clearly a precautionary measure (*ihtiyāt*).

This opinion was challenged by Ibn Idrīs. He advised the Shī'a not to follow this particular opinion of Tūsī, because Tūsī had not given the opinion with the specific intention of issuing a legal decision (*fatwā*), and had not discussed the full implications of the opinion in his *Nihāya*. In fact, as pointed out above in another context, Ibn Idrīs maintains that in several places in *Nihāya* Tūsī simply informs the reader about the situation without regarding this information as having the status of a legal opinion (which would necessarily require discussion of the problem in detail and documentation from authoritative sources like the Qur'ān, Sunna, or *ijmā'* to support the *fatwā*). This observation about *Nihāya* is corroborated by the fact that, according to Ibn Idrīs, there is an *ijmā'* of Imamite scholars to the effect that it is not permissible for the Shī'a to administer *hudūd,* nor is it intended for anyone but the Imams and the *hukkām* (judicial authorities) designated by them.[102] However, there was another *ijmā'* of Imamite scholars on the basis of the "investiture tradition" of 'Umar b. Hanzala, which was interpreted to include the generality of the Shī'ī scholarly elite as the indirectly appointed *hukkām.* This tradition enabled the Baghdad jurists of the classical age (Mufīd, Sharīf al-Murtadā, and Tūsī) to regard the Imamite jurist as the *khalīfat al-imām* (the "successor" of the Imam [in occultation]) whose authority among the Shī'a was comparable to the authority of the imam among the Sunnī community.

Tūsī's section on *al-qadā'* in *Mabsūt,* although written with due precaution, especially its detailed treatment, indicates recognition of the delegation of the Imam's juridical authority to the jurist who is referred to as the *khalīfat al-imām.* *Mabsūt,* Tūsī's final work on jurisprudence, reflected the cumulative experience of the Shī'a community in the eleventh century and the inner necessity to buttress

Imamite authority during the occultation so that it could function as a viable alternative to the de facto power of Imamite authority like that of the Būyids. Accordingly, the ambiguous usage of the title "Imam" in this whole section of juridical administration without any reference to the occultation and with full knowledge of the doctrinal Imamate (including the clause regarding the infallible knowledge of the Imam) signals the tacit recognition of the political imamate of someone other than the Imam himself during his concealment, which could be assumed by the ruling authority committed to uphold the Shar'ī vision of justice.

However, in subsequent works on *fiqh* this ambiguity was removed and the possibility of the existence of *al-sultān al-'ādil* (the just authority) besides the Imam *al-asl* as a fait accompli was considered legitimate. Perhaps the most explicit statement in this regard comes from Muhaqqiq:

> Acceptance of authority (*wilāya*) from *al-sultān al-'ādil* [the just authority] is permissible, and sometimes obligatory. It becomes obligatory [when the person] has been specifically appointed by the Imam *al-asl* [who is infallible], or when it is not possible to interdict evil and enjoin good except [by agreeing to hold the office from *al-sultān al-'ādil*]. But it is prohibited to accept [the position] from unjust authority when this is not the case.[103]

The difference between two categories of obligatoriness in accepting administrative position points to the fact that the two phrases *al-sultān al-'ādil* and *imām al-asl* have not been used synonymously by Muhaqqiq, and that the former clearly applies to *any* authority who upholds the Sharī'a law and rules justly. Moreover, in his next ruling Muhaqqiq speaks about the prohibition of accepting an office on behalf of *al-jā'ir*—unjust authority, the opposite of *al-'ādil*—which further characterizes political authority regardless of the fact of its not having been designated by divine authority.

As discussed in the previous chapter on the Imamite theory of political authority, it is plausible that during the period of the Complete Occultation—whether under *taqiyya* or under expediency created by realization of the existence of a Shī'ī political authority (other than that of the Imam) willing to consider the implementation of the Sharī'a law—the Imamite jurists began to apply the phrase *al-sultān al-'ādil* (the just authority) to mean any just Shī'ī authority committed to the promulgation of the Sharī'a.

There was a precedent for such a recognition among Sunnī scholars. Having witnessed the crumbling of the 'Abbāsid caliphate under the weight of the powerful de facto *sultān*s, Sunnī scholars had come to rationalize the concept of the *sultān* as being bound to govern according to the Sharī'a and thereby being bound to give effect to the general purposes of God for Islamic society. Shī'ī scholars saw a similar opportunity for God's purposes to be effected under the *sultān* who could be made responsible for the protection of the Shī'a community during the occultation.

Later on, during the Safavid and post-Safavid era, as we shall see in the next chapter, there was a clear indication in the *fiqh* writings of this period that *al-hākim*—that is, *al-imām al-'ādil*—was to be seen as the well-qualified *faqīh* during the *ghayba*.[104] It is in the extension of the application of the title *al-'ādil*—strictly reserved for *imām al-asl* to *hākim al-shar'*, the Imamite jurist—that one should reconstruct the inevitable and necessary growth, however gradual, of the *wilāya* of the Imamite jurists from *wilāyat al-qadā'* to *al-wilāyat al-ʿāmma*.

In this chapter I have concentrated on *wilāyat al-qadā'*, the authority to administer justice and its theologico-political implications. I turn next to *al-wilāyat al-ʿāmma*, the general authority invested in Imamite jurists under the aegis of *wilāyat al-faqīh*. But before I proceed to do so, some tentative conclusions would not be out of place.

Conclusions

Wilāyat al-qadā', the authority to administer justice, was perceived in Imamite jurisprudence as a constitutional right of the Imam, who alone could administer justice but who could also delegate this function to someone well qualified to exercise it on his behalf. This well-qualified person was the *rāwī* (transmitter) of the Imam's authoritative teachings, who was designated as *al-hākim* (judicial authority) over the Shīʿites by the Imam al-Sādiq.

During the Complete Occultation of the twelfth Imam the question was raised about the delegation of the Imam's judicial authority to an Imamite jurist to assume the position of *walī al-qadā'* (administrator of justice) because in practice he was not only the *rāwī* of the Imam's teachings, but also a functional imam, performing all the duties that the real Imam was to undertake. However, it was important for jurists to establish that the "permission" of the Imam was available for them to assume lawfully those functions of the Imam that carried politico-theological implications. To this end, the principle of *ijtihād* (independent reasoning) was evolved to give broader interpretation to the piece of documentation that was used in the first place as evidence for *al-hākim*'s position. At the same time, the qualifications of a person who could assume the *wilāya* were worked out in great detail, specifying the sort of training and strength of character required in *walī al-qadā'*.

The result of this intellectual exercise was to provide badly needed regulations governing the position of a judge and the method of his executing judgment on the basis of the sources of the Sharīʿa as adequately determined during the postoccultation period. Moreover, the juridical prescriptions that were formulated as a result of this intellectual exercise allowed the Shīʿī scholarly elite to lay down the firm foundation of their authority as the *khalīfat al-Imām* under the rationally and revelationally established duty of "enjoining the good and forbidding the evil" on which depended the entire administration of justice in Islamic society. "Enjoining the good and forbidding the evil" became the theologico-legal justification for the jurist to assume the *wilāyat al-qadā'* and for this juridical authority to become a legal basis for the legitimation of the all-comprehensive authority of the jurist that was emerging as an inner necessity to protect the interests of the Shīʿa community during the prolonged occultation.

Furthermore, the *wilāyat al-qadā'* and the authentication of the documentary evidence admitted to establish the comprehensive authority of the jurist both justified and validated the intellectual process by means of which scholars devised terminological and exegetical methods to settle the issue of the power that accrued to the "general" deputy of the Imam. The ability of the jurist to employ this methodology was contingent upon his rigorous training in the religious sciences; but his ability to exercise authority and to make binding judicial decisions in the community required him to be in possession of the Imamlike qualities. The combination of "sound knowledge" and "sound character," in addition to the fundamental "sound belief," allowed him to become part of the "apostolic" authority of the Imam and wield power in his name.

The coming together of all the qualifications of Imamite leadership, in other words, gave the jurist the necessary "permission" to undertake the *wilāyat al-qadā'*, even when he lacked the explicit designation of the Imam's "special" deputy for that purpose. More importantly, the *wilāyat al-qadā'* was the most relevant form of *wilāya* in the realm of the religiously ordained positions that carried definitive ramifications for the future growth of the politico-religious authority of the jurist.

Al-qadā', administration of justice, which was headed by the jurists, became one of the most important institutions in preserving the popular sense of common justice and, as such, at times when the central power of the Muslim rulers had disintegrated, the prestige of *al-qādī* or *al-hākim* became immeasurable. It became the only administrative institution committed to preserve the Sharī'a as the embodiment of divine justice. It is for this reason that *al-qadā'* should be regarded as the most fundamental aspect of the growing political power of the jurists, who (as the lawful administrators of *al-qadā'*) were regarded as the protectors of the people against the license or tyranny of those in power. The persistent Qur'ānic challenge of creating a just society, then, could be undertaken by these *hukkām*.

This challenge marked the transformation in *wilāyat al-qada'*, with all its theologico-political implications. It also forced the jurists to undertake the role of functional Imam, beyond their traditional role of the interpreters of Islamic revelation as the general deputy (*nā'ib al-'āmm*) of the Hidden Imam. The new role was carefully worked out in all its details in the form of *wilāyat al-faqīh*, which awaited its actualization at the proper time in history. In other words, *al-hākim* waited to assume the full responsibilities of *al-imām al-'ādil* (with whom his role became equated) as those in power in the Islamic empire persistently failed to realize divine justice on earth. This also helped to introduce the time when the idealized, Imamlike leadership of the jurist began to be seen as the only legitimate authority that can assume *al-niyābat al-'āmma* (the general deputyship) to further the welfare of the Imamite community.

5

The Comprehensive *wilāya* of the Jurists

In my study of the Imamite theory of political authority I have discussed the concept of *wilāya* and its ramifications for the Imamite classification of "just" (*'ādil*) and "unjust" (*jā'ir*) authority. The exercise of legal and moral authority (*saltana*) through the demand of obedience without proper designation by God in the form of *wilāya* (the faculty of that authority), which enables a person to assume authority in the Muslim *umma*, was, according to the Imamite theology, an act of injustice because it was tantamount to encroaching upon the right of an individual in whom *al-wilāyat al-ilāhiyya* (the divine sovereignty) had been invested by God. Accordingly, the question of *wilāya*, more particularly *al-wilāyat al-'āmma* (the all-comprehensive authority), in the case of Imamite jurists who were required to assume the exercise of authority affecting the religious-socio-political structure of the Imamite community, was of great consequence for the Imamite theory of political authority.

The fundamental difference between the *wilāya* of the Imams and that of the jurists was that, theoretically, there was an explicit designation (*nass*) for each of the Imams to undertake the *wilāya*. The *wilaya* of the jurist, however, depended on the recognition of the Shī'a. According to the tradition of 'Umar b. Hanzala, the Shī'a were responsible for seeking out the most qualified jurist to assume *wilāyat al-hukm*. As such, the *wilāya* of the Imams did not depend on the acknowledgment of the people, although assuming the office of the *imāma*, which was the manifest consequence of the *wilāya*, depended on the formal homage (*bay'a*) of the people. On the other hand, the *wilāya* of the jurist clearly depended on the acknowledgment by the Shī'a of the traits specified by the Imams as qualifying a particular Imamite scholar to assume the *wilāya* invested in him through a general designation of the Imam.

In addition to this fundamental difference in the *wilāya* of the Imams and that of the jurists, there were two important theological implications of the *wilāya* of the Imams, which were at the root of the discussion about the extent of the *wilāya* of the jurists and its theological ramifications. First, acknowledgment of the Imams' *wilāya* was necessary in order to follow them in all matters affecting spiritual as well material life. Secondly, acknowledgment of their *wilāya* involved recognition of their rights and belief in their Imamate and, as a consequence of this recognition, it meant obedience to their commands and interdictions. Both these theological implications of the *wilāya*, with obvious ramifications in the political leadership of the Imams, were determinant in defining the comprehensive authority of the jurists. It is to this *wilāya* of the jurists, which developed out of historical vicissitudes and was based on the theological underpinnings of the Imamite theory of political leadership, that I now turn.

The Kinds of *wilāyāt* Mentioned in the Juridical Works

The term *wilāya* occurs in Imamite jurisprudence in various contexts and in each one it means either "authority to act as a guardian" or "authority to administer or supervise a delegated task." Thus, in the section *Kitāb al-nikāh* of the work *Kanz al-'irfān*, al-Sīwurī (d. 826/1422) lists four kinds of *wilāya* in which a *walī* is given authority to act as a guardian or to administer a task:

(1) *wilāyat al-qarāba*. This type of *wilāya*, authority to act as a guardian, is based on relationship. This authority is given to a father and a paternal grandfather, excluding other relatives. A father and a grandfather, in Imamite jurisprudence, have the authority to exercise *wilāya* over minors and over the insane who continue to be so even after having attained the age of puberty. There is a difference of opinion, however, in regard to the exercise of authority over an adult virgin. The most probable opinion, according to Sīwurī, is that it is void, for there are traditions to this effect reported on the authority of the Imams al-Bāqir and al-Sādiq. This type of *wilāya* is arbitrary and *ijbāriyya* (coercive): a person over whom *wilāyat al-qarāba* is exercised (*al-mutawallā 'alayhi*) has no choice but to accept the guardianship of the father or grandfather.

(2) *wilāyat al-hākim*. This form of *wilāya* is authority to supervise all that is in the interest of a person (*al-mutawallā 'alayhi*) when that person is unable to exercise reasoning, because of intellectual immaturity or imperfect reasoning, and when that person does not already have a *walī*, or when that person's senses and opinions remain corrupt even after the attainment of one's majority. The *hākim*—that is, the regular administrator of justice—is given the authority to exercise this *wilāya*.

(3) *wilāyat al-mulk*. This *wilāya* accrues to a master who owns slaves. This *wilāya* is regarded as the most discretionary authority because it is exercised over slaves, both male and female, whether they have attained the age of maturity or not, and whether they are in possession of intellectual maturity or are mentally defective.

(4) *wilāyat al-'usūba*. According to al-Sīwurī, this *wilāya*, on the basis of being agnates, is gratuitously added by the Sunnites and is not valid among Imamites.[1]

Sīwurī's book deals with juristic decisions deduced from the Qur'ān only. Consequently, in his discussion on the forms of *wilāya* he does not take up the important question of the authority delegated to *hākim al-shar'* in certain types of *wilāya*. It is plausible that the discussion of forms of *wilāya* in the performance of certain religiously prescribed (*shar'ī*) obligations derives from the *ijtihād* of a jurist based on the extrapolation of the relevant *sunna*, not of the Qur'ān. Accordingly, from the early period of both Imamite and non-Imamite jurisprudence, forms of *wilāya* in those matters affecting public interest and protection of the rights of individuals incapable of protecting themselves was an important part of applied jurisprudence.

Thus, Tūsī takes up the question of a will or a testament (*wasiyya*) and the person who can delegate authority to act as his legatee (*wilāyat al-wasī*). The case Tūsī takes up in this particular section dealing with *bāb al-awsiyā'* (section on legatees) is the one in which the testament has already been made to a legatee who is about to die. The question is: Can he delegate another person, making him *his* legatee, to carry out the terms of the will in which he had the *wilāya?* Tūsī says that it is permissible for him to delegate his authority to another person for that purpose. However, other Imamite jurists, adds Tūsī, are of the opinion that a legatee delegating his authority to another legatee must necessarily delegate it to someone who can exercise discretionary control in such matters. Thus, when he dies it is the administrator of the affairs of the Muslims who, by virtue of his being invested with discretionary authority, should take upon himself to carry out the obligations specified in the testament in question. In the absence of such an authority, as in the case when the Imam is in occultation, it is the Imamite jurists who should administer the will insofar as it is possible. And if they cannot do so, then there is no blame on them.[2] Tūsī's delegation of the *wilāya* in the matter of the testament to the Imamite jurist recognizes the jurist's capacity as the tangible administrator of the affairs of the community in the absence of the Imam.

In *Tadhkira,* 'Allāma enumerates a number of *wilāyāt* in the section *Kitāb al-nikāh* and mentions possession of legal and moral power (*saltana*) as one of the factors that establish the *wilāya* of a person to undertake certain functions requiring the presence of a guardian, such as giving a girl in marriage. That person, according to 'Allāma, is *al-imām al-'ādil,* who can exercise the *wilāya* himself or can delegate it to one he judges suitable. This includes the dependable jurist who has all the necessary qualifications of being emulated (*al-iqtidā'*) and of assuming *wilāyat al-hukm* (authority to administer justice). However, adds 'Allāma, this *wilāya* does not include general authority (*wilāya 'āmma*), nor does he have *wilāya* over minors, or over those, whether male or female, who have attained the age of maturity. His *wilāya* in this particular function is established specifically with regard to the matter of a property left without any supervision. This is established by consensus (*ijmā'*), which allows him to become the guardian (*walī*) in marriage, because it pertains to the general welfare of the people.

After discussing *wilāya* through the possession of legal and moral power in matters affecting public welfare, 'Allāma elaborates on his equating of the *sultān* with the Imam or *hākim al-shar'* (the specifically legal authority) to whom the Imam has delegated his authority. The *wilāya* in marriage, according to 'Allāma, is not automatically delegated to the governor of a city, because *wilāya* in Imamite jurisprudence can be established only through the permission of the Imam or his deputy. This is a crucial difference between Imamite and non-Imamite jurisprudence. According to the Sunnites, who have depended upon a *hadīth* report related by Ahmad b. Hanbal, the governor of a region has the implied authority to perform the marriage of a person. But in another *hadīth* Ahmad b. Hanbal has contradicted this tradition, saying that this *wilāya* is vested in the *qādī,* not in the governor of a region. In any case, neither the governor nor the *qāḍī*, according to Imamite jurisprudence, can exercise any *wilāya* without proper deputization from *al-imām al-'ādil*.[3]

'Allāma's ruling can be taken to mean that the specific legal authority of *al-ḥākim* possesses the necessary deputization from the Imam, on the one hand, and that such deputization empowers the jurist to wield legal and moral authority to enforce his judicial decision in the interest of Islamic society, on the other. Moreover, from 'Allāma's recognition of the *wilāya* through the possession of legal and moral authority, it is clear that the well-qualified Imamite jurist in his capacity as a *sulṭān* (in the absence of the *sulṭān*—i.e., the Imam) has the *wilāya* to act as the guardian of a woman in the absence of any regular guardian, because, concludes 'Allāma, according to the Prophetic tradition, "the *sulṭān* is the guardian of the one who does not have a guardian, because he receives taxes (in his capacity as the ultimate receiver of revenue)."[4]

Both Ṭūsī and 'Allāma's discussion of the *wilāya*, although in different contexts (namely, *wasiyya* and *nikāḥ*, respectively), converge on a sensitive political issue affecting the exercise of authority in an administrative capacity. There is consensus in Imamite jurisprudence in all periods that the *wilāya* as exercised publicly needs proper deputization from *al-sulṭān al-'ādil*. This fact explains why the *wilāya* that has caused Imamite jurists to differ most often with each other in jurisprudence is the one that involves public recognition of authority, whether in the capacity of administrator of justice or of *waqf* (pious endowments), or any other function requiring the exercise of discretionary control. In other words, *wilāya* in the meaning of possession of legal and moral authority (*salṭana*) effected through deputization of a qualified individual (not contractually, as happens in the case of a *walī* appointed to carry out the terms of the *wasiyya*), is at the center of discussion in Imamite juridical writings.

It is in this context that *wilāyat al-qaḍā'* was discussed in the writings on jurisprudence, as demonstrated in the previous chapter. Whereas in the classical period of Imamite jurisprudence the *wilāya* of the jurists became well established in areas affecting the general welfare of the Imamite community, including its socio-religious as well as political structure, the question of the *wilāya* extended to enforce obedience was certainly prompted by later historical developments, such as the creation of the Imamite dynasties of the Safavids, the Qājārs, and the Pahlavīs. It was during this period of Imamite history that the classical *wilāyat al-qaḍā'* was interpreted to include *wilāyat al-niẓām* (authority to manage public order) and *wilāyat al-siyāsa* (authority to administer the government and hold political office in the very widest sense, including the use of force internally and relations with other regimes externally) on behalf of *al-sulṭān al-'ādil*.

As pointed out in the chapter 4, there were indications from the early period of Imamite jurisprudence that the specific *wilāya* of the jurists could be interpreted to include *al-wilāyat al-'āmma* (all-comprehensive authority), especially when the deputyship of the Imam was of a general category and dependent on personal qualifications of an individual rather than specific appointment. The absence of a specifically designated deputy during the Complete Occultation, instead of obstructing the development of the leadership of the jurist, actually reinforced the potential in the administration of juridical (*wilāyat al-ḥukm*) to become all-comprehensive.

Treatment of the question of the various forms of authority of a jurist, in

Imamite jurisprudence from the early days of the occultation, corroborates my observation that the absence of a specifically appointed deputy, on the one hand, made it necessary to take up the question of the qualifications of a jurist who could assume the *wilāya* of the Imam; and on the other hand, it brought out the need to determine the extent of that authority. Determination of the extent of *wilāya* in the general deputyship, where its forms were deduced through the interpretation of particular texts of the *hadīth*, was of the highest importance. In the case of "special deputyship" it was regarded as obvious that the designated person had the *wilāya* of the Imam in everything that the Imam performed when he was present. This position was construed from all the rulings regarding functions that are to be performed by the Imam or his specially designated deputy, including offensive *jihād*. The problem of *al-wilāyat al-'āmma* of a generally appointed jurist, as we shall see, was a complex one with theologico-political ramifications, because it was directly derived from the comprehensive authority of the Imam. Some theological implications relevant to the Imamate were discussed in my treatment of the *jihād*. At this juncture I need to take up the question of the Friday *salāt* (prayer) and the theologico-political implications of this duty in order to demonstrate the way that jurists, at different times in the history of Imamite jurisprudence, have ruled it obligatory or not, depending on their understanding of *al-wilāyat al-'āmma*.

The Friday Prayer *(salāt al-jum'a)* and its Prerequisites

I mentioned the Friday *salāt*[5] briefly in chapter 3 in reference to the invocation of the fourth Imam, 'Alī b. al-Husayn (d. 94/712–13), on the occasion of the Friday service, the convening of which is regarded as being the political right of the Imam in his capacity as the temporal ruler.[6] *Yawm al-jum'a* or simply *al-jum'a* is the weekly "day of assembly" to fulfil the religious obligation of *salāt* (prescribed worship), which takes the place of the daily midday (*zuhr*) worship, in accordance with the injunction about it in the Qur'ān, which says: "When you are called to pray on the day of assembly (*al-jum'a*), hasten to the praise of God and leave your business" (62:9). *Al-jum'a* is technically used to designate the particular worship on Friday, which is preceded by two sections of the sermon (*khutba*). That the *jum'a* ceremony carried political connotation and is closely related to the constitutional authority of the Imam as the head of the Muslim community becomes clear when one examines Imamite jurisprudence on the question of the validity of *al-jum'a* during the occultation of the twelfth Imam.

Imamite jurists have dealt with *al-jum'a* both in jurisprudence and in their exegesis of the Qur'ān, 62:9. The *jum'a* has been classified under the category of obligatory worship, binding on every male, adult, free, Muslim resident in a given locality. But the convening of the *jum'a* is dependent on the four following stipulations:

(1.) *Al-imām al-'ādil*, or, as Tūsī puts it, "*al-sultān al-'ādil*, or someone designated by him as his special deputy (*al-nā'ib al-khāss*)" must convene and perform the *jum'a*.

(2) There must also be present, according to Tūsī, seven believers where the service is held. The seven could include the Imam himself, his *qādī* (judge appointed by him), a prosecutor, a defending counsel, two witnesses, and the one who executes legal punishments (*hudūd*) in the presence of the Imam.

③ There must be a distance of at least two *farsakh*s (roughly six miles) between any two places where the service is held.

④ The sermon must be delivered before the performance of the prescribed worship.[7]

Some Imamite scholars mention six and others up to seven preconditions for the convening of the *jumʿa*. However, the variation in the number of stipulations is more a case of further elaboration of the above four conditions than of new stipulations regarding the *salāt*. Thus, some jurists have included the categories of Muslims on whom the *jumʿa* is incumbent as an additional stipulation, or the obligation of the performance of the *jumʿa* in congregation as a separate stipulation. Tūsī and others have regarded this latter stipulation as intrinsic to the primary signification of the *yawm al-jumʿa* ("the day of assembly").[8]

The most important precondition for this study is the first of the four stipulations given above. It was the prerequisite of the presence of *al-imām al-ʿādil* or his deputy that was directly relevant to the exercise of the *wilāya* with political connotation. Furthermore, it was precisely this political ramification of the *jumʿa* that led to different rulings regarding the legality of worship during the occultation of the Imam.

Before I embark on the imamate (i.e., the leadership of congregational prayer, which I spell with a lowercase *i*, in contradistinction with Imamate, the leadership of the Umma, as established in Twelver Shīʿī doctrine) of the *jumʿa,* it is well to compare the *jumʿa* with other daily canonical worship, especially the congregational aspect of the Friday ritual. Whereas the performance of all other daily prayers in congregation (*jamāʿa*) is "recommended" and even "preferred," the *jumʿa* must of necessity be performed in congregation. Ibn Bābūya relates a tradition reported on the authority of the Imam al-Bāqir, who said: "God has ordained thirty-five *salāt*s from one Friday to another, among which one has been required to be performed in congregation—namely, the *jumʿa.*"[9] This requirement evidently increases the importance of the Friday service, which should be performed only in the central mosque of a town or, as noted in the third stipulation above, in mosques that are more than three miles apart in a city.[10] Furthermore, attendance in the *jumʿa* as opposed to the daily worship, is a duty incumbent on every Muslim except for four classes of persons: slaves, women, minors, and the sick.[11] The requirement regarding a minimum number in attendance in the *jumʿa* (according to the Imamites, either five or seven) also indicates the importance attached to its public form, whereas the congregation in daily worship, which could be performed in all mosques, regardless of the distance between them, requires only two persons to make up the congregation: an *imām* (i.e., any qualified person who can lead the congregation in prayer [I spell the word with a lowercase *i*]) and a *maʾmūm* (who follows the imam).

The obligation with respect to the congregation, in addition to the compulsory participation of those legally qualified to attend the service, in the central mosque of a town, necessarily raises the issue of the imamate of such a congregation. Considering the religious-political background of early Islam, from its very inception the imamate of the *jumʿa* assumed great significance in jurisprudence, more particularly in the Imamite legal system, where the leadership of the *jumʿa* was closely bound to the doctrine of the Imamate. In this doctrine the Friday prayer was associated with the Imam's function as the politico-religious head of the government. Accordingly, the service stood as an important symbol touching on other duties of the Imam, which could be delegated to a qualified member of the community during the occultation. Significantly, if the *jumʿa* could be established as an act with all the essential elements (including the first stipulation, above) to render it legally valid, then a number of the Imam's other executive functions, which the occultation had postponed pending his return, could also become effective under the deputyship of the jurists.

Underlying the question of the validity of the *jumʿa* was, in fact, the legitimation of the jurist's prerogative to assume the leadership of the community in all its aspects. In other words, if the first precondition could be shown to be fulfilled, the *jumʿa* could resolve the legality of assuming the all-comprehensive authority of the Imam during the occultation. That the fulfillment of the first stipulation was of fundamental nature for the validity of the *jumʿa* will become obvious in the detailed treatment of the subject, which covers the endeavors of jurists to legitimize their deputyship as being the extension of the "apostolic" authority of the Hidden Imam.

The imamate of Congregational Worship

The subject of the general leadership of congregational worship has received still more attention in Imamite jurisprudence. Why? Because the imam who leads the faithful in prayer has been regarded as representing the Imam in his function as a religious leader. Accordingly, even for the short period of the duration of the prayer, the person leading the congregation is the Imam, the exemplary leader, whose actions must be followed carefully by everyone who joins the congregation and thereby accepts his leadership. It is for this reason that Imamite jurisprudence insists that besides being able to recite the Qur'ān correctly, and possessing the knowledge of at least the section of the Sharīʿa that deals with rulings on *salāt,* the *imām al-jamāʿa* should be *ʿādil*—that is, "a man from whom a thing occasioning doubt, or suspicion, or evil opinion has not appeared."[12]

I discussed the concept of *ʿadāla* (which may be rendered "moral probity" or "righteousness") in the previous chapter when I enumerated the qualifications of an Imamite jurist who can undertake *wilāyat al-qadāʾ*. Here it suffices to reiterate briefly that *ʿadāla* in the juridical texts is described as the aptitude that characterizes a person with justice and uprightness. It is a natural disposition of a person toward performing what is declared as obligatory and avoiding what is prohibited in the Sharīʿa. In other words, *imām ʿādil* is a person who is known for his habitual and evident conformity with the requirements of both moral and religious law.

The *ʿadāla* of the imam is so important that if a person performing the imamate of the congregation is known to commit grave sins, then to follow him in *salāt* renders the worship invalid (*bātil*) because it lacks an element essential for its legality. If a person commits grave sins, then in order to be reinstated as the leader of the congregational worship, in addition to possessing all other qualifications necessary for performing this function, he must repent for having fallen short of the requirements of moral goodness and the religious laws, and the members of the congregation should then assume that he still possesses the trait of being *ʿādil*.[13]

The question of *ʿadāla* in a person who becomes the leader of worship assumes great importance in all obligatory forms of worship, because leadership in *salāt* is regarded as part of *al-wilāyat al-sharʿiyya* (the authority to assume functions based on revelation). And because the *jum'a* is the only prayer that must be performed in congregation, the *imām al-jum'a* is accordingly required to be a man of singular piety, rectitude, and virtue to occupy the place of the Imam in his absence. The politico-religious nature of the imamate of the *jum'a* is further enhanced by the clause requiring that the *jum'a* be preceded by the sermon, *khutba*. In fact, the *khutba* is an integral part of the Friday service, without which the *jum'a* would not even be valid. The *jum'a*, which has a *salāt* of two *rak'as* (cycles), takes the place of the daily midday (*zuhr*), worship, which has four *rak'as*. The reduction from four to two *rak'as* on Friday, as implied in the explanation offered by Tūsī, is due to the importance of giving time to the *khutba*. Indeed, in the absence of the *khutba*, Tūsī rules, regular midday worship should be offered instead of the *salāt al-jum'a*.[14]

In explaining the reason for instituting two *rak'as* for the *jum'a*, Ibn Bābūya relates several factors recounted by Fadl b. al-Shādhān, a close associate of the eighth Imam, 'Alī al-Ridā. The latter said:

> First, on Friday people come from distant places to attend the service. As a result, God likes to lighten their burden by reducing the worship to the two *rak'as*. Secondly, the Imam detains the people in order to make them listen to the sermon. While the sermon is being delivered the people spend their time waiting to perform worship. Anyone who waits for the *salāt* has the status of the one who is actually performing it. Thus this waiting completes the *salāt* of the two *rak'as*. Third, *salāt* performed with the Imam is more complete and perfect because of his knowledge, excellence of character, religious comprehension, and moral probity (*'adāla*).

Ibn Bābūya then concludes the report by adding his own comments and says that the true reason why the *jum'a* has two *rak'as* is not merely because of the *khutba;* rather, it is the presence of the Imam, to whom the faithful have come to listen and who takes the opportunity to guide them to the right path and encourage them to worship God and be mindful of the divine commands and interdictions. "Friday," says Ibn Bābūya, "is the day of festival (*'id*), and like the prayer of the other two festivals [which have two *rak'as*] the *jum'a* also has two *rak'as*."[15]

The distinction between the imam who can lead the daily worship and the *imām al-jum'a* is implicitly drawn in a report cited by Tūsī in support of his ruling on the number of persons required for a valid *jum'a*, who must also be present when the *khutba* is delivered. A person asked one of the two Imams, al-Bāqir or al-Sādiq (some reports say one, some the other), whether the residents of a town could perform the *jum'a* in congregation. The Imam replied: "Yes, [they can. However,] they can pray four *rak'as* (of midday worship) [in congregation] if there is [no imam] who can deliver the *khutba*."[16] The Imam's response suggests that although there can be an imam who can lead the regular daily worship, the performance of the *jum'a* requires an imam who can also pronounce the *khutba*, because the *khutba* is one of the four stipulations on which depends the validity of the service.[17]

The inclusion of the *khutba* among the essential stipulations for its validity points to the public importance of the service, which provided the ruling sovereign a platform to assert his authority. In a distinct way, the *khutba* legitimized the authority of the person whose name was invoked in the sermon, and it set forth matters pertaining to the general direction adopted by the ruling authority. Because of this latter implication, the sermon was treated as an important statement of loyalty to the authority in power, who also reserved the right to appoint his deputy for that purpose.

From the early days of Islam, both the imamate and *khutba* of the Friday worship became a manifestation of the religious-political authority of the ruling sovereign, who himself led the service or appointed his deputy to perform this duty. As a consequence, the imamate of the *jum'a* was regarded as part of the constitutional prerogative of the sovereign, who sometimes delegated it to an official appointed to perform this duty on his behalf. It was possibly for this reason that the validity of the *jum'a* was held to depend on the *khutba*, which performed a crucial function of indicating the presence of the legitimate authority or his representative designated by him for that purpose. The fact that throughout Islamic history different ruling sovereigns reserved the right of appointing the *imām al-jum'a* in their respective capitals points to the major difference between this imam and the imam for the daily worship. This latter personage does not need any clear designation to perform his duty. Accordingly, it is plausible that it was the official designation by the Imam, not the *khutba*, that distinguished the imamate of the Friday service, whereas the importance of the *khutba* in some reports refers to its overall implications for the politico-theological recognition of the Imam. There are some reports that corroborate the fact that a person was appointed to deliver the *khutba* because he enjoyed the Imam's confidence to engage in performing this sensitive religious-political function. Acknowledgment of the ruling authority was more important in the ultimate validity of the *jum'a*, in which the *khutba* became the public form of that acknowledgment. It was for this reason that both the presence of the Imam and inclusion of the *khutba* were ruled as prerequisites without which the legality of the *jum'a* was vitiated and therefore null. However, it was the Imam whose presence was of fundamental nature in the validity of the Friday worship.[18]

Performance of the Friday Worship under *taqiyya*

Imamite jurists make frequent reference to the situation during the period of *taqiyya* (precautionary dissimulation) when the Shī'a are encouraged to assemble for the *jum'a* and perform the service with the two sections of the *khutba* "if there is no danger to their lives." Tūsī says:

> There can be no *jum'a* without a *khutba*. There is no objection if believers come together during the period of *taqiyya* and pray [the *jum'a*] with two [sections of] *khutba*, provided there is no detriment for them. But if they cannot pronounce the *khutba*, then it is permissible for them to pray in congregation; but they should offer four *rak'a*s [of the midday worship, instead of the *jum'a*].[19]

The period of *taqiyya* refers to both the period when the Imams were alive and the period of occultation, during each of which the Shī'ites were required to employ *taqiyya* in order to protect themselves against Sunnī majorities and Sunnī governments. It was most probably through the application of this principle that Imamites refrained from publicly performing those requirements of religion that would cause misunderstanding and enmity, and remained eager to identify themselves socially with the Muslim community at large.

There are reports that show, on the one hand, that the Imams avoided convening the *jumʿa,* at least publicly, for their adherents out of fear of arousing the animosity of the caliphal authorities; and, on the other hand, that they seem to have encouraged their followers to perform the *jumʿa* with the Muslim community at large, under *taqiyya.* Tūsī cites a *hadīth* report in which Zurāra, a close associate of Imam al-Sādiq, recounts a meeting with the Imam when the latter began to urge Zurāra and others to perform the *jumʿa.* Zurāra says that upon hearing the Imam they thought that he intended them to follow his lead in the Friday service. Thus he said to the Imam: "Do you wish to lead the *jumʿa* in which we all will attend?" The Imam replied: "No, I want to see you all [take care to perform the *jumʿa*]."[20] It is not explicit in the report whether the Imam was urging his disciples to perform the *jumʿa* with other Muslims, although it is evident that they were not performing the *jumʿa,* otherwise he would not have urged them to do so. However, it is plausible that the urging of the Imam implied that they should do so under *taqiyya.*

Similarly, it is not clear whether or not the Imams themselves performed the *jumʿa* separately with their adherents. But numerous traditions about the manner of performing *salāt,* and the various invocations recommended for recitation during this worship in the Imamite sources,[21] might support the view that these traditions go back to the period when Imam ʿAlī was invested with political authority and when, as the head of the government, he led the people in prayer and appointed officials with *wilāya* to convene and lead the *jumʿa.*

The practice of other Imams who were not invested with political authority is difficult to assess. On the one hand, we have the invocation of the fourth Imam, ʿAlī b. al-Husayn, and other utterances of Imams in which they considered the convening of the *jumʿa* the political right of the Imams, and which indicate that they did not participate in the *jumʿa* led by the appointees of the caliphs. On the other hand, we have accounts indicating that the Imams may have joined their Shīʿa and participated in the *jumʿa* under *taqiyya* to avoid any direct confrontation with the caliphal authorities, who were always suspicious of the political ambitions of Shīʿite leaders. It was also under *taqiyya,* it appears, that the Imams, in their reference to the imamate of *jumʿa,* employed general terms without referring to their own Imamate, to indicate the person who could convene and lead the worship. But when the qualities of any particular imam were enumerated, it was evident that these qualities reflect the special individual who alone could undertake the imamate of the *jumʿa.*[22]

Moreover, the position of the *imām al-jumʿa* was clearly seen as part of *al-wilāyat al-sharʿiyya* (the authority to assume functions based on revelation), which required explicit designation by *al-imām al-ʿādil.* The *jumʿa,* thus, became the most important symbol of *al-wilāyat al-sharʿiyya* in Twelver Shīʿism, especially in the absence of any other politically symbolic religious requirement, such as *jihād,* during the occultation. The performance of the *jumʿa* in Twelver Shīʿism, consequently, necessitated acknowledgment of the Imam's deputy for convening and validating the Friday service. In addition, it meant recognition of an authority that could exercise the *wilāya* of the Imam in all its aspects related to the life of Muslims. It was the acknowledgment of *al-wilāyat al-ʿāmma* with its constitutional ramifications in a jurist that made it problematic to regard the *jumʿa* as an obligation in the absence of the Imam or his special deputy.

Friday Worship during the Occultation of the Twelfth Imam

During the occultation, one of the main considerations in the juridical rulings regarding the *jumʿa* has been the nature of the *wilāya* of the Imam's general deputy, who, as we have seen, was invested with the *wilāyat al-hukm* (the authority to administer justice). Apparently, more than *al-hukm* it was the *jumʿa* that manifested the authority of the deputy, at least politically. The deputy, who could exercise *wilāya* in convoking or leading the Friday service, or in designating someone else to undertake that function, as the *sultāns* did, could also assume the position of *sultān al-zamān* (the ruling authority of the age), as discussed in chapter 3.

It was this overall implication of the politico-religious position of *al-wilāyat al-ʿāmma,* and its sensitive relationship with the *wilāya* of the Imams themselves, that gave rise to differences of opinion among the Imamites about the incumbency of the *jumʿa* during the occultation. The central issue in various legal opinions studied for this chapter makes it clear that the performance of the *jumʿa* was seen by a majority of Imamite jurists as the right of the Imam; and in the Imamite theory of political authority, this right could not be delegated either to the ruler professing the Imamite faith, or to the jurist in whom *wilāyat al-hukm* was invested.[23] Indeed, a number of Imamite scholars viewed the leadership of the *jumʿa* as an explicit extension of the Imam's "apostolic" authority to his general deputy. The consequences of this extension were clearly perceived in the danger of regarding an Imamite jurist as being the paradigmatic precedent for a deductively inferred ruling that would justify the wielding of the Imam's all-comprehensive authority by a qualified member of the Imamite scholarly elite. The *wilāya* to convene and lead the *jumʿa* provided the legal procedure through which the extension of the "apostolic" authority for a jurist could be validated in jurisprudence. The political implications of declaring the *jumʿa* legally valid under the leadership of a qualified jurist during the occultation included the issue of extending the Imam's "apostolic" authority to the jurist.

The Friday Worship as "Personally" or "Optionally" Obligatory during the Occultation

As seen above, one of the fundamental stipulations for convening the *jumʿa* is the existence of the temporal-religious authority, *al-imām al-ʿādil,* or someone designated by him as his special deputy for that purpose. It is important to keep in mind that in the Imamite conception of religious obligations (*al-takālīf al-sharʿiyya*), acknowledgment of the Imam is not only regarded as necessary to validate one's faith; it is also required in order to validate the area of religious obligation that relates to the temporal authority of the Imam. The performance of the *jumʿa,* as pointed out, entails the manifestation of the Imam's temporal-religious authorty as the Just Ruler of the Muslim community and the designation of the deputy to carry out this duty is among the Imam's constitutional rights. Consequently, in his or his deputy's absence, the *jumʿa,* according to the majority of the Imamite doctors, cannot become personally or individually binding; or, according to some others, even legally valid (*mashrūʿ*). The reason for its legal invalidity in the opinion of these scholars is that they have construed the position of *imām al-jumʿa* as being similar to that of a judge(*qāḍī*), who must be appointed and ordered by the Imam to exercise *wilāyat al-qaḍāʾ.*

This factor is also considered to be a decisive difference between the imam for daily worship and the *imām al-jumʿa:* the former does not need to be appointed, whereas the latter must be designated by the Imam. To this effect Tūsī says:

> One of the prerequisites for convening the *jumʿa* is the Imam or the one whom the Imam has ordered to perform [this duty], such as a judge, or a governor (*amīr*), or some such personage; and if it is convened without his appointing that person, then it is rendered invalid.[24]

In another place Tūsī mentions a ruling, in a *hadīth* of the Imams al-Bāqir and al-Sādiq, regarding the permission given to the residents of a town to hold the *jumʿa* service, provided there was an imam who could pronounce the *khutba;* otherwise the regular midday worship of four *rakʿas* should be conducted.[25] This second ruling, says Tūsī, implies a sort of permission of the Imam, which can be generalized and interpreted as the Imam's appointing a leader who can pronounce the *khutba* to pray with the residents of a town. Furthermore, there is consensus among Imamite jurists to the effect that the precondition for the *jumʿa* is the presence (*hudūr,* a term that implies possession of legal and moral authority—i.e., *saltana*) of either the Imam or his deputy who can execute his command. From the time of the Prophet on, contends Tūsī, no one except the caliphs, the governors, and those who were in charge of public worship, had convened the *jumʿa.* Hence, there is concurrence, for all periods, from the early days of Islam, that the *jumʿa* could be performed only by the one designated by the Imam to do so.[26] Accordingly, during the occultation of the Imam and the absence of his special representative, Tūsī's opinion explicitly establishes the *jumʿa* as not being a personally and individually obligatory duty. Nevertheless, it also implies that the *jumʿa* is optionally obligatory (meaning a choice between two or more obligations), a position close to saying that the permission of the Imam was available to perform either the *jumʿa* or the midday worship, contingent upon the fulfillment of the prerequisites, as Tūsī states in *Nihāya:*

> There is no harm for believers during the *taqiyya* to assemble, and pray [the *jumʿa*] with two sections of the *khutba,* provided there is no detriment for them; if they cannot pronounce the *khutba,* then they should offer the midday worship, four *rakʿas,* in congregation.[27]

In the same source Tūsī specifies the person who can undertake to perform this duty in the absence of the special deputy of the Imam, and who, it would appear, has the permission of the Imam to do so:

It is permissible for the jurists (*fuqahā'*) of the Imāmiyya [lit. *ahl al-haqq*, "belonging to the truth"] to assemble with the people and lead them in all prayers, including the *jum'a*, and the two festivals, and to deliver the two sections of the *khutba*, and pray with them the *salāt al-kusūf* [eclipse], as long as they [the jurists] do not fear any harm by doing so. If they are afraid that some harm may befall them, then it is not permissible for them to expose themselves to such a thing in any circumstances.[28]

In *Muqni'a*, Mufīd makes this permission available to "a reliable imam, who should possess the qualities described in the section dealing with the *jamā'a* (congregational prayer), and who can deliver for the people the two sections of the *khutba*."[29]

The judicial decisions of both Tūsī and Mufīd indicate that, so long as the occultation lasts, it is not obligatory to assemble the people on Friday specifically for the performance of the *jum'a*. If the appropriate imam is available to deliver the sermon, then there is permission both in convening and assembling for the Friday service after its having been announced. However, the permission in the legal decisions need not be construed as *'azīma* (an ordinance of obligation), because the clause about harm and the period of *taqiyya* indicate a *rukhsa* (an ordinance of indulgence), which may be applied if circumstances are appropriate.

That the performance of the *jum'a* is under the rubric of *rukhsa* in Imamite jurisprudence is further established in Muhaqqiq's judicial decision:

When neither the Imam nor the one appointed for leading the worship is present, and if it is possible to assemble and deliver the two sections of the *khutba*, some are of the opinion that it is recommended to perform the *jum'a*; whereas others maintain that it is not permissible to do so. The former opinion is more sound.[30]

Further elaboration of the nature of obligatoriness (*wujūb*) is necessary here to comprehend the meaning of "personally and individually obligatory" (*wājib 'aynī*) when used in the context of the *jum'a*. If the *jum'a*—because all the preconditions are fulfilled—is declared personally and individually obligatory, then this obligatoriness will have two significations: first, it would apply to those for whom the worship is obligatory in the Sharī'a (*al-mukallafūn*), in that they would be required to attend the service, personally and individually; secondly, it would apply to the Imam or his deputy who would be required to convene it, if the other prerequisites are fulfilled.

Some Imamite jurists have interpreted Tūsī and Mufīd's permission to the Imamite jurists to undertake the imamate of the *jum'a* during the occultation as fulfilling all the necessary preconditions of the *jum'a*; and hence, on the basis of these rulings, they have maintained the *jum'a* to be personally and individually obligatory in the sense set forth above. In other words, the qualified and reliable Imamite jurist must convene and the Shī'a must attend and perform the *jum'a*, as an individual and personal obligation in both cases. But it is evident that such an incumbency cannot be adequately deduced from the texts of the judicial decisions themselves, because, according to Abū al-Makārim Hamza b. Zuhra (d. 585/1189–

90), who commented on Tūsī's opinion in his *Ghunya*, there is consensus among the Imamites that the *jumʿa* is incumbent only when the precondition concerning *al-imām al-ʿādil* or the one appointed by him to act on his behalf is fulfilled.[31]

'Allāma in his *Tadhkira*, although concurring with Muhaqqiq's opinion that the *jumʿa* is recommended, raises the fundamental question: Must the Imamite jurist undertake to convene worship during the occultation, if it is possible to congregate and to deliver the two sections of the sermon? It is interesting that *Tadhkira* of 'Allāma is in the area of comparative law and it normally cites opinions of different schools of Islamic legal thought. But when it comes to this section, because of the absence of the *sultān*, the problem is seen as so peculiar to the Imamites that no other schools are mentioned in connection with the imamate of the jurists in the *jumʿa*.

'Allāma begins his response by declaring that all the Imamite jurists agree that the *jumʿa* is not incumbent during the occultation, because the preconditions remain unfulfilled, even though there apparently is permission from the Imam to pray the *jumʿa*. As for its being *recommended*, there is a difference of opinion—although, adds 'Allāma, most scholars seem to favor this ruling, because of the several traditions in which the Imam has encouraged his close associates to perform this worship. 'Allāma then proceeds to mention the opinions of Sallār al-Daylamī and Ibn Idrīs al-Hillī, who have maintained the impermissibility of the *jumʿa* because of the four preconditions that cannot be disregarded without well-established evidence. We shall see their full reasoning for this opinion below. Briefly stated, in the context of 'Allāma's *Tadhkira*, both these jurists are hesitant to recognize the Imamite jurist as the deputy (*nāʾib*) of the Imam in matters of the *jumʿa* during the occultation.

'Allāma then takes up the case where an unjust (*jāʾir*) ruler deputizes a righteous (*ʿādil*) person, meaning an Imamite jurist, to conduct the *jumʿa*. 'Allāma rules that it is recommended to come together and convene the service under his imamate. However, he cautions, this does not do away with the prerequisite concerning the presence of the Imam or his deputy.[32]

On the basis of the legal decisions mentioned above, it should be noted that, on the one hand, Imamite jurists speak of consensus regarding the precondition for the Imam or his deputy to validate the *jumʿa;* on the other hand, they simultaneously seem to suggest, with understandable caution, that the righteous jurist (*al-faqīh al-ʿādil*) can function as a substitute for the Imam and his special deputy, and so convene the *jumʿa*. Their caution in regarding the *jumʿa* as essentially an optional duty during the occultation is obvious. Mufīd's use of "reliable Imam" and 'Allāma's use of an unspecified "righteous person" in their opinions could be construed as rulings that do not require the *faqīh* to be present in order to render the *jumʿa* optionally valid during the occultation.

However, as we move away from this classical period, the opinions of Imamite jurists tend to regard the presence of *al-faqīh* (necessarily the Imamite jurist) as necessary for the validity of the *jumʿa*, even if it be conducted as an optionally incumbent duty. Thus, Abū ʿAlī al-Tūsī (Tūsī's grandnephew), according to Sabzawārī, argued in support of the opinion of those scholars who adopted the view that among the conditions that must be fulfilled to render the *jumʿa* valid is the presence of the *faqīh*. The reason, according to Abū ʿAlī al-Tūsī, is that the permission of the Imam is a necessary stipulation to validate the *jumʿa*. When the Imam is present he can convene it himself or he can deputize someone else to do so. But during the occultation, the *faqīh,* who as a deputy of the Imam in general has the Imam's permission to represent him, can substitute for the Imam and for any deputy who might be designated for a specific purpose. To support his opinion Abū 'Alī al-Tūsī provides three kinds of documentary evidence: first, the practice of the Prophet and the caliphs, who used to appoint the imam of the *jumʿa* as they appointed *al-qāḍī;* secondly, the tradition of the Imam al-Sādiq reported by Muhammad b. Muslim al-Thaqafī, in which the Imam specifies the seven categories of persons among believers on whom *jumʿa* was incumbent, including the imam of the *jumʿa;* and thirdly, the consensus of the Imamites as reported by Muhaqqiq, 'Allāma, and Shahīd I in his *Dhikrā.*[33]

Shahīd I is more explicit than 'Allāma in regarding the Imamite jurist as the deputy of the Imam during the occultation in the convening of the *jumʿa.* His opinion, to be discussed presently, has been commented upon by Shahīd II in considerable detail. Most of the Imamite jurists after Shahīd I followed him in the main point of his opinion, emphasizing the fact that the imamate of the *jumʿa* is an official position that requires proper designation by the Imam, who should appoint a special deputy for this purpose.[34] Consequently, during the occultation, there should be a deputy of the Imam, whether through special or general designation, such as the *faqīh,* in order for worship to become personally obligatory, at times when it is possible to assemble and perform the *jumʿa.*

It is important to remember that both Shahīd I and Shahīd II, who wrote their works in the Sunnī environment of Syria, regarded the performance of the *jumʿa* as one of the most decisive religious duties that would ensure recognition of the Imamite school among the other Islamic legal schools. Moreover, by the time Shahīd II wrote his commentaries on the earlier works of Shahīd I and Muhaqqiq al-Hillī, the position of the Imamite jurist as the general deputy of the twelfth Imam had been further strengthened by the judicial decisions of Muhaqqiq al-Karakī, writing under the reign of the Safavid Tahmāsp. In Shahīd II's commentary on the question of the *jumʿa,* as we shall see presently, the deputyship of the jurist is taken as a well-established judicial ruling and discussion centers on the extent of this deputyship.

In his *Lumʿa,* Shahīd I gives the following ruling on the question of the *jumʿa:*

> The *jumʿa* cannot be convened except by the Imam, or his deputy, even if it is a *faqīh,* when it is possible to assemble during the occultation (*ghayba*) [of the twelfth Imam].[35]

Commenting on this ruling, Shahīd II in his *Rawda* makes it explicit that the 'Imam's deputy," whether specially designated or in a general capacity, should be a well-qualified *faqīh* who can issue legal decisions (*fatwās*). This stipulation concerning the *faqīh* who can take the place of the Imam during the occultation, elaborates Shahīd II, is based on the fact that he is designated by the Imam in a general way, as reported in the instruction received from the twelfth Imam to have recourse to persons who relate the Imamite teachings (*ruwāt*) and in other traditions in which Imams have appointed jurists as judges and administrators of justice. In short, Shahīd II continues, there is agreement among the Imamites that when the Imam is present the *jumʿa* must be convened by him, or by his specially designated deputy for that purpose. If the *jum'a* is convened by anyone besides these two, then the worship becomes legally invalid.

Shahīd II then takes up Shahīd I's judicial decision in his *Durūs* regarding the incumbency (*wujūb*) of the *jum'a*, given the presence of the *imām al-jum'a*, the *faqīh*, because the presence of the *faqīh*, according to this ruling, fulfills the stipulation regarding the permission of the Imam. This latter clause is recognized, in most cases, on the basis of the consensus of Imamite jurists. However, in his *Bayān* according to Shahīd II, Shahīd I has declared the incumbency of the *jum'a* even if it is not led by the *faqīh* himself, because of the unrestricted nature of the evidence provided by the Qur'ān (62:9) for its incumbency. Furthermore, the prerequisite regarding the Imam, or the one designated by him, applies specifically to the period when the Imam is present, or when the alternative precondition regarding the designated deputy can be fulfilled. The generality of the injunction regarding the *jum'a* is evidenced in the Qur'ān and the Sunna, which remain unchallenged even when a particular condition remains unfulfilled.

Shahīd I is not alone in maintaining such an opinion, because, according to Shahīd II, there are other jurists who also contribute to such an opinion, provided there is a possibility to assemble and that the other stipulations, such as the sermon and the righteousness of the *imām al-jum'a,* are observed. The sense one derives from these opinions, says Shahīd II, points to the *permission* to convene the *jum'a* during the occultation, and to its being simply *recommended* on account of the consensus that denies its being personally and individually incumbent. Yet other jurists, following Shahīd I, have regarded *jum'a* as optionally incumbent and have ruled accordingly.

Some jurists, points out Shahīd II, have obscured their arguments because of their regarding the status of worship as recommended, whereas emphasis should be on the recommended status of the person who might want to exercise his option between the midday and the Friday worship. However, they all agree on making the Imam or his deputy a prerequisite for its incumbency on the basis of the consensus of the earlier jurists—and then they proceed to discuss the period of occultation and disagree in their rulings. In other words, they misconstrue the purpose of the aforementioned consensus, that the *jum'a* without the jurist should not be permissible during the occultation. According to them, the reason why the *jum'a* is not personally obligatory during the occultation is because the precondition regarding the presence of the deputy of the Imam must be fulfilled for the worship to be valid. This being so, when the *faqīh* is present, argues Shahīd II, these jurists must rule the status of the *jum'a* as personally incumbent.

On the basis of the ruling of jurists who have not recognized the *faqīh* as the deputy of the Imam, some other jurists have declared it impossible to convene the *jum'a* during the occultation, because the stipulation regarding the deputy is unfulfilled. Shahīd II refutes their opinion, stating very clearly that the opinion of these jurists is weakened by the argument that there is a possibility for the fulfillment of the condition regarding the deputy—namely, by the presence of the *faqīh*. Moreover, insistence on fulfilling the precondition for the deputy is unacceptable, because it is known that there is no explicit textual proof to that effect. The error of these jurists, according to Shahīd II, is that they have claimed an *ijmā'* in support of their opinion regarding the absence of personal and individual incumbency in the *jum'a;* their case would have been very strong if they had ruled for optional incumency with the preponderance in favor of the *jum'a*. The inference of Shahīd I and others regarding the "possibility of assembly" is meant to convey the sense of 'assembly with an *imām 'ādil*," because the assembly had almost never taken place during the time of the Imams. This is the reason why Shahīd I and others who followed him have not been content with the *jum'a* excluding the regular well-established midday worship, in spite of all that is reported in connection with their opinions upholding the *jum'a* during the occultation.[36]

Shahīd II in *Masālik al-afhām*, his commentary on Muhaqqiq's *Sharā'i'*, is most explicit regarding the deputyship of the Imamite jurist in the *jum'a*, whereas for the other two occasions of festival worship (marking the end of the fasting of Ramadān, and the feast of sacrifice during the *hajj*) he does not regard it necessary for the *faqīh* to convene these public rituals.[37] It is reasonable to suggest that Shahīd I's argument in favor of the *faqīh*s as convenors of the *jum'a*, not merely "out of respect" for their position in the Imamite community but as fulfilling the essential prerequisite as deputies of the twelfth Imam, certainly marks an important advance in legitimizing the all-comprehensive authority of the jurist.

In his *Dhikrā*, Shahīd I concludes emphatically that jurists have the permission of the Imam available in the *sahīh* ("sound") *hadīth* of Zurāra, in which Imam al-Sādiq urged his close associates to pray the *jum'a*. The permission of Imam al-Sādiq given to Zurāra is similar to the permission of the Imam of the age to the jurists, who have undertaken much greater responsibilities than convening the *jum'a* during the occultation. Indeed, the *faqīh* is both the *nā'ib* of *al-sultān al-'ādil* and possessed of the *idhn al-imām* (the permission of the Imam) to lead the *jum'a*.[38] It is significant that Shahid I seeks documentation for his juridical extrapolation in the permission given to one of the most prominent disciples who formed the early group of *ashāb al-ijmā'* (the associates who participated in the concurrence with the Imams al-Bāqir and al-Sādiq).

This method of documentation corroborates my observation in chapter 1 that the most decisive method of deriving an authoritative juridical prescription was to establish a chain of transmission (*isnād*) that could be traced back to the only valid precedent provided by the Imam. Moreover, the documentation legitimized the extension of the "apostolic" authority of the Imam to the contemporary jurist (in this case, Shahīd I) through Zurāra, who was the actual link to the source of paradigmatic authority. By the time Shahīd I and Shahīd II wrote their juridical works, the "apostolic" link between the *ashāb al-ijmā'* and subsequent jurists was taken as a duly established fact. This point is corroborated by their judicial decision that regarded the jurist as fulfilling the essential prerequisite as the deputy of the twelfth Imam in functions that required specific designation. Accordingly, it is reasonable to assert that it was a matter of further extrapolation in the documentation that legitimized the subsequent equation of the *faqīh* with *al-sultān al-'ādil* in the juridical works of Safavid and post-Safavid scholars.

Undoubtedly, Shahīd II deduced his opinion from Shahīd I's conclusion. But there was much deeper concern in the minds of the Shī'ī jurists living in Sunnī majorities during this period. Indications of this situation can be discerned in juridical works earlier than those of Shahīd I and II—that is, in 'Allāma's long discussion on the question of the right of jurists to convene the *jum'a*.

During the eighth–ninth/fourteenth–fifteenth centuries there seems to have been much criticism of the Imamite community by Sunnī jurists for not regarding the *jum'a* as obligatory during the occultation of the Imam. The duty, explicitly ordained in the Qur'ān without any preconditions, was regarded by the Sunnites as the most important act of public worship to express one's loyalty to Islam and to the political authority who ruled as the protector of Islam. On both grounds, the Imamite minority was viewed as deviating from the established Islamic practice: deviation from the principle of *jamā'a* (community of believers), which recognized the *umma* as under the *sultān's* authority, and deviation from the well-established practice based on the expressly ordained obligation in the Qur'ān. Jurists like 'Allāma, Shahīd I, and Shahīd II, who, as discussed in the Introduction, were conducting their affairs under Sunnī dominance, were faced with a situation in which the authority of Imamite scholars needed to be reaffirmed in such a way that their long-established concurrence requiring the presence of the Imam or his deputy for the *jum'a* could be maintained in changing circumstances.

Theoretically, it was true that jurists were designated as the "general" deputies of the Imam, lacking special appointment for specific functions. But for the general welfare of the Imamite community, such a theoretical distinction between "special" and "general" deputy, particularly when "special deputyship" had been terminated early on, could not be permitted to make a difference in the actual role of the "general deputy." Nevertheless, "general deputyship" needed some theoretical basis for the exercise of comprehensive authority in the absence of the Imam, and there is no other place than the *jum'a* in the whole of jurisprudence where the *faqīh* could assume theoretically what was already available to him in practice—namely, the leadership of the Imamite community.

Although *jihād* in jurisprudence could have provided this theoretical base for the legitimation and extension of the *wilāya* of the *faqīh,* in the absence of offensive *jihād* during the occultation the question of waging it was postponed to the return of the twelfth Imam. But such a postponement in the matter of the *jum'a* was detrimental to the religious standing of the Imamite community in the context of the *umma.* Moreover, the jurists did not lack the Imamlike qualifications, either in personal moral development or academic preparation, in such a way as to hamper their assuming the *wilāya.*

But there were some Imamite jurists, however few, who regarded such equation of the "general deputyship" with the "special" one, and thus declaring jurists permitted to assume that *wilāya* of the Imam in the *jum'a,* as innovative and contrary to the well-established doctrine of the *wilāya* of the Imams among the *ahl al-bayt.*

These jurists argued against the *wilāya* of the jurists in the *jum'a.* They maintained that even during the lifetime of the Imams, prior to the occultation, *jum'a* could not have been personally and individually obligatory, because the Imams were living in *taqiyya* and were unable to convene the *jum'a.* The Imams needed to be reinforced by the people before they could execute their duty of assuming the leadership of the Friday service. As a result, it was the people who should be blamed for the inability of the Imam to shoulder his obligation of convening the *jum'a,* as expressly ordained in the Qur'ān.

It is also evident from some authentic reports that even the close associates of the Imam were not performing the *jum'a.* There is an allusion to this fact in the *hadīth* where the Imam al-Sādiq urged his associates to perform the *jum'a,* saying: "I want to see you all [take care to perform the *jum'a*]." Moreover, there are reports that served to remind the Shīʿa during the *taqiyya* that the Imams were unable to assume the leadership of the service because of fear of their enemies, who had encroached upon their right to do so. Hence, concluded these jurists, the *jumʿa* cannot be absolutely incumbent unless the stipulated prerequisites are fulfilled. The reports that allude to "permission" to perform the *jumʿa* are mostly in the form of *rukhsa*, meaning one "can" perform, rather than *ʿazīma*—that is, one "must" perform—the duty. That there is permission of the Imam for the *faqīh* to convene the *jumʿa* as an optionally incumbent duty (*wājib takhyīrī*) is open to debate, and I shall consider that matter in the following section.

The Friday Prayer as "Prohibited" *(harām)* during the Occultation

A careful scrutiny of judicial decisions, both in favor of or against the "personal and individual incumbency" of the *jumʿa* reveals that the doctrinal difficulty concerning the Imamate is at the root of different opinions formulated by Imamite jurists. Without exception, the jurists had recognized the prerequisite regarding the "presence" or "permission" of the Imam or of the one appointed by him. By "presence" (*hudūr*) the jurists meant *wilāyat al-tasarruf*—that is, possession of legal and moral authority to exercise discretionary control over the affairs of the community. It also signified *saltana*—that is, "the power to enforce or exact obedience"—to further God's cause on earth. By "permission" (*idhn*) they meant either *idhn khāss* (special permission) given to a particular individual, at a particular time, to assume functions accruing to the Imam; or *idhn ʿāmm* (general permission for the same purpose) given to all qualified believers, at all times.

The difficulty lay in the fact that during the occultation the jurists were left with two options in their rulings about the *jumʿa* (if they did not adhere to "optional incumbency," as explained above): either to declare the *jumʿa* to be permissible (*jāʾiz*) as a personally incumbent or recommended act of worship, or to consider it prohibited (*harām*). In either case the crux of the problem had to do with the definition of the prerequisite concerning the imamate of the *jumʿa* with its theologico-political implications for the *wilāyat al-tasarruf* (discretionary control over the affairs of the community). *Wilāyat al-tasarruf* was regarded as the manifestation of the legal and moral authority of the Imam and his all-comprehensive authority, which the qualified jurist could exercise if the legality of his right to convene and lead the *jumʿa* could be substantiated. In other words, the validity of the *jumʿa* during the occultation could be linked with the legitimate claim to *saltana*. The political implications of such a judicial decision cannot be overemphasized, for the decision would lead to the investiture of divine sovereignty in persons other than the Imams. As pointed out earlier, a number of Imamite jurists interpreted the legal validity of the *jumʿa* in the absence of the Just Ruler as an explicit acknowledgment of the extension of the Imam's "apostolic" authority to his general deputy, the jurist.

The Imamite doctrine of the Imamate explicitly considers the *jumʿa* as the political religious function of the rightly designated Imam. This is stated in very clear terms by Mufīd in his section of *Irshād* dealing with the Imamate of the twelfth Imam:

Among the rational proofs [with the aid of which] the intellect, with appropriate argumentation, establishes the necessity of an infallible Imam, who should be independent of his subjects in matters pertaining to religious injunctions and other sciences [is the one that requires] that he should exist in every age. This is so because it is impossible for believers [*mukallafūn*, those on whom duties of the Sharī'a are imposed] to be without a ruler (*sultān*) whose presence would draw them closer to righteousness (*salāh*) and would keep them away from sinful deviation (*fasād*). It is well established that every imperfect being needs someone who can discipline him so that he will refrain from evil acts. . . . He should also be the one who will protect Islamic territory, and will assemble the people in order to convene the *jum'a* and the *'īds* [festivals].³⁹

The Imam is *al-sultān al-'ādil*, the Just Ruler, and he alone is empowered to delegate an official function to his deputy. So, if it could be established that jurists can serve as the deputy of the Imam, then the question regarding the *jum'a* and other areas of the deputyship would be resolved. But in the absence of any explicit designation of jurists to assume all forms of *wilāya*, especially the *wilāya* to convene the *jum'a*, the question of the deputyship of the jurists would inevitably go through much discussion and debate as the need for them to assume the leadership became imperative. The classical solutions offered to the question of *wilāya* and *saltana*, under *taqiyya*, as indicated in the terms *sultān* or *imām 'ādil*, were too ambiguous to determine the extent of the *wilāya* of the jurists. These titles were sometimes used without the definite article *al-*, so as to hide the true identity of the one to whom the jurists referred in their jurisprudence.

The designation of the Imam as *sultān 'ādil*, when the Imams (with the exception of 'Alī b. Abī Tālib for a short period) were never fully invested with de facto power, carried some ambiguity, and was hence open to interpretation, especially when this title was used interchangeably with a less political one—namely, *imām 'ādil*. The latter title was generally used, at least in the Imamite jurisprudence, for those who led the daily worship in congregation. Thus, explaining the prerequisites for the *jum'a* in his *Nihāya*, Tūsī uses *imām 'ādil* or "the one appointed by him (*imām 'ādil*) to pray with the people," whereas in his commentary on the verse of the *jum'a* (62:9) in *Tibyān*, he explicitly employs the title *sultān 'ādil;* and in his book on comparative law, *Khilāf*, he simply says *al-imām*.⁴⁰ That this Imam, whose presence Tūsī makes a precondition for the *jum'a*, was the one invested with full power is corroborated by the fact that in the latter work, after mentioning the requirement regarding "the one appointed by *imām 'ādil*," he adds that the one appointed by the Imam to convene and lead the *jum'a* might in fact be a *qādī* or *amīr*, or other such government official normally designated by the ruling sovereign. In a widely quoted tradition, Imam 'Alī is reported to have said: "The administration of justice (*al-hukūma*), the execution of legal punishments (*al-hudūd*), and the *jum'a* cannot [any of these three] be valid without the Imam or the one substituting for the Imam."⁴¹ Thus, in Tūsī's usage, *sultān* or *imām 'ādil* or merely *al-imām* were one and the same—namely, the Imam invested with full legal-moral authority.

It was the exercise of authority—*saltana*—in the case of *sultān 'ādil*, understood in the sense derived from Tūsī's usage, that gave rise to the opinion that it was only the "presence" (*hudūr*—i.e., *saltana*) of this Imam or his special deputy that could render the performance of the *jum'a* legally valid. In the absence of such an authority at any given time, the *jum'a*, according to some Imamite jurists, was prohibited ending the establishment of the rule of justice and equity by the twelfth Imam. Although they affirmed the *wilāyat al-hukm* and *al-qaḍā'* of the Imamite jurists (who, in accordance with the permission given by the Imam al-Ṣādiq, could sit as judges and give legal opinions during the occultation), they could not concede the delegation of the Imam's all-comprehensive authority (*al-wilāyat al-ʿāmma*) to them. To assume such duties as fell under the Imam's *al-wilāyat al-ʿāmma*, jurists were in need of a special permission from him. This permission, they argue, is a precondition that must be fulfilled at all times. As such, there is no difference between the period when the Imams were present or when the Imam is in concealment.

Among the jurists in the classical period of Imamite jurisprudence who declared the *jumʿa* prohibited during the occultation was Sallār al-Daylamī (d. 448 or 463/1056 or 1071), who was Ṭūsī's contemporary. He is regarded as being the first Imamite jurist to have given this legal decision.[42] In his *Marāsim*, in the section on *al-'amr bi al-maʿrūf wa al-nahy ʿan al-munkar* (enjoining the good and prohibiting the evil), where Sallār discusses the functions that jurists can undertake, he explicitly mentions that the *fuqahā'* of the Imamites are allowed to assume the Imamate of the two festivals and the *istisqā'* (prayer for rain), which are also followed by a two-part sermon; but they are not permitted to lead the *jumʿa*, the performance of which depends on the presence of the *imām al-asl*—meaning, the original Imam— or of the one substituting for him.[43] This denial of the right to convene the *jumʿa* to Imamite jurists is consistent with the doctrine of the Imamate, which never conceded the delegation of the political authority of the Imam either to a ruler professing the Imamite faith or to the body of religious authority. The Imamite jurists, according to Sallār al-Daylamī, are ordinary *mukallafūn* on whom the duties of the Sharīʿa are imposed, and hence they are like other believers who cannot assume the political authority of the Imam in his absence. The logical outcome of such a stance concerning the jurists was to declare the convening of the *jumʿa* during the occultation as *bidʿa* (innovation). The innovation was obviously felt to lie in its connection with the delegation of the Imam's political authority to the jurists.

Sallār al-Daylamī's opinion was accepted by another prominent Imamite scholar, Ibn Idrīs al-Hillī (d. 598/1201–2), the leader of the Imamite school of jurisprudence at Hilla, Iraq, whose disagreement with Ṭūsī in the matter of *wilāyat al-hukm* we have seen in the previous chapter. In fact, Ibn Idrīs's name is always mentioned with that of Sallār al-Daylamī in regard to the question of the *jumʿa* and their refusal to recognize the jurists as qualified to convene this worship during the occultation. Ibn Idrīs takes up the question of the *jumʿa* in *Sarā'ir* and disagrees with Ṭūsī who allows the jurists to convene the *jumʿa*. He then cites Sallār's opinion and says:

That the four *rakʿas* [cycles] [of the midday worship] are obligatory is certain; [whereas] those who consider the two *rakʿas* [of the *jumʿa*] to be sufficient [at that time] are in need of a proof. It is improper to revert from that which is based on certainty (*ʿilm*) to that which is founded on supposition (*zann*). [In addition,] "single traditions" (*akhbār al-āhād*) do not make anything binding, whether [in the form of] knowledge or practical rules.[44]

It is significant that, although Ibn Idrīs through his rational methodology has conceded to Imamite jurists the prerogative of formulating judicial decisions to effect God's purposes, he has consistently doubted the authenticity of any opinion that could be contrued as delegating or conferring the Imam's constitutional authority to a member of the Shī'a. His rejection of the "single" tradition, generating conjecture rather than certainty, is based on his awareness of the implications derived from some of the documentary evidence, such as the permission of the Imam al-Sādiq to Zurāra, as seen above in reference to Shahīd I's opinion, which could be extrapolated to reinforce the jurist's constitutional status.

It can be concluded from 'Allāma's criticism (below) that Ibn Idrīs did not reject *all* the "single" traditions as "unsound"; otherwise he would not have been able to formulate the majority of his deductively inferred judicial decisions. Rather, he rejected those "single" traditions that were utilized as documentation to consolidate the all-comprehensive authority of the jurist. As we shall see below, Ibn Idrīs's concern in this regard was not totally unfounded. Others had in fact depended on some of these "single" traditions for documentary evidence to argue for the all-comprehensive authority of the jurist. Consequently, according to Ibn Idrīs, it is not permissible to shorten the midday worship on Friday: permission to do so is not available with certainty.

'Allāma in his *Mukhtalaf* undertook to discuss systematically the question: Could the Imamite jurists convene the *jum'a* in the light of Sallār's and Ibn Idrīs's opinions? 'Allāma takes up the discussion by stating that the crux of the problem lies in determining the status of this worship during the occultation. Then he proceeds to cite those who have not conceded to the jurists the right of convening the *jum'a*. First on the list is Sharīf al-Murtadā. Sharīf al-Murtadā's judicial decision in his treatise *al-Masā'il al-maiyafāriqiya* is that the *jum'a* cannot be convened without *al-imām al-'ādil* or his deputy. "If they are not present, then one should pray the four *rak'a*s of the regular midday worship." The second name on 'Allāma's list is that of Sallār al-Daylamī whose opinion (cited above) he quotes in full with an addendum to the effect that Ibn Idrīs also accepts this view.

Then 'Allāma turns to discuss Tūsī's contradictory rulings in his *Nihāya* and *Khilāf*. In the former work Tūsī permits believers to assemble and jurists to lead and pray the *jum'a* with the sermon, provided there is no harm to them in so doing. But in the latter work Tūsī recounts the precondition for the *jum'a*— namely, the presence of the Imam or the one appointed by the Imam for that purpose from among the *qādī, amīr,* and other such official ranks. Moreover, according to 'Allāma, Tūsī also maintains that, on the basis of the permission given by the Imam to the inhabitants of a city or a village (if the required number of believers for the *jum'a* can assemble) to pray the *jum'a,* this permission can be extended to other periods, and Tūsī accordingly deduces that it is available for believers during the occultation. This interpretation is not permitted by Ibn Idrīs or Sallār al-Daylamī, but 'Allāma contends that Tūsī's interpretation is probably valid in this regard.

Then 'Allāma gives his own views on the issue. He begins with the verse on the *jum'a* (62:9) and argues that the text of the verse points to the generality of the ordainment, without any preconditions attached. In fact, the fundamental point in all the traditions on the subject is the absence of preconditions, and because it takes the place of the midday worship, there is no additional requirement for it. 'Allāma affirms that there is an *ijmā'* against Ibn Idrīs's counterargument regarding the fulfillment of the precondition. Moreover, says 'Allāma:

We maintain the requirement is fulfilled because a trustworthy jurist is designated by the Imam. It is for this reason that his rulings are effective and it is
obligatory to assist him in implementing legal punishment (*hudūd*) and administering justice (*qadāʾ*) among the people. Also, certainty supports what we have
mentioned; and in the case of "single" traditions, even if they lead to conjecture,
a decision made by using them produces certainty. [If this were not the case,]
then much that has been reported by him [Ibn Idrīs himself] would become null
and void.[45]

'Allāma's critique of Ibn Idrīs's ruling corroborates the point that in the matter
of the *jumʿa* the *wilāya* of the jurists was a fundamental issue if the *jumʿa* was to be
considered as obligatory or prohibited in the absence of prior deputization of the
jurist during the occultation. Undoubtedly, as explained by the prominent Safavid
jurist, Fādil al-Hindī (d. 1137/1724), what could prove the *jumʿa* to be individually
and decisively obligatory was the "explicit permission" of the Imam for the jurists,
and this was not available. This permission was a precondition to the legality of the
jumʿa during the occultation, and because it was not available, says Fādil al-Hindī
(according to Sharīf al-Murtadā, Sallār al-Daylamī, and Ibn Idrīs), it is not valid or
legal to perform the *jumʿa*.[46] "Explicit permission" implied the exercise of the
Imam's discretionary authority (his *saltana*), as the leader of the community without any restriction. By assuming the leadership of the Friday worship without
"explicit permission" from the Imam, necessary for undertaking the exercise of
authority on the basis of *al-wilāyat al-sharʿiyya* (authority derived by virtue of divine
designation), the jurists were claiming the "special" deputyship that had been
terminated in the year A.D. 941. Significantly, even Sharīf al-Murtadā, who held
high official positions in the 'Abbāsid juridical administration, could not regard the
constitutional aspect of the fait accompli authority of an Imāmī jurist as delegated.
This matter was indeed crucial in light of the theory of Imamite political authority
during the occultation, with practical implications for the leadership of the community. It was for this reason that judicial decisions of Sharīf al-Murtadā, Sallār, and
Ibn Idrīs in the matter of the *jumʿa* were taken seriously by the jurists. They were
fully aware of their ramifications for the leadership of the Imamite community
during the occultation.

As indicated earlier, the development of the *wilāya* of the jurists was an
inevitable corollary to the theory of the Imamate of the Hidden Imam who had
deputized them to undertake certain functions in his absence, including functions
with political ramifications, such as *wilāyat al-hukm*. That the *wilāyat al-hukm* had
the potential to become *al-wilāyat al-ʿāmma* can be construed from the rulings
regarding the incumbency of the *jumʿa*. From early on, the general deputyship of
the Imam was deemed possible in a person who, because of his qualifications and
the best possible disposition, could undertake the deputyship of the Imam and
represent him in Friday service.[47] Thus even when Sharīf al-Murtadā and those who
followed him in denying this *wilāya* to the jurists argued against the validity of the
jumʿa during the occultation because the jurists had not been authorized by the
Imam, they could not deny the *wilāya* in other obviously official positions like *al-
hukm* or *al-hudūd*, also requiring official appointment.

Fādil al-Hindī, who tends to agree with Sharīf al-Murtadā, explains that administration of justice and the giving of legal decisions on matters pertaining to the

Sharī'a, for scholars like Sharīf al-Murtadā, were far more significant for the welfare of the Shī'a. As a consequence, they did not object to the exercise of this
authority by jurists. On the other hand, if the *jum'a* was not performed, no sinful
deviation or corruption resulted among the people. This is the difference between
the exercise of two types of *wilāya*, and the *jum'a* certainly requires the *wilāya* to be
exercised by the Imam or the one delegated by him.[48]

It is plausible that those scholars who ruled that the Imamite jurists did not
have the permission of the Imam to convene the *jum'a* were also implicitly denying
the extension of the *wilāyat al-hukm* to *al-wilāyat al-'āmma*, whereas those who
ruled that it was permissible for the jurists to convene and lead the *jum'a* were
implying the delegation of *al-wilāyat al-'āmma* to well-qualified, righteous jurists.
Of course, as our sources reveal, the majority belonged to the latter group, albeit
with variations in the interpretation of *al-wilāyat al-'āmma*. The Imamite juridical
sources from the Safavid and post-Safavid era demonstrate, in some cases, a shift
from implicit to explicit delegation of *al-wilāyat al-'āmma* to the jurists.

In his commentary on *Qawā'id* of 'Allāma, Muhaqqiq al-Karakī (d. 937 or 941/
1530 or 1534), the Shaykh al-Islam under Tahmāsp Shah, takes up the question of
the *jum'a*. He points out the hesitation on the part of 'Allāma in his other work,
Muntahā, to concede permission to the jurists in view of Tūsī's opinion in *Khilāf*
and the opinions of Sharīf al-Murtadā, Sallār, and Ibn Idrīs to the same effect.
Their argument, according to Karakī, is based on an important consideration—
namely, that had there existed evidence to support the permission, then it would
have also supported individual, personal incumbency of the *jum'a*. In the absence of
such a ruling by anyone, the conclusion regarding its invalidity is sound.

Karakī responds by stating in the clearest terms that an argument based on the
absence of the precondition regarding the presence of the Imam or his deputy is
void:

> The reliable, well-qualified jurist who can issue legal decisions is designated by
> the Imam. Accordingly, his rulings are effective and it is obligatory to assist him
> in the administration of *al-hudūd* and *al-qadā'* among the people. It is not proper
> to say that the jurist is designated for administration of justice and for giving
> legal decisions only, and that the Friday service is a matter outside the scope of
> these two responsibilities. Such an opinion is extremely weak, because the jurist
> has been appointed as *al-hākim* by the Imams, which is well documented in the
> traditions.

That the jurist has been recognized as having the permission of the Imam for
the *jum'a* is, according to Karakī, well known among the Imamites, even more so
among the *muta'akhkhirūn* (i.e., from 'Allāma on, including his contemporaries),
which is apparent from Shahīd I's opinion in his *Dhikrā*.[49]

Karakī then proceeds to discuss the "permission" (*al-jawāz*) for the jurist on
the basis of 'Allāma's opinion in *Qawā'id*, which has been interpreted by other
jurists to mean "an act that becomes commendable (*al-mustahabb*)." According to
Karakī, giving the "permission" cannot mean simply that an act is neutral in status
in relation to ultimate reward or punishment, whether one performs it or not. Thus,
when the "permission" is applied to the *jum'a*, it connotes something more than
merely neutrality. The *jum'a* is a form of worship, and as such it is something more

than merely a commendable act with neutral status. Thus, the meaning of "permission" is general, in the sense that it signifies general permission in performing an obligatory act. Accordingly, "permission" in the context of the *jum'a* does not connote *al-istihbāb* (commendable), which is actually categorized as a *mandūb* (recommended) act, because the *jum'a* is supposed to replace the regularly incumbent obligation of the midday worship.

Moreover, there is an *ijmā'* to the effect that midday worship is not legal when the *jum'a* becomes obligatory. No recommended act can substitute for an incumbent duty. Rather, argues Karakī, "permission" in the context of the *jum'a* means that it is the more excellent of the incumbent acts one could perform at the same time. There is no contradiction in maintaining that the *jum'a* is both definitely recommended and optionally incumbent. It can be asked: What is the better legal expression for the *jum'a* during the occultation, "commendable" or "permissible"? Karakī responds "permissible," because the controversy centers around the prohibition of the *jum'a* and its legality, not around its being "recommended" or "optionally incumbent."[50]

Karakī's judicial decision indicates his confidence in the opinion that regards the *jum'a* as being legally valid during the occultation when a qualified jurist can undertake to convene and lead the service. It was probably this confidence that led him to conclude that the jurist possessed all the necessary qualifications for assuming the position of *al-imām al-'ādil* (the Just Ruler) when the twelfth Imam is in concealment.

Another prominent jurist of the Safavid era who has greatly contributed to our understanding of the question of the *jum'a* and its relationship to *al-wilāyat al-'āmma* of the Imamite jurist is Sabzawārī (d. 1090/1679). His *Dhakhīra* contains a detailed discussion of the differences of opinion among the Imamites. Sabzawārī also cites some sources that have not been discussed in the context of the *jum'a* before by anybody and they reflect the Imamite concern over the way the Sunnites viewed their perceived deviation from the well-established practice among Muslims. Thus, Sabzawari cites 'Imād al-Dīn al-Tabarsī's *Nahj al-'irfān ilā hidāyat al-īmān,* where the author (after mentioning differences of opinion among Muslims about the stipulations of the *jum'a*) says:

> "The Imamites have for the most part declared the *jum'a* to be incumbent, even more than the Sunnites; in spite of this the Sunnites have condemned them for having abondoned it; whereas [the actual situtation is that] they do not permit the imamate of a corrupt, sinful individual or [the imamate] of one who opposes the correct belief." This statement [says Sabzawārī] indicates that the Imamites consider the *jum'a* to be individually and personally incumbent without any reference to the precondition regarding the Imam or his deputy, because such a condition could not be fulfilled during the occultation. [According to al-Tabarsī] the Imamites even hold it as incumbent more than do the Sunnites. The Sunnites do not make a precondition except that it should be within the city [i.e., a settled area, where there is a suitable mosque], as the Hanafīs say.[51]

Sabzawārī gives his own ruling on the *jum'a* in his other work, *Kifāyat al-ahkām.* Here he contends that the plausible opinion is that the *jum'a* is individually and personally incumbent, without any precondition about the leadership of the jurist. But, as a precautionary measure, if there is a jurist in a city, then no one else should convene it. This precautionary measure also applies in the case of a jurist who is more learned, for the more learned should be given preponderance over any others to convene the worship. Furthermore, no one should undertake to lead the *jumʿa* without the qualified jurist's permission, who should also see to it that the general prescribed conditions are fulfilled in order to render the *jumʿa* valid.[52]

From Sabzawārī's discussion it can be maintained that gradually the performance of the *jumʿa* was being viewed less as the function of the Imam and more as an obligation prescribed by the Qurʾān without any precondition. This was also the way the Sunnites had come to accept the Friday worship in the absence of "rightly guided" caliphs. In Sunnism, the absence of strict adherence to the ideals set forth in the doctrine of the Imamate, which underlay the judicial decisions on the *jumʿa* in Imamite jurisprudence, allowed Sunnī scholars to rationalize the authority of sinful rulers, who disregarded Islamic norms, and work toward the autonomy of religious observances whose fulfillment was, in theory, dependent upon the ruling authority. In Shīʿīsm, it seems that, more so under the Safavid rule than at any other time, such an autonomy of religious observances was contemplated by jurists like Tabarsī, Sabzawārī, and all those scholars who can be designated as belonging to the Akhbārī school of Imamite jurisprudence. However, among the jurists who followed the Usūlī methodology in jurisprudence, this process of autonomy was contemplated under the leadership of the jurists who could become *al-imām al-ʿādil* in the absence of the twelfth Imam.

During the Safavid era there were other jurists besides Sabzawārī who did not insist on the precondition about the Imam or his deputy for the *jumʿa*. The most prominent jurist during the later part of the Safavid rule, Muhammad Bāqir Majlisī (d. 1111/1699–1700), also maintained a similar opinion. Majlisī upheld the individual and personal incumbency of the *jumʿa* during all the epochs of Imamite history. As for the imamate of the *jumʿa*, all that was required for convening and leading the *jumʿa* was a person whose righteousness (*ʿadāla*) was recognized in the congregation and who was knowledgeable about the rulings pertaining to worship (*salāt*) in general, whether as a *mujtahid* himself or as one who followed the rulings of another *mujtahid* (i.e., a *muqallid*).[53] He argues that the precondition about the presence of the Imam or his deputy was not mentioned by the early traditionists like Kulaynī or Ibn Bābūya. In fact, says Majlisī, Kulaynī not only did not relate any tradition about this precondition, he also provided no report of the particular tradition related on the authority of Muhammad b. Muslim al-Thaqafī in which Imam al-Bāqir told him: "The *jumʿa* is incumbent upon seven: the Imam, . . ." The latter tradition, adds Majlisī, led to the delusion concerning the prerequisite for the Imam in the *jumʿa*.[54]

Shaykh Muhammad Hasan al-Najafī (d. 1266/1848–49), the leading Imamite jurist during the Qājār period, believed that the jurists who had declared the observance of the *jumʿa* individually and personally incumbent during the occultation, without requiring believers to take any precautionary measure in performing the regular midday worship, had ruled differently in their previously expressed opinion on this issue. This change of their opinion, says Najafī, was because of their "love for leadership (*riʾāsa*) and power (*saltana*) in Iranian lands (*bilād al-ʿajam*)."[55] There is little doubt that the declaration that the *jumʿa* was individually and personally incumbent had ramifications for the leadership of the community. Najafī's statement about the "love for leadership" points to the political ambitions of scholars like Majlisī who had no scruples in assuming the Imam's theologico-political function as the imam of the Friday service, which led to the denial of the precondition in the *hadīth*.

Majlisī's *Bihār* is more inclined toward the Akhbārī trend in Imamite jurisprudence (although, as pointed out in the Introduction and in the previous chapter, official Akhbārī-Uṣūlī debate is usually placed in the Qājār era). In the judicial decision on the status of *jumʿa* in *Bihār*, a particular line of thought can be detected regarding the *jumʿa* and the prerequisite of the Imam or his deputy during the occultation. This can be attributed to the Akhbārī trend in Imamite jurisprudence. The performance of the *jumʿa* was viewed as the fulfillment of the obligation ordained in the Qur'ān without any preconditions. The Akhbārī trend goes back to the early period of Imamite legal history when the great traditionist Ibn Bābūya, in his *Muqniʿ*, maintained an opinion similar to that of Majlisī regarding the individual and personal incumbency of the *jumʿa* without any preconditions about the Imam or his deputy.[56]

Another prominent traditionist during the Safavid era was Mullā Muhsin Fayd al-Kāshānī. In his work *Mafātīh* he argues against the precondition about the Imam or his deputy in the *jumʿa* and considers Sallār al-Daylamī and Ibn Idrīs's opinion regarding *imām al-asl* as an unsubstantiated assertion. As for those who maintain that there is an *ijmāʿ* among the Imamites on the precondition regarding the presence of the general deputy of the Imam (that is, the well-qualified jurist who could give legal decisions for the *jumʿa* to become obligatory), they need to clarify their assertion. If their assertion means that the precondition regarding the general deputy is for the sake of seeking his opinion in the matter of the performance of the *jumʿa* (especially in view of the difference of opinion among jurists on this question and their own abstention from its performance in consequence), then the assertion is reasonable; otherwise, such a precondition has no textual support in Imamite sources, nor does it have any rational proof.[57] The reference to rational proof is meant to be a rebuttal of those theologians who have enumerated the *jumʿa* among the constitutional functions of the Imam as the political head of the Muslim community. The use of such rational proof, according to the traditionists, is questionable when the revelational proof establishes the opposite in the case of the Friday worship.

This Akhbārī trend in ruling the *jumʿa* as an individually and personally incumbent duty reached its logical conclusion during the Qājār period when, according to al-Shaykh Jaʿfar al-Kabīr, Kāshif al-Ghīṭāʾ (the contemporary of Fath ʿAlī Shāh), anyone who denied its incumbency was charged with apostasy (*al-ridda*).[58] The problem, according to Shaykh Jaʿfar, was that some jurists had refused to recognize the precondition about *al-sulṭān al-ʿādil* (designated by God, together with the Prophet and the Imams, as able to exercise discretionary control in the managing of the affairs of the Muslims) for the *jumʿa* to be declared definitely incumbent (*wājib hatmī*). The appointment of a deputy who can function as an occasional imam to lead the *jumʿa* is dependent on the consolidation of the Imam's authority. During the occultation such reinforcement of the Imam is impossible, and hence the designation of the imam for the *jumʿa* likewise cannot be regarded as definitely incumbent on the Imams. This is well established in the authentic practice of the Imams.

In addition, says Shaykh Jaʿfar, the imamate of the *jumʿa* is among the official

positions established by revelation (*al-manāsib al-sharʿiyya*) that require the permission of the Prophet or the Imam. This conviction has ultimately become well
established among Imamite scholars, both "ancient" and "modern"—except for a
few, adds Shaykh Jaʿfar, who have insisted otherwise. This consensus, he says,
whether reported or not, can be ascertained by thorough investigation of Imamite
sources throughout the ages, and it has actually attained the level of *al-mutawātir*
(the consensus reported uninterruptedly by so many trustworthy scholars that excludes doubt as to its truth).[59]

Shaykh Jaʿfar's claim to *mutawātir* indicates the necessity to maintain the perpetuity of a precedent for its eligibility as an authoritative documentation to deduce
juridical decisions. However, the *mutawātir* he is reporting is not of the tradition
narrated on the authority of the Imams; rather, he is relating the *mutawātir* of the
consensus (*ijmāʿ*) reported in the works of the jurists.

It is relevant to note the way the jurists—who, as a group, were entrusted with
the Imamite teachings—form a link in the *ijmāʿ*. This type of consensus is differentiated from the consensus that was arrived at during the lifetime of the Imams.
Technically, the former type is classified as *manqūlī*—that is, "reported" consensus
that can be attained through investigation of the works of the jurists in all periods of
Imamite jurisprudence. During the occultation, "reported" consensus at times attained the level of *mutawātir* and served as a valid documentation for a juridical
prescription. It is probably for this reason that in some instances the consensus that
was claimed by both the proponents and the opponents of a particular judicial
decision was a "reported" type recorded in the works of early Imamite scholars.
The use of contradictory "reported" consensus of Imamites indicates both the
individual nature of the juridical works produced by these scholars, and the flexibility in decisions based upon this documentation. Furthermore, the method of resolving the contrariety in *hadīth* reports also became applicable to "reported" *ijmāʿ*,
which came to be regarded as part of the Sunna in the later works of *usūl al-fiqh*
(theoretical basis of Islamic law).

The validity of the *jumʿa* as depending on the Prophet or the Imam or a person
specifically appointed by them in regard to their "presence" (*hudūr*) (that is, having
the power and authority [*bast al-kalima*] to perform it) was undisputed by Usūlī
jurists during the Qājār period. This is well demonstrated by al-Wahīd al-
Bihbahānī, Shaykh Jaʿfar's teacher.

In his critique of the Akhbārī position that the *jumʿa* was definitely incumbent
during the occultation as authoritatively proven by the Qurʾān, the Sunna, and the
ijmāʿ, Bihbahānī declares that the *jumʿa* is among the *manāsib al-saltana* (official
positions based on investiture with power), which require the Imam's special permission. From the time of Ibn Bābūya until the time of Shahīd II, Imamites were in
agreement that a prerequisite to convene the *jumʿa* was the special permission of
the Imam given to a specific person in order for the worship to become valid in
general, or for it to become individually, personally, and definitively incumbent.
But, says Bihbahānī, for the period immediately preceding Ibn Bābūya we have no
information regarding the stance and the terminology employed by jurists, although
we know of the existence of well-informed jurists, close to the period immediately
before Ibn Bābūya. We also have information about those who were connected
with and were contemporaries of the Imam al-Sādiq, and those who lived in the
succeeding period, because there exists a body of consensus that goes back to the

period of al-Sādiq and to the period preceding it—that is, linking up with the Prophet's own time. This consensus was in regard to the prerequisite of the Imam or his special deputy. In spite of all this, says Bihbahānī, the Akhbārīs maintain that the *jum'a* is definitely incumbent on the basis of consensus, the Qur'ān, and the *mutawātir* traditions, without any precondition. Disregarding the application of precautionary measure (*ihtiyāt*) in a matter where there is also an opinion that maintains its illegality is, according to Bihbahānī, equivalent to disregarding the *akhbār*—the traditions that support or challenge its incumbency.

As for the role of the jurist in the *jum'a*, it is supported by the traditions that speak about their *wilāya* (meaning their authority to administer justice). These traditions demonstrate permission for jurists, who are the Imam's general deputies in matters that only the Imams used to undertake. However, the performance of the *jum'a* is not dependent on the legal decision of the *faqīh*, and regardless of whether a person is a *mujtahid* or not, he still has to perform the *jum'a*, other things being equal. But whether the *jum'a* suffices on Friday in place of the regular midday worship is dependent on the judicial decision of the jurist.[60]

In this overall view of the *jum'a*, Bihbahānī indicates a tacit recognition of the fact that the legally valid status of the *jum'a* had implications for the Imam's constitutional right, and, consequently, in the absence of the Imam or his general deputy's political investiture, it is the obligation of a *mujtahid* to rule whether performance of the *jum'a* could be regarded as individually and personally incumbent, and hence sufficient for the fulfillment of the duty of worship on Friday. In other words, it was the personal *ijtihād* of a jurist that could resolve the fundamental question about the nature of his deputyship in the Shī'ī community. Bihbahānī's juridical prescription in this connection was going to serve as the general concern shared by Usūlī jurists at this time in defining their deputyship in explicit terms under the Qājār rulers.

It is important to summarize the Akhbārī position in this connection because its proponents have asserted that the precondition regarding the presence of *al-sultān*, which has been interpreted as *al-imām*, is not based on the traditions of the Imams but on the *ijtihād* of the Usūlīs and their conjectural opinion (*al-ra'y*). Bahrānī, the Akhbārī jurist, takes up the question of the precondition in the *jum'a* and his discussion in *Hadā'iq* is centered on the following questions:

1. Is it a precondition for the *jum'a* that there be present *al-imām al-ma'sūm* (the infallible Imam) or his deputy?

2. Does this precondition relate to the convening of the *jum'a* or to its being regarded as incumbent?

3. Is this precondition specifically for the period of the *hudūr* ("presence" of the Imam where he could be invested with political power) or does it extend to the occultation?

4. Is the deputy mentioned in the precondition a "special" or a "general" functionary, who could also include the *faqīh* during the occultation?

5. Can any imam of the congregation lead the *jum'a* or does it have to be the *faqīh*?

6. On the basis of its condition, is the incumbency in the *jum'a* individual and personal (*'aynī*) or is it optional (*takhyīrī*)?

The Akhbārī jurists regard the *jum'a* as incumbent, as long as the conditions are met. These conditions differ from those that pertain to the regular daily worship, but they do not include, as a prerequisite, the presence of the Imam or his deputy, or his permission, because the ostensible sense derived from the ordainment of the *jum'a* in the Qur'ān and the Sunna does not support such a prerequisite. As Bahrānī states, there is no difference of opinion among the Imamites that the *jum'a* is individually and personally incumbent when the Imam or his special deputy is present. But there is a difference of opinion touching the period of occultation, when the specific permission is not available. He then systematically discusses different opinions throughout Imamite jurisprudence and treats the question of *al-sultān al-'ādil* in *Man lā yahduruh al faqīh* of Ibn Bābūya (who is referred to as "our master, the chief of the *muhaddithūn*") and in Shahīd II's *Risāla* on the *jum'a*. Bahrānī argues that both of these jurists have spoken of the Imam who is other than *al-sultān al-'ādil;* rather, in their usage, he is merely the imam for the daily worship, with whom one can participate in the *jum'a* also. Further on, he says that those Imamite scholars who have argued that *al-sultān* is the Imam as a precondition of the *jum'a,* have done so in the same way as "they have preferred *al-ijtihād* [independent reasoning] [to *al-akhbār*— i.e., "traditions"], basing their opinions on *al-ra'y.*" As a result, says Bahrānī, they have come to assert this as a precondition for the *jum'a* to be incumbent and valid. Thereafter, they have differed among themselves as to the applicability of the condition. Some say it applies both to the period when the Imam is "present" (i.e., invested with political power) on earth and when he is in occultation. Consequently, they have ruled the *jum'a* as suspended at virtually all times in history. Others say it applies only to the period when the Imam is present and hence, during the occultation, the condition for the presence of the Imam is waived.

This group is divided into those who consider the *jum'a* definitely (*hatmiyyan*) incumbent, to which most of the "ancient" Akhbaris have adhered; and others who have regarded the *jum'a* as definitely recommended and optionally incumbent (*mustahabba 'aynan, wājiba takhyīran*). This is maintained by a few, but prominent, jurists like Tūsī, and their importance has led some Akhbārīs to agree with them. There are some jurists, who, according to Bahrānī, have argued that the general permission is a substitute for the special permission during the occultation. Accordingly, they regard the presence of a *faqīh* as a condition for the validity of the *jum'a,* because he is the deputy of the Imam in general, and authorized by him to execute rulings. This opinion has been adopted by Kāshānī in his treatise on the *jum'a,* on the basis of an *ijmā'* claimed by this group, but one for which they have provided no documentation from either the Qur'ān, the Sunna, or the *khabar*.

Majlisī, reports Bahrānī, has argued for the definite incumbency of the *jum'a* in his *Bihār,* insisting that just as there is no precondition regarding daily worship or *zakāt* (alms-levy) to the effect that there should exist an Imam and that he should be present and that his permission should be available, so too is the case with the *jum'a.* Majlisī criticizes these jurists for their claim to the *ijmā'* in this regard and in other questions, because *ijmā',* in order to become authoritative and binding, must include the Imam's opinion. There is no doubt, according to Majlisī, that during the occultation one has no access to the *ijmā',* even if one had knowledge of all the opinions of the Imamite scholars and their differences throughout Islamic lands. Such an alleged *ijmā'* cannot be used as evidence to promulgate any law. It is claimed that the Imam would reveal his opposition if the *ijmā'* were erroneous; but how can one accept this argument, asks Majlisī, in the face of the variations in the traditions? Majlisī's criticism of the consensus pertains to the "reported" type of *ijmā'*, to which I made reference earlier. His remark about the variations in the *ijmā'* and their similarity with the *hadīth* reports is consistent with the way both *ijmā'* and *hadīth* were transmitted and, at times, attained the level of *al-mutawātir* (uninterrputedly reported), and at other times remained "single" (*al-āhād*).

Then Bahrānī names his teacher, al-Sayyid Majīd b. Hāshim al-Sādiqī Bahrānī, as one of those who used to perform the *jum'a* in Shiraz and adds that he himself had prayed with him. The fact that this prominent Sayyid performed the *jum'a*, according to Bahrānī, is evidence for its definite incumbency and the alleged *ijmā'* does not establish its opposite. Also, in Mashhad, Amīr Muhammad Ja'far and Amīr Muhammad Mu'iz al-Dīn Muhammad used to convene the *jum'a* as an individually and personally incumbent duty. And, adds Bahrānī, there used to be heated debate in Isfahan between al-Shaykh 'Abd Allāh Sālih Bahrānī, the Akhbārī jurist, and Fādil al-Hindī, who maintained that the *jum'a* during the occultation was prohibited.[61]

It is significant that the Akhbārī jurists like Bahrānī maintained the permissibility for the imam of the daily worship to assume the imamate of the *jum'a* precisely because they denied the validity of the *ijmā'* to the effect that the presence of the Imamite jurist was necessary to legalize the *jum'a* during the occultation, and that there was a difference between the imamate of the *jum'a* and that of the daily regular worship. It would be reasonable to surmise on the basis of this ruling that they did not regard the leadership of the *jum'a* as among the official positions of the one in authority, because of their denial of the precondition of the Imam in the first place. This could be regarded as one of the fundamental differences between Usūlī and Akhbārī jurists, which had ramifications in their whole respective perceptions of the position of the deputy of the Imam in the Imamite community during the absence of the Imam. The Usūlī jurists, through their exegetical extrapolation of the basic "investiture traditions" were able to argue for the comprehensive authority of the jurist in his capacity as the generally appointed deputy of the Imam. But the Akhbārī jurists rejected both the *usūlī* method as well as the prerogative of the jurist to go beyond the ostensible sense of the precedents preserved in the traditions (*akhbār*) from the Imams in order to deductively infer the all-comprehensive authority of the jurist. It is important to remember that even among the Usūlīs the opinion that the *jum'a* had to be convened by a *faqīh* was not accepted unanimously. The opinion that the *faqīh* should convene the *jum'a* was common among those Imamites who maintained the optional and contingent incumbency of the service, which implied a choice between two incumbencies.

This issue has been taken up in detail by another eminent Usūlī jurist during the Qājār era, Najafī (d. 1266/1849). He believes that the optional incumbency of the *jum'a* had been particularized by Muhaqqiq al-Karakī and others who followed him by making it necessary to be convened by *al-mujtahid,* although there is nothing in the sources to require this. Moreover, he goes on, although Muhaqqiq al-Karakī has ruled to this effect authoritatively on the basis of the interpretation of the particular texts in 'Allāma and Shahīd I's writings, the apparent sense to be derived from these texts is that it is not necessary that the *faqīh* lead the worship. Thus, maintains Najafī, the opinion is established in the explicit textual evidence even before the designation of the jurists is mentioned in traditions like the one reported by 'Umar b. Hanzala, in which jurists are designated to exercise *wilāyat al-hukm.* This delegation of the *wilāya* occurred partly because the Imams were unable to assume authority themselves.

There is a *mursal* tradition (which has not been fully authenticated, because of a defective chain of transmission) regarding the *faqīh,* says Najafī, in which the Prophet prayed, saying: "O God, have mercy on my successors (*khulafā'ī*)!" He was asked who these were. The Prophet said: "Those who will come after me and will report my tradition and custom." The tradition is *mursal,* emphasizes Najafī, and there is a possibility that it points toward the Imams or the one designated by them; it certainly does not provide evidence for the designation of the *faqīh.*

But what does point toward this precondition regarding the *faqīh* is *al-ijmāʿ,* strengthened by evidence that proves this condition to be necessary if the *jumʿa* is to be individually and personally incumbent, which in Najafī's opinion is more appropriate to maintain. Alternatively, one should maintain its legality only during the period of the Imams. Consequently, the *faqīh's* appointment as the Imam's deputy during the occultation does not change the *jumʿa* from being optionally incumbent on anyone other than the *faqīh,* and this is because of the generality of the permission. The documentation for its being optionally incumbent is provided by the nature of the deputyship (*niyāba*), which is something other than "permission" (*al-idhn*), because the latter is given to a specific person, whereas the deputyship is general and can be applied to anyone who is qualified as a deputy (*nāʾib*). This is the essence of Shahīd I's statement in *Dhikrā* regarding the difference between *niyāba* and *al-idhn.* This difference, says Najafī, is probably correct because *al-takhyīr* ("option") supports the view that "deputyship" is open to anyone who qualifies, unlike "permission," which is a specific designation, and as such is regarded as one of the nine conditions required in a *nāʾib* (deputy).[62]

It is evident from the above argument that the *niyāba,* or deputyship of the jurist, was almost a settled question, even when the specific permission in the *jumʿa* was insisted upon for the *faqīh* as a precondition for the validity of the *jumʿa.* Nevertheless, even when the *jumʿa* was ruled optionally incumbent, there was an implicit acknowledgment that there was a general deputy of the Imam, with *wilāyat al-hukm,* who could assume the imamate of the *jumʿa.* Such an acknowledgment of the deputyship of jurists in the *jumʿa* had repercussions in the interpretation of their *wilāya* in the context of *al-wilāyat al-ʿāmma.*

The *jumʿa,* as indicated earlier, was the only religious duty with theologico-political implications, which, if determined as a valid legal activity during the occultation, could provide the indispensable legitimacy for the jurist to assume the all-comprehensive authority of the Imam. In other words, the convener of the Friday service as a valid legal activity would have fulfilled all the essential stipulations, including the one requiring the presence of *al-sultān al-ʿādil,* the Just Ruler, to render it thus. The convener would have assumed the function of the Imam and the official position of the Imam's deputy.

It is to this position of the deputy to whom *al-wilāyat al-ʿamma* (the comprehensive authority of the jurist) accrues that I now turn.

Al-wilāyat al-ʿāmma and the Problem of its Legitimation

In my discussion of the *wilāya* so far, it has become evident that the legitimation of the authority of Imamite jurists in the area of religious-juridical functions of the Imam has not caused any major revision in the theory of political authority in the doctrine of the Imamate. The authority of jurists in areas affecting the religious and social welfare of the Imamite community, as I have argued in this study, has been the inevitable consequence of the prolonged occultation of the Imam. Furthermore, I have demonstrated that Imamite jurisprudence has preserved ample documentation, which (in some cases explicitly and in others implicitly) justifies as much as legitimizes the assumption of the Imam's religious-juridical authority by Imamite jurists, who have, from the early days of the Imamite history, provided necessary leadership as functional imams of the Shī'a.

As regards classical as well as "premodern" Imamite jurisprudence, it may be confidently asserted that the assumption of direct political authority by jurists has not been discussed at any point in the sections of jurisprudence examined so far. However, as I have pointed out on several occasions, the assumption of the authority of *sultān al-zamān,* to establish a just rule based on the precepts of the Sharī'a by jurists, was implied in many rulings that had ramifications for political authority. In other words, *al-wilāyat al-'āmma* of the jurist was implied in several rulings, but was never discussed as part of the *wilāyat al-faqīh* until much later in the premodern *fiqh* when the assumption of such an authority by jurists became a probability.

As I have argued throughout this study, there was nothing in Imamite jurisprudence that could be construed as compromising the theological Imamate if the jurist assumed *al-wilāyat al-'āmma* as the deputy of the Imam; this is especially so, when, as I have shown with substantial documentary evidence, the existence of a just ruler (*sultān 'ādil*) who promulgated the Sharī'a during the occultation was taken as a reality after the establishment of the Imāmite Būyid temporal authority. Just as there was nothing in the juridical works that declared it impossible when the Imam was in occultation for any Imamite to assume temporal authority, even through use of force, so there is no reason to believe that Imamite jurists who exercised different forms of *wilāya* in the name of the Hidden Imam, were barred from assuming *al-siyāsa* on behalf of *al-sultān al-'ādil* (i.e., the Just Ruler, the Imam himself).

There is evidence in the *fiqh* writings of all the major schools of legal thought in Islam that the exercise of *al-wilāyat al-'āmma* accrued to the *sultān* essentially because he received taxes. In my discussion about those who can exercise *wilāya* I have mentioned 'Allāma's opinion regarding the *wilāya* of a *sultān* who, according to the Prophet's tradition, becomes "guardian of the one who does not have a guardian," this being so because, says 'Allāma, "he received taxes." It is significant that the jurists in the Imamite system, like the *sultān*s in the Sunnite system, were, according to the majority of scholars, entitled to receive the *zakāt* and other religious taxes in the absence of the Imam. Thus, says 'Allāma, it is better to disburse the *zakāt* on "openly visible" goods (*al-amwāl al-zāhira,* which include crops, fruits, camels, cattle, and so on) to *al-imām al-'ādil.* This opinion, he says, is maintained by Imams al-Bāqir, al-Shu'bī, al-Awzā'ī, and Ahmad b. Hanbal, because, according to them, the Imam knows best how this should be spent. During the occultation, it is recommended to deliver the *zakāt* to the trusted jurist among the Imamites.[63] Shahīd II, in his commentary on Muhaqqiq's *Sharā'i',* explains why the *zakāt* should be given to *al-faqīh al-ma'mūn* (the trusted jurist):

> This applies to the jurist in general on the basis of the *wilāya* invested in him because of his qualifications to give legal decisions and his reliability because of his being exempt from using any tricks to get his hands on the wealth of others.

[If he were to do such a thing] then that would detract from his status [as the trusted one]. One must bear in mind that he is appointed in the interest of the public, and if he were to be dishonest there would occur harm to those who were entitled to receive the *zakāt*.[64]

The ruling that the *zakāt* should be delivered to the Imam or, when he is in occultation, to the trusted jurist, conveys an important consideration—namely, that the Imam or the jurist is invested with *al-wilāyat al-ʿāmma*. Mufīd, Abū Salāh al-Halabī, and al-Qāḍī Ibn al-Barrāj, who have all ruled and made it incumbent to carry the *zakāt* to the reliable *faqīh*, have in fact required all other forms of dues like *khums* (the fifth) and *sadaqāt* (other forms of benevolent charity) to be taken to the *faqīh*, because he is the deputy of the Imam in this and other related matters during the occultation. Such an incumbency, according to historical precedent, can apply only when the Imam or his deputy is invested with political authority. According to Qāḍī Nuʿmān, who cites several traditions in *Daʿāʾim al-islām*,[65] the practice of the Prophet and the caliphs was to collect taxes from the people, and no one else besides them was authorized to distribute it. These traditions demonstrate that those who were invested with *al-wilāyat al-ʿāmma* administered the financial structure of the community.

But, as in other areas of jurisprudence where such political implications of the *wilāya* could be deduced, so in the disbursement of the *zakāt* also some jurists were hesitant to require the *zakāt* to be delivered to the Imamite jurist as an incumbent duty, especially during the *taqiyya* and the occultation when such an activity could arouse the suspicion of the de facto ruling authority. Nevertheless, even these scholars did not deny that the jurist had a better insight as to how the *zakāt* should be spent. Moreover, if the jurist specifically asks for it, then, they ruled, it was obligatory to deliver the *zakāt* to him. ʿAllāma hesitated in concurring with the idea that the *zakāt* must be delivered to the Imam or the jurist, because the rule applies specifically only when the Imam asks for it, and with the appearance of the Imam all Islamic laws will become clear.

Commenting on this last part of the argument, Najafī (writing during the Qājār period) says that if it is ruled obligatory to implement all the laws of the Sharīʿa during the occultation, it is impossible to maintain that the implementation could take place by asking a well-qualified jurist to execute the laws. The reason is that the deputyship of the jurist is of a general type which does not include the responsibility of implementing the Sharīʿa. This general sense of deputyship is implied in the ruling of Shahīd I in the context of the collection of the *zakāt:*

It is said that it is obligatory to give *zakāt* to the jurist during the occultation if he asks for it himself or through his agent, because he is the deputy of the Imam, just as the collector of the taxes (*al-sāʿī*) is. Rather, however, it is more appropriate to state that his deputyship (*niyāba*) on behalf of the Imam is applicable in all those matters in which the Imam himself has authority; whereas the collector is the agent of the Imam only in a particular function.[66]

Najafī then proceeds to say that the evidence supporting the authority of the jurist in particular is the tradition regarding the "future contingencies" in which the "investiture" (*nasb*) of the jurist is specifically mentioned. This tradition, commonly known as *riwāyat al-nasb* (the "investiture tradition" among the Qājār jurists), is reported on the authority of the twelfth Imam. According to Najafī, it places the jurist among the *ulū al-'amr* (those who possess authority through investiture), obedience to whom is incumbent on Shīʿites. It is well established, says Najafī, that the jurist has been specifically designated to handle all the taxes ordained in the Sharīʿa. The claim that his *wilāya* extends only to matters of juridical decisions is disproven by the fact that his authority includes matters that are not confined to legal issues, such as guarding the property of a minor, an insane or absent person, and so on. The *wilāya* of a jurist is mentioned on various occasions and his being designated among the *ulū al-'amr* clearly points to the urgent need of the community in more than merely juridical matters.[67]

There is an important *hadīth* report related by the prominent associate of the Imam ʿAlī al-Ridā, Fadl b. Shādhān, who had once asked the Imam to explain the reason why those who can exercise authority, the *ulū al-'amr*, were appointed and were required to be obeyed by the people. The Imam replied that no community can survive without those in whom authority is invested to manage the affairs of the people. The *ulū al-'amr*, he says, have been appointed and their obedience required by God so that they can defend Muslims against their enemies, can collect taxes and dispose of them in accordance with the law, can convene and lead the *jumʿa*, can protect the people from being oppressed and the religion of God from being distorted by sinful deviators.[68]

The response of the Imam certainly points to the general character of the need for those who can exercise authority justly at *any* time in history. Accordingly, the designation by the twelfth Imam of the jurists as *ulū al-'amr* has been regarded as consistent with the rational argument for the necessity of the Imamate at all times in Imamite theology, and its implication for the theological Imamate has not been overlooked by any Imamite scholar up to the present day.

A critical problem is that there are many *hadīth* reports that emphatically declare Imams from among the *ahl al-bayt* to be the only *ulū al-'amr* to whom obedience is incumbent. In fact, in the exegesis of the Qur'ānic verse in which obedience to the *ulū al-'amr* has been ordained (4:59), Imamite exegetes have maintained unanimously that the *ulū al-'amr* are the infallible Imams only, and they cite many traditions to support their interpretation. However, the general application of the *ulū al-'amr* has been upheld by all Sunnī exegetes, on the basis of both the lexical explanation of the phrase and its historical usage in relation to those who held power among Muslims, whether a caliph or a *sultān*.

It is this apparent contradiction of the general and restricted application of the *ulū al-'amr* among the Imamites that has raised doubts about jurists being part of the *ulū al-'amr* who should be obeyed as the Imams were to be obeyed. If the implications of the designation of the jurists by the twelfth Imam as one of the *ulū al-'amr* is accepted, then there is no problem in accepting also *al-wilāyat al-'āmma* for the jurist. But such an interpretation is based on juristic deduction rather than on the actual use of the phrase by the twelfth Imam; at least in the various traditions examined for this study and discussed below, I have not come across any specific text that applies *ulū al-'amr* to jurists. It is by the aid of demonstrative jurisprudence, where exegetical extrapolation of "investiture traditions" was achieved, that Imamite scholars like Najafī (whose opinion on *zakāt* uses the phrase for jurists on the basis of the "investiture tradition") have deduced such signification from the

Imam's general designation of jurists in his letter regarding the "future contingencies" affecting the Shī'a.

In addition, it is through the *ijtihād* of the jurists that the section on *wilāyat al-faqīh* has taken up issues that, in the Sunnī *fiqh,* did not arise at all, because of the presence of *ulū al-'amr* for Sunnī Muslims throughout history. In some respects, the development of *al-wilāyat al-'āmma* in Imamite Shī'īsm corresponds to the development of the authority of the *sultāns* who exercised full de facto authority while the divinely ordained caliphs remained as the legitimizers of their authority without themselves being invested with discretionary control over the affairs of the empire.

It is not without careful inference that Tūsī, as discussed in chapter 3, described the Imamite *faqīh* as the *khalīfat al-imām* and compared his position to that of the Imam among the Sunnites.[69] In other words, the jurist among the Imamites could exercise the authority of the *sultān* among the Sunnīs, who held the position of those invested with authority (*ulū al-'amr*). This was the implication of al-Sayyid Muhammad Kāzim al-Yazdī's ruling on *zakāt,* to the effect that the jurist received *zakāt* in his capacity as the person who possesses *al-wilāyat al-'āmma.*[70] It is, then, in this sense valid to equate *al-wilāyat al-'āmma* with the *saltana* of *ulū al-'amr* in the Muslim public order.

There is evidence for such an equation in the following rule of Najafī:

. . . *Al-wilāya,* which includes holding office for administering justice (*al-qadā'*), organizing [the affairs of the people] (*al-nizām*), [holding] political office (*al-siyāsa*), collecting taxes, managing the affairs of minors, and so on, on behalf of *al-sultān al-'ādil* preeminently, [more so] when in assuming it there is also assistance to uphold probity and piety, and service to the Imam. Sometimes it is individually and personally incumbent, when it is declared so by the Imam himself. This is obedience to God, more particularly when the obligation to forbid evil and enjoin good is dependent on [the exercise of the *wilāya*]. When the assumption of the latter duty becomes confirmed in one person, then it is incumbent on that person to assume this responsibility. Some have argued that *wilāya* is among the premises of the power (*al-qudra*) that is actually the condition for its incumbency. As such, it is incumbent neither to acquire it nor to accept it, because of the absence of definite obligation in relation to the *wilāya* [during the occultation]. This argument can be challenged on the ground that the definite obligation to enjoin good requires that all that is necessary in the form of premises to fulfil it must be satisfied, and that this obligation is not omitted except in cases of incapacity. Thus, [possession of] *wilāya* is part of these premises, after which one has the power to carry out the obligation [of enjoining good]. [The *wilāya* of] the deputy is connected to [the *wilāya* of] *al-sultān al-'ādil* [i.e., the Imam] during the occultation, on the basis of his having *wilāya* in some areas. [And, according to Najafī's teacher, Muhammad Bāqir b. Muhammad Akmal al-Wahīd al-Bihbahānī] if the *faqīh* who is appointed by the Imam on the basis of general permission is appointed a *sultān* or *hākim* for the people of Islam, there will be no unjust rulers, as was the case with the Children of Israel. This is so, because *hākim al-shar'* and *al-'urf* are [in such a case] both appointed by the divine will (*al-shar'*).[71]

Although Najafī does not fully agree with his teacher's assertion that *al-faqīh* is more capable of establishing the rule of justice than is any other member of the Shī'a, what emerges from the above detailed opinion stated in regard to the generality of the obligation of "enjoining the good and forbidding the evil" and the *wilāya* being a precondition in its fulfilment, corroborates my observation in chapter 3 that "enjoining the good and forbidding the evil" provides moral-religious justification for the existence of government in Islam. Moreover, *al-faqīh* in this context as well as in the general context of Imāmī Shīʿīsm should be understood as the most qualified person in the community to assume the *wilāya* in the best interests of the community. Accordingly, "general permission" given to *al-faqīh* supports the sort of *wilāya* that accrues to a "philosopher-king" by virtue of his "sound belief," "sound knowledge," and "sound character." Hence, the statement in the above opinion that if such a person, being appointed by the Imam on the basis of the "general permission," becomes a *sultān* or *hākim,* "there will be no unjust rulers, as was the case with the Children of Israel," who were ruled by their well-qualified scholars pending the return of the Messiah.

It is not difficult to discern *moral* grounds for any qualified member of the Shīʿa to undertake to establish a rule of justice, based on the purely moral requirement of the Qurʾān—namely, "enjoining good and forbidding evil"—pending the return of the Mahdī at the end of time. If the duty of "enjoining good and forbidding evil" is based on reasoning initially and confirmed in revelation subsequently, as argued by Tūsī and others, and if the duty is dependent on the assuming of the *wilāya,* then it is possible to argue rationally the right of *any* qualified individual to assume the responsibilities of *al-wilāyat al-ʿāmma* in the absence of the Imam, for the moral well-being of society.

It is significant that Imamite theologians, like Muslim philosophers, have maintained consistently that the necessity of the Imamate is established on the basis of the intelligence that rules that the well-being of society is dependent upon a well-qualified leader (*imām*). In the classical period, the inability of the Imam to stand up to his obligation of leading the community during the occultation was blamed on the people who failed to acknowledge and reinforce the Imam. But in later periods (when the Imamite temporal authority had come into existence without the Imam at the head of the government), emphasis was laid on the persons among the Shīʿa qualified to exercise authority in the name of the actual holder of that authority, the Imam, in the general interest of Islam and Muslims. That such a person ought to be the *faqīh* seems to have resulted from the criticism by the Imamite jurists of the failure of the actual wielders of political authority in the Imamite state to further the establishment of just public order.

In other words, it was the question of adequate justice, and who could further the cause of justice, that brought about discussion of *al-wilāyat al-ʿāmma.* Thus, in Imamite jurisprudence, the decisive factor of assuming the *wilāya* has centered on the ultimate purpose of having a government, rather than the form that *wilāya* ought to take. It is for this reason that Imamite jurisprudence speaks about *al-sultān al-ʿādil,* the Just Ruler, who governs in accordance with the divine laws, without treating the actual form of *saltana* (exercise of authority) that can attain the goal of creating an adequately just order.

It is plausible on the basis of our sources that those Imamite jurists who thought it permissible for the *faqīh* to assume *al-wilāyat al-ʿāmma* did not conceive the *faqīh* in any other form than the caliph, he being of course *khalīfat al-imām,* who ruled in accordance with "sound knowledge," being confirmed in "sound belief" and consolidated by "sound character," as well attested among the populace. On the other hand, those who opposed *al-wilāyat al-'āmma* of the *faqīh* could not conceive of the *faqīh* in any other form except that of the Imam, who ruled with absolute power invested by divine designation (*nass*) and protected from being or becoming corrupt by infallibility (*'isma.*)

Among these latter scholars, we ought to consider in some detail the discussion on *al-wilāyat al-'āmma* by al-Shaykh Murtadā al-Ansārī (d. 1281/1864). His personal eminence and highly technical discussion of the topic have greatly influenced those present-day jurists who have opposed granting the right of assuming *al-wilāyat al-'āmma* to the *faqīh* during the occultation of the twelfth Imam.

Al-wilāyat al-'āmma and al-Shaykh al-Ansārī's Interpretation

Al-Shaykh al-Ansārī is considered the "seal of the *mujtahids*" and is perhaps the most eminent Imamite jurist in the history of Imamite jurisprudence in the "modern" period. He was the younger contemporary of Najafī who, as the leader of the Imamites, was instrumental in achieving Ansārī's recognition as one of the most profound *mutjahids*. It is impossible to assess his contribution to Imamite jurisprudence in this study. However, it is sufficient to point out that his works have remained unsurpassed. They are, to mention two of them, *Farā'id al-usūl,* or as it is generally known, *al-Rasā'il,* on *usūl al-fiqh;* and one in applied *fiqh, Kitāb al-matājir* or *al-Makāsib,* where his ability to utilize the principles of law to deduce sound opinions is demonstrated with remarkable intellectual percipience.

In the latter work Ansārī has demonstrated the most original aspect of his method of deducing judicial decisions by determining the ordinary usage of the terms employed by the people in the context of social relationships. He makes continual reference to *al-'urf* in the sense of "ordinary language," as it is understood in analytical philosophy, in the *mu'āmalāt* (social transactions) section of applied jurisprudence, showing the importance of analyzing terms and understanding their conventional, customary significations before any juridical prescription can be given. Accordingly, Ansārī introduces *al-'urf* as a minor term of a syllogism in order to clarify and dissolve a problem.

Ansari demonstrates his application of *al-'urf* in his analysis of the phrase *ulū al-'amr* (those invested with authority). The phrase is discussed in the context of the discretionary control that can be exercised by various persons, including "those invested with authority," in managing the affairs of the Muslim community. He explains the use of the term in revelation and then embarks upon the sense in which people ordinarily understaand the phrase. Following this discussion he says that the ostensible sense derived from the term in its conventional usage (*'urfan*) is that "it denotes a person to whom it is obligatory to have recourse in all those matters for which there is no single person responsible."[72] He then concludes that *al-'urf* expects "those invested with authority" to assume political responsibility as part of the general matters for which the position of *ulū al-'amr* is responsible. Accordingly, anyone who is invested with authority to join the *ulū al-'amr* can be said to possess discretionary control over the property and lives of the people as established by *al-'urf*.

It is interesting that Ansari's discussion on *wilāya* is placed in the middle of his *Makāsib,* in the *Kitāb al-bay',* where he is engaged in treating the categories of persons who can exercise discretionary control (*tasarruf*) over the goods that belong to minors and other such persons who are legally incompetent to exercise that authority on their own. One of those who can exercise *wilāyat al-tasarruf,* acting according to his own discretion in the disposal of goods belonging to others, is *al-ḥākim*—a jurist who has all the necessary qualifications to issue a legal opinion (*al-faqīh al-jāmiʿ li-sharāʾit al-fatwā*). From the reference in the text it appears that Ansārī was asked to explain *wilāyat al-tasarruf* of jurists, and he undertook to explain it in detailed fashion, appending to it a general discussion on *wilāya.* But before he does so, he sets forth the problem that has caused jurists to give different opinions on the authority of the jurist (*wilāyat al-faqīh*).

A well-qualified jurist, according to Ansārī, has three functions:

1. *al-iftāʾ*

This is the function of issuing legal decisions in those matters in which the average man (*al-ʿāmmī*) needs guidance in the performance of the duties (*takālīf*) imposed on him by the Sharīʿa. The purpose of this function is to extract subsidiary legal ordinances through the study of secondary juristic decisions and derivative issues. There is no objection to this, nor is there any disagreement among Imamite scholars in establishing this position for the *faqīh,* except that there are those who do not consider *taqlīd* (following the rulings of a *mujtahid*) permissible for the ordinary person.

Although Ansārī refers his readers to details on this later issue in his treatment of *ijtihād* and *taqlīd* in other places in his work, it is important to point out that the treatment of the legality of *taqlīd* in Imamite jurisprudence has been taken up subsequent to the permissibility of *taqlīd* in the matter of fundamentals of the faith (*uṣūl al-dīn*).

From the theological works of both Sunnī and Shīʿī scholars during the classical period, it is evident that among the Muslims two groups—namely, the Hashwiyya and the Taʿlīmiyya (the Bātinī Ismāʿīlites)—were criticized for upholding *taqlīd* in the sense of "uncritical faith" or "blind adherence" in the matter of their belief, thereby prohibiting discursive inquiry and discussion in order to arrive at knowledge of the truth. The Hashwiyya belong to the *ashāb al-hadīth* (people relying on *hadīth*), who accepted anthropomorphic traditions without criticism and interpreted them literally. The Taʿlimiyya unquestioningly recognized the Imam as the final authority in all matters of dogma, in opposition to the free inquiry encouraged by the Imams in the matter of faith. Thus, Ghazālī in his work *al-Mustasfā* refers to these two groups as those who maintained that the only way to the knowledge of the truth was *taqlīd;* whereas, he says, the Muʿtazilites considered discursive inquiry into every item of evidence as necessary to arrive at the truth.[73]

The Imamites were also concerned with the question of *taqlīd* in the conventional meaning of the term, which connotes unquestioning acceptance of an opinion in the matter of *uṣūl al-dīn* as indicated by a section in Kulaynī's *Kāfī* dealing with the principles of faith. In his "Book on the excellence of knowledge," Kulaynī proposes to show that the Imams encouraged discursive inquiry into the matter of faith, and of the practice based on it, by requiring the Shīʿa to ask questions before accepting an opinion.[74] In addition, in another section of the same "book," Kulaynī cites a tradition on the authority of al-Sādiq. The Imam was asked about the passage of the

Qurʾān (9:31): "They [i.e., Jews and Christians] have taken as lords besides God their rabbis and their monks."[75] The Imams observed that Jews and Christians "*followed* them (*ittabaʿūhum*) when they declared the unlawful as lawful and the lawful as unlawful.*" In other words, al-Sādiq's comment implied criticism of these two communities for having followed their rabbis and monks unquestioningly even in the most erroneous of their distortions of the two monotheistic religions.

In the theological works of Mufīd and his disciples one finds even more explicit condemnation of *taqlīd* in the *uṣūl al-dīn*. In *Awāʾil al-maqālāt* and *al-Fuṣūl al-mukhtāra*, in his discussion on the necessity of divine guidance through the prophethood, Mufīd argues that although human reasoning (*al-ʿaql*) needs particular guidance through revelation (*al-samʿ*) for securing necessary confidence in the faith and for arriving at rational understanding about the goodness of religious-moral obligation (*al-taklīf*), the believer's faith must rely on rational proof.[76] Mufīd was fully aware that not all members of the Shīʿa were capable of attaining their understanding of the faith through rational discourse, and for this reason he was not willing to condemn them all to infidelity (*kufr*), although he criticized them for being *muqallidūn*.[77] Whether theologians like Mufīd or traditionists like Kulaynī, all Imamite scholars were opposed to *taqlīd* in the *uṣūl al-dīn*, as indicated by numerous traditions in the "Book of intelligence and ignorance" of *Kāfī*, and by frequent references to the interpretation of Qurʾānic passages like the following:

> And when it is said unto them: "Follow that which God hath revealed." They say: "We follow (*nattabiʿū*) that wherein we found our fathers." What! Even though their fathers were wholly unintelligent and had no guidance? [2:17].

This passage censures "following" any authority (even "our fathers") in *uṣūl al-dīn*. However, there are passages in the Qurʾān that direct believers to a form of *taqlīd* of those who possess "sound knowledge in religion," such as the following passage, which requires some believers to undertake to equip themselves with religious knowledge in order to warn community members to "take heed to themselves":

> Of every group of them, a party only should go forth [for *jihād*] so that [among those who are left behind there should be some]) who should [undertake to] gain knowledge (*yatafaqqahū*) in religion, in order for them to warn their folk when they consult them, so that they take heed to themselves [9:122].

Two important considerations have been inferred by Imamite jurists from the above passage to support the incumbency of *taqlīd* in matters pertaining to religious practice (*furūʿ al-dīn*). First, there is the necessity of referring to person with sound knowledge, because this passage clearly makes such a person responsible for putting the believers on their guard by warning them about their obligations, whether in the realm of religion or morality. Secondly, whether they commit themselves to follow his legal decisions or not, it is incumbent upon this knowledgeable person to issue legal decisions and to warn believers about fulfilling their religious and moral obligations—a warning coupled with the threat of due punishment for anyone who neglects obligatory acts and performs prohibited ones.

There is a misleading connotation in the lexical meaning of the term *taqlīd* ("girding someone's neck"), which has been alluded to in various opinions that onsider *taqlīd* "blind adherence" and as such impermissible in religion. The techni-al sense of the term, as elucidated in juridical texts, denotes a "commitment" *iltizām*) to accept and act in accordance with the rulings of the Sharīʿa as deduced by he independent reasoning (*ijtihād*) of a well-qualified, righteous jurist (*mujtahid*). Moreover, in its conventional signification, corroborated by rational proof, *taqlīd* means "confidence" or "trust" (*istinād*) in the rulings of another, someone authorita-ive, so as to base one's practice on them.[78] In this conventional usage (*al-ʿurf*), the person who is adopting the rulings of a jurist with trust in their correctness, and without investigating the reasons that led the jurist to make his decisions, is putting the responsibility for his actions on the *muqallad* (the jurist thus "made responsi-ble"). This is one of the specialized senses of the meaning of *qilāda* ("hanging a certain object around the neck of animals to be slain in sacrifice"), which figuratively denotes the meaning of *taqlīd* as "fastening the responsibility for one's religious acts on the neck of a *mujtahid*."[79]

It is significant that although there are numerous traditions requiring the Shīʿa to refer to learned authorities in order to obtain a ruling over any question in the Sharīʿa, the word *taqlīd* appears in only one tradition, which has not been fully authenticated—it is *mursal* (i.e., has a defective chain of transmission). Neverthe-less, rational proof has been cited as the most important piece of evidence, juridi-cally, to support the legality of *taqlīd* in a matter pertaining to a believer's religious-moral life. In fact, the Imamite jurists during the Safavid and post-Safavid era have argued emphatically for believers to employ either *ijtihād*, *ihtiyāt* (precautionary measure—i.e., taking the "safe," prudent line), or *taqlīd*, in descending order of preference, in their observance of acts prescribed by the Sharīʿa. They have, more-over, regarded *ijtihād* and *taqlīd* as an optionally incumbent duty that has been imposed "representatively" (*bi al-kifāya*) in accordance with the Qurʾānic injunc-tion in 9:122. This means that if *ijtihād* is undertaken by a sufficient number of believers, then others are relieved of undertaking this responsibility; instead, they can follow the course of *taqlīd* individually and personally to fulfil their obligations. On the basis of both rational and revelational proofs, these jurists have contended, it is important for believers to feel confident in their religious observances, and that confidence can be attained through *ijtihād*, *ihtiyāt*, or *taqlīd*.[80]

The question of *marjaʿ al taqlīd* (the highest juridical authority to be referred to in case of religious difficulty) is the corollary of the rational necessity to consult those who are specialists in matters of the Sharīʿa. Those who have maintained the rational necessity of *taqlīd*, have also argued rationally for the necessity of following the most learned (*al-aʿlam*) among the scholars, who is the point of reference—*al-marjaʿ*—for all the Shīʿa. However, the question of *aʿlamiyya* ("learnedness") in *taqlīd* appears for the first time in the writings of the Safavid jurists. In fact, Muhaqqiq al-Karakī has claimed an *ijmāʿ* of the Imamites on this requisite for *taqlīd*. This *ijmāʿ*, however, has been challenged by others because it is a *manqūlī* ("reported") *ijmāʿ*, which, according to some Usūlī jurists, cannot be used as a proof to make a legal decision.[81]

Ansārī, in his discussion on the need to unify Imamite applied *fiqh* and to centralize the leadership of the jurist, laid great emphasis on the question of *aʿlamiyya* (being the most learned) in applying the principles of *fiqh* for a *mujtahid* to deduce legal decisions inferentially. However, the Akhbārī *ʿulamāʾ*, who rejected the Usūlī methodology in jurisprudence, also rejected the necessity of *taqlīd* in a *mujtahid*, because they believe such a requirement has been deduced from juristic supposition (*al-zann*), and hence cannot become binding. Moreover, they believe that if one does not know a ruling in the case of a particular problem, then one can follow the precautionary measure (*al-ihtiyāt*) and avoid acting on that problem until doubt is removed.[82]

It is plausible that the Akhbārī jurists sensed a further growth in the prestige and authority of the Usūlī jurist as the *marjaʿ al-taqlīd*, whose leadership was being institutionalized and centralized by requiring all the Shīʿa, as an incumbent duty, to accept his authority and pay allegiance to him by declaring *taqlīd* to him. Thus, in the exercise of *wilāyat al-iftāʾ* for jurists, on which there was consensus among the Imamites from the early days on, one can notice development in the office of *marjaʿ al-taqlīd* and the sense of loyalty being created between *marjaʿ* and *muqallid* (the one who accepts the *marjaʿ* through *taqlīd*). This sense of loyalty to the *marjaʿ*, which was justified and formalized through a juridical prescription requiring the ordinary person (*al-ʿāmmī*) to declare his intention to follow the most learned *mujtahid* through *taqlīd*, was the culmination of a lengthy historical process initiated by the scholarly elite to emancipate believers and organize them as an autonomous religious community under Islamic law. The religious autonomy of the community, through the *mujtahid-muqallid* relationship, consolidated the position of the jurist within the community, whose members depended upon the *mujtahid* not only for religious prescriptions but for total guidance in realizing the ideal Islamic society.

The *mujtahid* who occupied the position of *marjaʿiyya* (authoritative reference) attained a unique status: he was the conscience of the community. And regardless of the fact that he lacked a specific investiture from the Imam to become his "special" deputy (*al-nāʾib al-khāss*), for the average member of the Shīʿa the *marjaʿ al-taqlīd* was the "special" deputy of the Hidden Imam. Accordingly, the institution of the *marjaʿ al-taqlīd* was the logical outcome of the juridical requirement of *taqlīd* through which the *mujtahid* legally established the fait accompli of his discretionary control over the affairs of the Shīʿa.

2. al-hukūma

This is the authority to judge matters in dispute and other related matters in the administration of justice. This position is, according to Ansārī, well established for the *faqīh*, without any disagreement in the judicial decisions of jurists or any dispute in the interpretation of the explicit textual proof (*nass*).

I have already dealt with the function of *al-hukūma* in chapter 4, where I examined the opinions as well as the texts that have been used as documentation. Hence I now move on to the third function of the Imamite jurist.

3. wilāyat al-tasarruf

This is the authority to freely dispose of goods and lives. This *wilāya* receives detailed treatment in Ansārī's *Makāsib*. It has to do with those individuals who have the authority to act freely in the disposal of the property of others. Ansārī begins his discussion with a general introduction to *wilāyat al-tasarruf*.

This *wilāya*, says *Ansārī*, can be conceived with two objectives. First, it can be conceived with the objective of giving total discretion to the *walī* ("guardian") in such a way that he possesses the right of disposal, and the exercise of this right by someone other than himself depends on his permission. The source of this form of discretion is his competence, which is the reason (*"sabab"*) for allowing him the right of disposal.

Secondly, this *wilāya* can be conceived with the objective of giving the *walī* the right of disposal that would normally be dependent on someone else's permission. This *wilāya* entails lack of discretion for someone other than the *walī* to undertake disposal, and the dependence of someone else's right of disposal on the *walī*'s permission, even when the *walī* cannot act independently in disposal. The source of this right is his competence, which is a condition (*"shart"*) for allowing the right of disposal to any person other than a *walī*.

Between the two objectives of *wilāya* there is a point of semblance—namely, that the *walī*'s permission is necessary for someone else to exercise the right of disposal, whether in the form of deputization (*istināba*)—for example, when *al-hākim* appoints his agent; or in the form of delegation (*tafwīd*) and designation (*tawliyya*)—for example, when an administrator is appointed for endowments (*awqāf*) by *al-hākim;* or in the form of consent (*ridā*)—for example, when *al-hākim* designates someone to lead the funeral prayer for a dead person who does not have a *walī*.[83]

The First Kind of *wilāya*: Total Discretion of the *walī*

Having described the two objectives of the *wilāya,* Ansari proceeds to determine whether the first kind of *wilāya* is applicable to the Imamite jurist. On the basis of one of the Practical Principles (*al-usūl al-'amaliyya*) applied to deduce a law in any matter in which one has doubt and there does not exist any other form of evidence, Ansārī rules that such a right to disopose of anything cannot exist for anyone. The Practical Principle applied by Ansārī to deduce his opinion is known as *istishāb 'adam* (absence of *istishāb* or "link"). (*Istishāb* is one of the important principles of juridical practice; it seeks a link that would establish the validity of a given injunction.) In the case of *wilāya* of the first kind, one can ask the question: "Until now such a right did not exist; now we are in doubt, and we ask, has it come into being?" The principle of *istishāb 'adam* is applied and it is concluded: "Even now such a right does not exist."[84]

However, excepted from this conclusion are, according to Ansārī, the Prophet and the Imams, as proven on the basis of four sources of the Sharī'a. First, says Ansārī, the Qur'ān states that the Prophet has more right (*awlā*) over believers than they have over themselves (33:6). Secondly, the Prophetic tradition reported by Ayyūb b. 'Atiyya declares that the Prophet is more entitled to a believer's self than is the believer. In the Ghadīr tradition the Prophet asked: "Am I not more entitled to you than you to yourselves?" The people replied: "Yes." Then the Prophet declared: "Whoever I am the master of 'Alī is the master (*mawlā*) of also." There are also traditions that dictate obedience to the Imams, and declare disobedience to them equal to disobedience to God. In addition, there are traditions—such as the *maqbūla* (accepted tradition) of 'Umar b. Hanzala, the *mashhūra* (well-known tradition) of Abū Khadīja, and the rescript received from the twelfth Imam—in which *al-hukūma* ("judgment") of the Imamite jurist and his authority over the people have been justified. Thirdly, there is also *ijmā'* on this *wilāya* as belonging to the Prophet and the Imams. Finally, *al-'aql* ("reason") also establishes that this form of *wilāya* be invested in them. As for autonomous rational proof, reason requires that one show gratitude to one's benefactor. Once one knows that the Prophet and the Imams are benefactors, it is evident that they should be obeyed. As for nonautonomous rational proof, it is contained in the ruling on paternity (*ubuwwa*) and the necessity of obedience to one's father. In general, the Imamate necessitates obedience to the Imam by subjects; and, as a matter of fact, it is even more appropriate to obey the Imam because the position of the Imam is clearly greater than the position of one's father.

All the above proofs from the four sources of the Sharī'a refute the conjecture that there is no obligation to demonstrate obedience to him in matters pertaining to social conventions (*al-'urfiyya*) and the exercise of authority (*saltana*) over the property and lives of the people. In short, says Ansārī, what is derived from these four proofs of the Sharī'a, after careful examination and reflection, is that the Imam has absolute authority (*saltana mutlaqa*) over the people, by reason of his designation by God, and his discretionary control is effective among the people absolutely. This is the meaning of *wilāya* in the first sense.[85]

The Second Kind of *wilāya:*
Lack of Discretion over Others without Permission

This type of *wilāya* is dependent on the Imams' permission. That is, the right of disposal vis-à-vis others depends on the Imams' permission. Although contrary to the Practical Principle stating that such a right does not exist for anyone, this *wilāya* has support in the reported traditions, which make it specifically obligatory to refer to the Imams and instance the lack of discretion vested in anyone other than they. The reason is its relation to the desired general welfare in the eyes of the divine lawgiver, which has not been assigned to any particular individual in Muslim society, but has been made representatively obligatory on Muslim society, as in impementing statutory penalties (*al-hudūd*) or different forms of discretionary penalties (*al-ta'zirāt*), exercising discretionary control over the property of those who are legally incompetent (*al-qāsirin*) to do so, requiring the people to pay taxes, and so on.

In this regard it is sufficient to take note of that which corroborates the authority of those who have the permission of the Imams—namely, that they form part of the *ulū al-'amr* (those who possess authority) and that they are administrators (*wulāt*) of the affairs of Muslims. The title *ulū al-'amr*, in its conventional usage, is applied to persons to whom it is obligatory to refer in all those matters for which there is no specific person responsible in Islamic law. The rescript of the twelfth Imam requiring the Shī'a to refer to them in future contingencies (*al-hawādith al-wāqi'a*) corroborates the fact that they are the *hujja* ("irrefutable manifestation"— i.e., "competent authority") representing the Imam, whereas the Imam is the *hujja* of God. Furthermore, the rescript confirms that it is the Imam himself who is the ultimate point of reference. Ansārī then cites *'Ilal al-sharā'i'* of Ibn Bābūya, where there is a report on the authority of Fadl. b. Shādhān, who heard the Imam 'Alī al-

Ridā explaining the reason why the Imam is needed. After mentioning several other reasons for the need of the Imam, ʿAlī al-Ridā says:

> We do not find any group (*firaq*) of persons or any nation (*milla*) among the nations who survived and lived on earth without a custodian (*qayyim*) and a leader (*raʾīs*), [sent to them] because of their need in the matter of religion and this world. It is unthinkable in the wisdom of the Wise One (*al-hakīm*) to leave the people without providing that which they need and without which their lives would not be upright.[86]

I have discussed this *hadīth* earlier in the context of the rational proof of the necessity for having a leader. Ansārī, although he regards this as textual proof to support his explanation of the *ulū al-ʾamr*, suggests its rational appeal because of the generality of the terms *qayyim* and *raʾīs* as much as those like *firaq* and *milla*, also used in the *hadīth*. He then points out that this tradition is in addition to all those traditions which speak about the *wilāya* of appointed persons in the administration of legal penalties, different forms of discretionary penalties, and just rule generally (*al-hukūmāt*). These are the functions of the Imam of the Muslims. As for the funeral prayer, the *sultān* appointed by God has more right to its performance as the *walī* than anyone else.

There are many more traditions, recalls Ansārī, that can be discovered by an investigator to ascertain the need of appointed persons in the administration of the affairs of the community. There is no objection to maintaining the position that there is no permission to administer freely a number of things connected with public welfare in the absence of the permission and approval of the Imams. However, there is nothing to necessitate abstention from exercising free discretion in administering all the affairs of Muslims until the permission of the Imam becomes available. Indeed, says Ansārī, the matters that the masses refer to their leaders are provided for, because there are the *ulū al-ʾamr,* in whom authority has been invested to administer Muslim affairs. These *ulū al-ʾamr* are the ultimate reference point for future contingencies.

When it is doubtful whether certain matters should be referred to the *ulū al-ʾamr,* such matters should be studied in the light of the general principles established through the essential sources of the Sharīʿa and their application, to determine whether discretionary control is allowed or forbidden in such cases. If, from essential sources like the Qurʾān, the Sunna, and so on, one cannot find documentation to remove doubt in these cases, then the Practical Principles should be applied. However, in cases where the permission of the Imam or his special deputy is obtainable, then it is not permissible to apply the Practical Principles, because these are redundant if one can refer to a well-established proof. Reference to the Practical Principles applies to those cases where one cannot find such proofs.

It is obvious that Ansārī regards the "special" deputies of the twelfth Imam during the Short Occultation (873–941) as those invested with authority (*ulū al-ʾamr*) to carry on the function of guiding the community in all those matters in which they had the Imam's permission. However, he regards the jurists as the *ulū al-ʾamr* during the Complete Occultation (from 941 onward), who should be consulted by the Shīʿa on the ground of their being invested as the "general" deputies of the Imam. Ansārī refers to the twelfth Imam's rescript as an important piece of documentary evidence to infer the "permission" of the Imam applicable to the jurists to administer the affairs of the community as *ulū al-'amr* in certain specific areas. But he does not regard this investiture as authorizing the assumption of the all-comprehensive *wilāya* by the jurists, as will become evident in the following discussion. Nevertheless, by establishing the limited authority of the jurist he reinforced the centrality of the juridical authority that had been envisioned in the position of the *marja'iyya* (authoritative reference).

After this introduction on the two meanings of *wilāya*, Ansārī, without offering his own critique, proceeds to embark upon the question of *wilāyat al-faqīh* and to discuss whether the *faqīh* has *wilāya* in any of the two meanings discussed above.[87]

The *wilāya* of the Imamite Jurist: Is it of the First or the Second Kind?

Ansārī's elucidation of the *wilāya* in the introduction was meant to prepare the investigator for his judicial decision in the section that follows: "Thus, we say. . . . " According to him, *wilāya* in the first meaning, that of "total discretion of the *walī*" to act freely in the affairs of the Muslims, has not been established. However, says Ansārī, some jurists, relying on traditions that narrate the position of the *'ulamā'*, have conjectured this to be so for the *fuqahā'*. Among these traditions are:

① "The *'ulamā'* are the heirs of the prophets."[88] This tradition, according to those who have interpreted it in the first signification of the *wilāya*, indicates that the *'ulamā'* (i.e., the *fuqahā'*) are the heirs of the prophets, not only in their knowledge but also in their authority as the *walī* of their community. On the other hand, others have interpreted the tradition as informing the people of the excellence of religious knowledge and encouraging them to acquire it. As such, the tradition has nothing to do with the *wilāya* of the jurists. Moreover, even when the *hadīth* uses the term *waratha* ("heirs"), it has the restricted sense of being heirs to their knowledge, not their authority, which would require divine designation, for the prophets' authority is derived from God. In other words, the *'ulamā'* are heirs of the prophets as having acquired religious knowledge, not because of having inherited the right to exercise their authority as the divinely appointed *walī*.

② "The prophets do not leave inheritance in the form of wealth (lit. *dinār* and *dirham*); rather, they leave behind their teachings (*ahādīth;* a variant reads *al-'ilm,* "knowledge"). Anyone who acquires a part of it has gained a great amount of it." This *hadīth* is actually the continuation of the first *hadīth* and Kulaynī mentions it as a tradition related on the authority of the Prophet through al-Sādiq. The implications of the tradition, accordingly, are similar to those discussed above.

③ "The *'ulamā'* are the trustees (*umanā'*) of the prophets." This tradition is reported on the authority of the Prophet, who in another similar tradition is reported to have said that the *'ulamā'* are the trustees of the prophets as long as they do not enter the world. On being asked to elaborate on what he meant, the Prophet replied: "In following the *sultān.*" And he added, "When they do so, be on guard against them for your religion."[89] In still another tradition the Prophet declared the *'ulamā'* to be the trusted ones of God in respect to God's law, and their position to be similar to that of the prophets among the Children of Israel.

The purpose of all these and similar traditions, according to those who uphold the *wilāya* of the jurist in its first signification, is that when the Prophet went on to declare them to be even superior to the prophets of the Israelites and the trustees of the prophets, they are responsible for bringing about the establishment of just public order on earth, because that was the purpose of God in appointing prophets in history. However, other scholars differentiate between the position of a prophet and that of a jurist, however exalted the latter's position might be, because whereas it is possible for the latter to "enter the world," and become sinfully deviant, such a thing is impossible in a prophet who must be obeyed in all that he teaches.

In the above three traditions that were cited to corroborate the claim to the *wilāya* similar to that of the Prophets and the Imams, it is possible to discern the extension of their "apostolic" authority to those who derived legitimacy by being the knowledgeable persons who could effect God's purposes for humanity. The *'ulamā'* became the "heirs" and the "trustees" of the prophets by continuing to make the will of God known to the community through their interpretation, and the community acknowledged their tangible authority derived from their "sound knowledge" and obeyed their religious prescriptions. In a distinct way the *'ulamā'* formed the link with the source of the theologico-political authority of the Prophet from generation to generation, providing the crucial sense of continuity in the Prophetic paradigm that was recognized as the only valid basis for juridical prescription.

④ The Prophet said three times; "O God, have mercy on my successors (*khulafā'*)." He was asked who these were and he replied: "Those who will come after me and relate my traditions (*ahādīth*) and my customs (*sunna*)."[90] This *hadīth* uses the key term for the assumption of authority in Muslim public order—namely, *khalīfa;* and it elaborates on the qualification of the *khalīfa* as the *rāwī* ("reporter") of the prophetic traditions and the custom, thereby establishing an inescapable connection between the *khilāfa* ("caliphate") and the *faqāha* (knowledge of jurisprudence). This connection has been regarded as evidence in support of the similarity between jurist's *wilāya* and the Imam's discretionary power. If the *faqīh* can assume the *juridical* authority of the Prophet by disseminating his teachings as his successor, argue the supporters of discretionary power for the jurist, then, in the same capacity, he can also assume the Prophet's *constitutional* authority. However, like other traditions seen above, this too has been interpreted by other scholars as being an encouragement from the Prophet to spread his teachings among the people so that the people may become worthy of his prayer of mercy for them. Moreover, it also makes an important point for those who want to become the *khalīfa* of the Prophet—namely, that the true caliphate is in teaching and spreading Islam, not in wielding power. The use of the plural noun, *khulafā'* ("successors"), in the text, say these scholars, indicates that there could be many individuals to ensure the stability and smooth running of the government.

It is important to note the connection between the *khalīfa* of the Prophet and the phrase *khalīfat al-imām,* which was used for the Imamite jurists, as discussed in chapter 3. According to the classical usage of this phrase, the jurist as the *khalīfat al-imām* could assume authority similar to that of the caliph among the Sunnites.

⑤ The Imam al-Sādiq, according to the *maqbūla* ("accepted tradition") of 'Umar b. Hanzala said: "I have appointed him a *hākim* over you."[91] I have discussed in detail in chapter 4 the long-credited *hadīth* of Ibn Hanzala. This is one of the three decisive *hadīth* reports that have been used as a proof for *al-wilāyat al-*

'*āmma* of the jurists. More importantly, it is in this tradition that one can discern the right that a Shī'ī can exercise in determining a jurist who can be regarded as a *hākim*. The question of the *a'lamiyya* (most learned) jurist can be inferred from the long discourse that the close associate had with the Imam. Undoubtedly, as pointed out in the context of *wilāyat al-hukm*, the purpose of the *hadīth* was to provide the Shī'ites with sufficient ways of determining their juridical authority in the absence of the Imam's constitutional authority. However, whether the *hadīth* supports any wider implications of the *hākim*'s authority is a matter of independent reasoning (*ijtihād*), which can reinforce the limited or comprehensive *wilāya*, depending on how the position of *al-hākim* is interpreted. There is consensus among the Imamites that *al-hākim*, in accordance with the Qur'ānic usage, is a juridical position, and unlike its premodern and modern usage, which carries the meaning of "ruler" or "sovereign," its classical usage was limited to *al-qādī*—that is, "administrator of justice" or "judge." It is in its implication for *al-wilāyat al-'āmma* during the absence of the political authority of the Imam that scholars have differed and *ijtihād* has been applied.

It is probably correct to say that *al-wilāyat al-'āmma* of the jurists has depended on rational exegesis rather than on revelational proof, as corroborated by the inferential deduction of *al-hākim*'s position from this report.

6. According to the *mashhūra* (generally well-known) tradition of Abū Khadīja, the Imam al-Sādiq said: "I appoint him as a judge (*qādī*) over you." This is the second of the three most important *hadīth* reports used as essential proof in support of *al-wilāyat al-'āmma* of the jurist. The implications of this tradition are not different from that of the one related on the authority of 'Umar b. Hanzala, except that in the latter tradition the term *al-hākim* has admitted a much broader political signification of "guardianship" of the Imamite community in the absence of the twelfth Imam. Furthermore, there is a technical difference in the two traditions. Both are categorized as *khabar al-wāhid*, meaning a "single" tradition that has been transmitted by just one source, but 'Umar b. Hanzala's tradition has been credited in practice in virtue of its contents, whereas Abū Khadīja's tradition has become merely well-known among the people, even though the chain of its transmission cannot be traced back to more than one individual or to none at all. Accordingly, it is Ibn Hanzala's tradition that has been cited as documentation in the exegetical deduction of the jurists, as we have seen in this study on various occasions.

The significant word that appears in both traditions, and which is regarded as decisive in establishing the *wilāya* of the jurist, is the verb *ja'altu*, meaning "I have appointed [the jurist]." As such, *ja'l* has been taken in the sense of *nasb*, "investiture (with authority)." Nevertheless, the investiture, although general in nature, necessarily becomes effective in a jurist whose qualifications to assume the position of *al-hākim* or *al-qādī*, as detailed in the previous chapter, are well established.

7. In the rescript of the twelfth Imam, he responds to the question regarding future contingencies: Who should be referred to when problems arise in religion? The Imam directs his Shī'a to "refer in those matters to the transmitters (*ruwāt*) of our *hadīth*, because they are my proof to you and I am the proof of God."[92] This is the third piece of documentation enlisted to support *al-wilāyat al-'āmma* of the jurist. The key element in the rescript is the phrase *al-hawādith al-wāqi'a* ("the future contingencies"). As we have seen before in the context of *zakāt*, the phrase has been interpreted to include a wide range of socio-politico-religious problems

that may be encountered by the Shīʿa, and whose solutions should be sought from the learned jurist. Consequently, if a well-qualified jurist is the proof of the Imam— that is, representing him as a competent authority in one area during the occultation—then he is also his proof in other areas affecting the lives of his followers. This is a rational arugment based on the function of the Imam in society and the continuing needs of society.

Clearly, *ijtihād* has been employed to deductively infer *al-wilayāt al-ʿāmma* of the well-qualified jurist as the proof representing the Hidden Imam in *all* matters affecting the socio-politico-religious welfare of the Shīʿa. More importantly, without methodological advances in the theoretical basis of jurisprudence—more particularly, in the area of the Practical Principles (*al-uṣūl al-ʿamaliyya*) to derive juridical prescriptions to guide the socio-political affairs of the community—the jurists during the Qājār period would not have been able to increase their practical control of the social and financial structure of the community. The Practical Principles, in addition to the linguistic and semantic discussions in the *uṣūl*, furnished the essential premise to infer deductively relevant signification from the phrase "future contingencies" in the rescript, to include anything that affected the welfare of the Shīʿa when the Imam was in concealment. The entire development of the juridical authority that led to the institution of the *marjaʿ al-taqlīd* in Ansārī's discussion of *ijtihād* and *taqlīd* can be attributed to the methodological clarification of the unresolved ambiguities in the documentation that was designated as "investiture traditions."

I have expanded sufficiently on what Ansārī had in mind in his brief statement about some jurists who, in his opinion, have attributed the "total discretion of the *walī*" (*wilāya* in its first signification) to the jurists, basing themselves on the traditions mentioned above. But the truth, says Ansārī, is that after careful consideration of the context and the wording of the above documentation, it becomes necessary to assert authoritatively that these traditions elucidate the status of the ʿulamāʾ and their function vis-à-vis the rulings of the Sharīʿa; they do not, as with the Prophet and the Imams, confer on the ʿulamāʾ a greater entitlement than belongs to the people generally in the matter of property. Thus, rules Ansārī, if the *faqīh* asks for the *zakāt* and the *khums* from the Shīʿa, there is no proof to substantiate its legal obligation to give them to him. Nevertheless, if it is proven through a Sharʿī process that the legality of the *zakāt* and the *khums* depends on giving it for distribution to the qualified *faqīh,* under any circumstances or after being asked to do so, and if the *faqīh* issues a judicial decision to this effect, then it is obligatory to follow this decision. This ruling applies to the one who has begun to follow the rulings of this *faqīh* through *taqlīd* or has chosen to do so after having reflected at length on following this *faqīh.*

Thus, the functions of the Prophet (like collecting and distributing taxes) are denied to the jurist, including those functions reported in the traditions as being specifically assigned to the Prophet by virtue of his being a prophet, the messenger of divine revelation. If the Prophet's functions were generalized, they could be extended to all members of society, because the *faqīh* does not have legal and moral authority (*saltana*) over the property and the lives of the people except in a few cases; in most cases he lacks authority.

After this discussion, Ansārī gives his opinion that it is impossible to establish a proof to substantiate the obligation of obedience to the jurist similar to that for the

Imam. In other words, the jurist lacks full investiture to assume *al-wilāyat al-'āmma*.[93] It is evident that Ansārī recognizes the lack of *saltana* in the jurist's investiture, which would have given him the right to demand obedience in all matters related to the Islamic public order.

As for the second kind of *wilāya* (lack of discretion for others without permission)—that is, the authority that depends on the permission of the Imam in all cases in which the Imam's permission is known to be necessary (keeping in mind that there are very many cases where this is not known precisely)—it is important to recount the factors that could determine precisely the extent of this *wilāya*. According to Ansārī, it includes all that is good (*ma'rūf*), of which God has apprised human beings, intending its actualization in human society. The good can be attained by making it an obligation of a specific person, such as the supervision undertaken by a father in regard to the property of his minor son, or by creating a special function in society, such as the giving of a legal decision (*al-iftā'*), the administration of justice, and all that can be accomplished through this kind of function, like "enjoining the good," and so on. There is no problem in understanding the *wilāya* in these matters, even when it is not known precisely whether one has the permission of the Imam or not. It is possible, however, that there might be a precondition attached to some of these functions—that they should be performed under the supervision of the *faqīh*. In that case, it is necessary to refer to him. Moreover, if the *faqīh* knows that there is authoritative evidence permitting him to assume responsibility in these matters—because of their not being dependent on the special control of the Imam or his specifically designated deputy—then he can undertake these matters directly or as a deputy, if it is necessary to assume them through deputization; otherwise he should not assume them.

Furthermore, when something is good, then it is inconceivable that it should be dependent on the Imam's supervision, either directly or through his deputy; and preclusion from it when the Imam is absent is analogous to preclusion from all other blessings denied to believers because of his being in occultation. The source of this preclusion is the doubt about the unconditional existence of authoritative evidence or the existence of a special person, such as the Imam, for the realization of total good.

There is a fundamental recognition of the rational need for the establishment of a just public order in Ansārī's estimation of the basic function of society. But that rationally established duty is not contingent upon the jurist's assumption of the *wilāya* in all spheres related to the creation of just public order. Moreover, there is a generalness in this rational obligation that is not dependent on the existence of the Imam or his deputy; rather, any qualified member of the Shī'a who possesses the *saltana* could undertake to bring about the good that God wants for human society unconditionally. Accordingly, Ansārī maintains that the second kind of *wilāya* should be assumed by the jurist only in those areas where it is explicitly required of him to do so. Such a function must be assumed by the jurist in virtue of his being qualified to do so, not because of any prerequisite specified in the "investiture traditions." The "investiture" by the Imams in all the three essential records of delegation referred to any member of the Shī'a community who was acknowledged to be qualified to undertake a specific function for the welfare of society.

The full implications of Ansārī's rational argument to deduce the general application of the "investiture traditions" for the actualization of total good for humanity would become adequately institutionalized in the constitutionalization of the *wilāya* in modern Iranian experience. The Iranian constitutional revolution of 1906 and the Islamic revolution of 1978–79 were specific instances demonstrating how Ansārī's open-ended judicial decisions regarding responsibility for the entire Shīʿī community to put into effect all the good that God intended for human welfare provided the necessary legal procedure to combine the *saltana* (legal-moral authority to exact obedience) with the *wilāya* (the faculty that gave the right to demand obedience) pending the return of the twelfth Imam.

The obligation to refer to the *faqīh*, says Ansārī, is inferred from all the above-mentioned documentation, which confirms the appointment of the *faqīh* as a *hākim*, as stated in the "credited" tradition of ʿUmar b. Hanzala. Like all the other *hukkām*, he was *ostensibly* appointed during the time of the Prophet and the companions, with the requirement that the people refer to them in all matters stated as demanding his supervision and control.

However, states Ansārī, what is conventionally understood is that when the *sultān* (the ruling authority) appoints a *hākim*, he is to be referred to in general matters as desired by the *sultān*. Furthermore, the tradition that says "the management of affairs is in the hands of the ʿulamāʾ who are the trustees of God in His matters, both lawful and prohibited," also corroborates that the ʿulamāʾ, as *hukkām*, must be referred to in matters of the general good of the people.

Finally, the rescript of the twelfth Imam, as reported in several early sources (like Ibn Bābūya's *Kamāl al-dīn*, Tūsī's *Kitāb al-ghayba*, and Tabarsī's *Ihtijāj*), which was written in reply to the question submitted by Ishāq b. Yaʿqūb through ʿUthmān al-ʿAmrī, deals with difficulties in the future. The reply of the Imam was that the Shīʿites should refer to the transmitters (*ruwāt*) of the traditions of the Imam for guidance in the case of future contingencies, "for they are my proof to you as I am the proof of God." The purport of the term *al-hawādith* ("contingencies") in the rescript, explains Ansārī, is general, and includes all matters that should be referred to the *ruwāt*. These include matters that are customarily (ʿurfan), rationally (ʿaqlan), and legally (sharʿan) referred to the leader (*al-raʾīs*), such as supervision of the property or goods belonging to legally incompetent persons (minors, absentees, dead, or insane).

However, to say that these *hawādith* refer only to *sharʿī* matters, says Ansārī, is farfetched on several grounds. First, the apparent sense of the deputization of the *faqīh* with regard to future occurrences is that the *faqīh* should guide the Shīʿa directly or as a deputy of the Imam, and not that he should be referred to regarding the rulings of the Imam in those matters. Secondly, there is the argument that the Imam provides in the rescript by declaring the *faqīh* to be the proof of the Imam and the Imam the proof of God. His being the proof of the Imam refers to those matters in which competent authority is the jurist's own sound opinion and insight. This was indeed the position of the *wulāt* of the Imam who represented him, and it is not so because God made it incumbent on the *faqīh* to assume this responsibility after the occultation of the Imam, otherwise it would have been appropriate to say that the *ruwāt* were the proofs of God, just as they have been described in a tradition as the "trusted ones of God" in His lawful and prohibitive injunctions."

In addition, the necessity to refer to the jurists in *sharʿī* matters is self-evident from the early days of Islamic history. This was undoubtedly known by Ishāq b. Yaʿqūb, who wrote the letter to the Imam seeking his guidance about future contingencies—quite apart from the necessity of asking for the opinion and insight of a jurist in matters affecting general welfare. Accordingly, it is possible that the Imam was entrusting Ishāq to a person or persons among his thoroughly reliable followers at that time. Thus, it is obvious that the word *al-hawādith* ("future occurrences") is not limited to seeking the jurist's ruling in any matter that has given rise to doubt or dispute.

One must also consider the explicit relationship between this kind of rescript and the manifest generalities (*al-'umūmāt al-zāhira*) that are part of the permission given by the lawgiver regarding whatever promotes the public interest. This permission is included in the Prophetic tradition, "All good is charity (*sadaqa*)." In another tradition the Prophet says: "Helping the weak is the best charity"; and there are other such traditions. This is a rational argument employed by Ansārī to demonstrate that the rescript applies to these generalities. The "future occurrences" in the rescript include all general matters affecting public welfare. These matters can be determined and ruled upon rationally as being necessary for the well-being of society, in accordance with the instruction in the rescript, by referring to the "transmitters" of the Imam's teachings. These individuals are the ones who in conventional usage are known as *ulū al-'amr*—that is, those in whom authority has been invested and who should be referred to during the occultation.

It is plausible that Ansārī's rational argument led to the logical necessity of upholding the all-comprehensive *wilāya* of the jurist in all matters conventionally referred to "the leader" (*al-ra'īs*). However, Ansārī is hesitant to draw such a conclusion for two reasons. First, because there was a Shī'ī *sultān al-waqt* who was the actual wielder of power in the Shī'ī public order and who appointed *al-hākim* (the administrator of juridical authority—i.e., the jurist) to perform a specific official function. Secondly, because there was not sufficiently authenticated documentation that could be extrapolated to deduce legally valid all-comprehensive authority for the jurist. This latter point is taken up in the following section; the former point is tacitly acknowledged above and again at the end of the section on *wilāya*.

However, there is a conflict between the rational necessity of fostering the general good and the nonlegality of this general good when an issue arises in the absence of the stated opinion of the one invested with authority (*walī al-'amr*). The entire matter of the general good implied in "future occurrences" is, according to Ansārī, not without problems, even though many jurists have given rulings in this area. The general good lacks the ostensible sense of the *al-hawādith*, because what is undubitably established on the basis of the evidence in the texts of the communications from the Imam is the *wilāyat al-faqīh* in matters that can be legally enacted in society—and this in such a way that if the *faqīh* is not available, it is incumbent upon the people to undertake it representatively (*kifāyatan*). Nevertheless, there is no doubt about the legality of the assumption of the *wilāya* by someone other than the Imam in assigning legal penalties, appointing replacements for a father or grandfather in giving a minor into marriage, entering into or canceling a business contract having to do with goods belonging to an absent person, and so on. The proofs that establish the legality of the *wilāya* for someone other than the Imam do not restrict it to the *faqīh;* rather, says Ansārī, it is necessary for the *faqīh* to establish this restriction from other proofs. On the other hand, the *wilāya* in this regard is well established for the Imam for the reasons stated earlier in the discussion of the first kind of *wilāya,* where it was demonstrated that the Imams have more right than others to assume total discretion over the affairs of the community. It is clear, rules Ansārī, that establishing the general deputyship of the Imamite jurist on behalf of the Imam in any way similar to the *wilāya* of the Imam is impossible.

Ansārī thus concludes that the rescript of the twelfth Imam, although in some sense portending the general deputyship of the Imamite jurist, lacks sufficient evidence to establish total discretion of the jurist over the lives and property of believers. The authoritativeness of the tradition commonly cited, apparently by the contemporaries of Ansārī, to support the general deputyship of the jurist—"The *sultān* is the *walī* of the one who does not have a *walī*"—according to Ansārī, needs further evidence. Its weak transmission and ambiguity have been admitted by prominent Imamite scholars like Aghā Jamāl al-Khwānsārī in his discussion on *khums,* and Muhaqqiq al-Karakī in his treatise dealing with the legality of a jurist's receiving rent on certain classes of land *(anfāl)* taken from enemies of the Shīʿite state (land that would belong to the Imam when he appeared) because of doubt surrounding the generality of deputyship. And the tradition has other weaknesses militating against its use as evidence for a ruling. Moreover, it is not general in application; rather, it means that the *sultān* is the *walī* for someone who needs a guardian, according to particular needs. For instance, a minor whose father has died, or an insane person who has reached maturity but cannot take charge of his own affairs, needs a guardian and that guardian is the *sultān.* However, admits Ansārī, what results from this tradition is not what is understood in the rescript—namely, permission to do whatever is in the interest of the people. This tradition establishes the legality of the *faqīh*'s deputyship where it is not established by the rescript. Thus, it is permitted for him to undertake that which is in the general interest of the Imamites.

In the final analysis, according to Ansārī, there is nothing wrong with doing something beneficial for the people. This is maintained in view of what Ansārī interprets this tradition to mean—that is, that the *faqīh* is the *walī* (guardian) of the one who needs a guardian, and he is the one who will undertake what is in his interest; but not that he has to be the *walī,* of necessity, in accordance with his discretionary authority over that person in the past. Indeed, concludes Ansārī, God has appointed a *walī* for the one who needs a *walī,* and it is necessary that such a *walī* be the *sultān* (i.e., the one who has power).[95] Without the power to demand obedience the *walī* cannot undertake to protect the interests of those who need a guardian. Obviously, Ansārī recognizes the interdependence of the *wilāya* and *saltana* in the "enjoining of good and forbidding of evil" in human society and realizes that the *saltana* has been invested in the Shīʿī *sultān,* not in the jurist.

Al-wilāyat al-ʿāmma and Islamic Government

Are the Shīʿa, in general, in need of the *walī* when the *walī* designated by divine authority is in occultation? Those who might have acted as the *walī* of the Shīʿa, protecting their rights and administering justice (namely, the *sultāns*) proved to be unrighteous and unjust long before the Shīʿa entered modern times. The unjust behavior of the *sultāns* merely came more sharply to a head in modern times.

With the establishment of the position of the *marjaʿ al-taqlīd* in the Qājār and post-Qājār era, and the consolidation of the socio-religious authority of the Imamite jurist (*al-mujtahid*) as the *nā'ib al-imām* ("deputy of the Imam") who, unlike the *sultān*, truly represented the Shī'a at the popular level and on many occasions proved to be a guardian (*walī*) against the encroachment of the *sultān*, the *wilāya* of the jurist went through a period of further affirmation in jurisprudence. Although the limitations on *al-wilāyat al-'āmma* of a jurist, as discussed by Ansārī, remained very much the same, the jurists in the late nineteenth and twentieth centuries saw the possibility of further inferential extrapolation through *ijtihād* in this matter, as Ansārī had alluded to in his *Makāsib*.

It is important to note that in all the works on jurisprudence during this period (post-Ansārī) the method of argumentation and the documentary evidence provided to support one or another view of *al-wilāyat al-'āmma* have remained almost the same. What draws the attention of an investigator is the wider use of rational proof, sometimes of an autonomous type derived from a purely intellectual foundation, with no reference to inferences extrapolated from the textual proof that was prominent in Ansārī's work.

At the turn of the century, there appeared a work entitled *Bulghat al-faqīh* (The competence of the jurist), a collection of treatises given in the form of lectures by al-Sayyid Muhammad Āl Bahr al-'Ulūm (d. 1326/1908). One of the treatises deals with different forms of *wilāya*. The author expounds in detailed fashion what was implied or alluded to in Ansārī's *Makāsib*. This treatise, in my opinion, remains unsurpassed in the details and systematic discussion it has on *wilāya* and the way in which a jurist can undertake it. The author does not distinguish between political and religious spheres of human activity and maintains that the *wilāya* of the jurist is of a comprehensive ('*āmma*) nature, which cannot be limited only to matters of God-human relationship (*al-'ibādāt*); rather, it includes person-to-person relationship (*al-mu'āmalāt*) as well.

The constitutional revolution of Iran in 1906 played an important role in the political awareness of the Imamite jurists. It provided them with a vision of what has become a common phrase among jurists in modern times, "the Islamic government" (*al-dawlat* or *al-hukūmat al-islāmiyya*). There is no direct discussion on Islamic government or its form in jurisprudence, but what is implied there is very obvious. Islamic government is necessarily a government in which the administrators are, to use Imam 'Alī b. Abī Tālib's statement, "those who deserve to exercise authority because they are most knowledgeable regarding the injunctions of God."[96] To put it differently, it is a government headed by *al-sultān al-'ādil*, the Just Ruler, who rules according to the norms provided by Islamic revelation. So even when modern systems of public order provided a quite different model of government, in Imamite jurisprudence it was *al-sultān al-'ādil*, legitimately invested with *wilāyat al-tasarruf*, who alone could establish the Islamic rule of justice.

There was no doubt in the minds of jurists that *al-sultān al-'ādil* as such was the Imam in occultation. He was divinely and explicitly designated to exercise this authority, whereas the authority of the jurists depended not only on the Imam's designation but also on the acknowledgment of the Shī'a. It was in this latter requirement that tension lay. On the other hand, to use modern constitutional language, sovereignty was invested in the people who, however indirectly, could exercise that sovereignty in their recognition of the *walī al-faqīh*. On the other hand, it was the right of the Imam to appoint his deputy and impose him on the populace. However, for jurists writing in the twentieth century, as for those who wrote before them, it was a well-established fact that the deputyship of the Imam was of the "general" nature during the Complete Occultation and, as such, the recognitionn of the a'lam (the most learned), most pious, and so on, was dependent on the Shī'a, in accordance with the instruction in Imam al-Sādiq's response to 'Umar b. Hanzala's questions. Accordingly, if at any time al-wilāyat al-'amma of a jurist was to be established, it was impossible to impose it on the Shī'a on merely theological grounds. The moral grounds for such an imposition were also lacking, unless it was proven with certainty that for the good estate of the Shī'a it was only a jurist who was fit to exercise the *wilāya*.

It was precisely on this issue that the major Imamite jurists differed when it came to the institutionalization of *al-wilāyat al-'āmma* of a jurist by means of constitutional requirement. In part, this was because it is certain that Imamite jurisprudence did not view the *fuqahā'* as forming a class—a "clergy"—whose ranks were closed to the laity, as was the case with medieval Christianity. A jurist was like any other *mukallaf* (a believer subject to the Sharī'a), and hence, had no special right or privilege in Islamic jurisprudence. What accrued to the *faqīh* as a *faqīh* was open to any member of the Shī'a who could prove himself in the three areas of "sound belief," "sound knowledge," and "sound character." As a result, there was no way that Imamite jurisprudence could uphold *al-wilāyat al-'āmma* for a particular class or individual during the Complete Occultation. Moreover, in the absence of the infallible Imam, whose own *wilāya* was absolute and protected by God from becoming arrogant or deviant, there was no assurance that the *wilāya* of a fallible individual could not lead to the creation of another unjust rule on earth. Imamite jurisprudence has persistently reminded Imamites of the injustice inherent in power as such, and the weakness of human beings in the exercise of power. The consequence of all these tensions in the Imamite *fiqh* was an argument for constitutional *wilāyat al-faqīh*, in the modern political experience of the Shī'a in Iran, which addressed some of these underlying concerns of Imamite jurisprudence.

It is reasonable to say that *ijtihād*—independent reasoning by a jurist—has shown its ability to deal with questions entangled in the long history of Imamite Shī'īsm in the absence of the Hidden Imam, especially in the area of *al-wilāyat al-'āmma* and its relationship with the modern national state. For Sunnī jurists it has been a relatively less trying experience in modern times to invoke the Qur'ānic principle of *shūra* (consultative government), designated to legitimize their different forms of government, especially in the absence of a universal Muslim caliph. But for Shī'ī Imāmī jurists, dealing with the exercise of power in the name of the living, but concealed Imam, especially in face of the failure of the ruling Imamite authority to establish an adequately just order, it has required much precaution and exertion of *ijtihād* to guide the Shī'a in its political life. One can understand and appreciate the caution applied by some of them in any concession of *al-wilāyat al-'āmma* to the jurist. What is at the root of the problem is not so much the possibility that a particular jurist assume the authority under the aegis of *al-wilāyat al-'āmma* as the institutionalization of the *faqīh*'s authority in the Imamite community generally.

Some prominent jurists have understandably pointed out that Islamic government is not necessarily government by a jurist; it is government by any qualified member of the Shī'a who has the confidence of the Shī'a public to assume leader-

ship, regardless of his being the *faqīh* or not. The ultimate responsibility for ensuring just government, pending the establishment of the Imam's just order, lay with
the entire community.[97]

Institutionalization of the *wilāya* of a jurist through a modern legislative process was made possible by the pronouncement of the Āyatullāh Khumaynī.[98] Although not very different in content from the customary formulation examined in
this chapter, it laid down the groundwork for rational inferences to institutionalize
al-wilāyat al-'āmma in the Iranian constitution. The most important aspect of this
constitution is acknowledgment of the right of the Shī'a in determining the *walī al-
faqīh*, as an inseparable right given by God to humanity to decide its social destiny.[99] The second most important aspect is the affirmation of the position of the
faqīh during the occultation of the Imam by declaring him the *walī al-'amr* and
leader of the community by virtue of his being righteous, godly, fully aware of the
events of his time, courageous, possessed of the qualities of an administrator and
supervisor, and one who has been recognized and accepted by the majority of the
people.[100] The latter clause has replaced the traditional insistence upon the
a'lamiyya (being the most learned), which had become a prerequisite for the *marja'
al-taqlīd*.

Both these aspects of the constitution point to the *ijtihād* of the jurists who, on
the one hand, elaborated on the already established position of the *walī al-'amr* (the
one put in authority through investiture) in Imamite tradition and conventional
usage (*al-'urf*), as demonstrated by Ansārī. On the other hand, the jurists were
convinced that the elements of majority recognition would ensure the centralization
of the position of the *walī al-'amr*, and would put an end to the factionalism that had
marred the institution of *marja' al-taqlīd*.

One of the problems facing the Imamite community during the last century and
a half has been the factionalism caused by the recognition of several *mujtahid*s as
marja' al-taqlīd.[101] This development in the politicization of *al-wilāyat al-'āmma*
marks the distinction between *al-walī al-faqīh* and *marja' al-taqlīd*. Although the
former can also be the *marja' al-taqlīd*, if he combines in himself the preconditions
for assuming that position, the latter cannot be *al-walī al-faqīh* unless he is invested
with *al-wilāyat al-'āmma* through the political process recognized as valid by the
constitution. The position of *marja' al-taqlīd* was seen as a position limited to the
religious guidance of the Shī'a, although there were arguments for the comprehensive authority of the *marja'* by such a politically quietistic jurist as the late Āyatullāh
Burūjirdī (d. 1961), who believed in the inseparability of the religious and secular
realms of human life and contended that if the *marja'* was a point of reference in
one realm, he could also be the competent authority in the other.[102]

However, the separtion and distinction of the two positions is the legitimation
of the existing situation in religious leadership of Iran rather than anything necessitated by juristic considerations. *Mujtahid* and *āyatullāh* (the sign of God [on earth])
were, up to the recent past, positions of accomplishment and the titles were reserved for those who deserved to be so addressed. But, as happened under the
Sunnī caliphate, these positions in present-day Iran have been conferred on individuals who have been recognized by the government as deserving them. This
development could have negative ramifications for the quality of a *mujtahid* in the
future, especially if the process of *ijtihād* were to become uniform under a
government-sponsored body of religious scholarship.

There has been no other period in Imamite Shī'ism when fundamental official steps were taken to organize Shī'ites nationally on the basis of *al-wilāyat al-'āmma* of a jurist. Such a move threatens to give rise to the tensions commonly experienced by Sunnī Muslims who, on the one hand, are committed to the universal vision of Islam and, on the other, are asked to be loyal to their national identity as created in modern times. If *al-wilāyat al-'āmma* of a jurist becomes well established among the Imāmī Shī'ites, what should be the obligation of those Shī'ites who do not share that national identity with Iran, and hence loyalty to that nation? This question remains to be addressed by those Imamite jurists who have supported the particularization of *al-wilāyat al-'āmma* in the national constitution of Iran.

Conclusions

Although Ansārī's detailed discussion of the *wilāya* of the jurist is unprecedented in Imamite jurisprudence, it has been overshadowed by the development of the constitutionalized *al-wilāyat al-'āmma* in literature published since 1980. The subject of *al-wilāyat al-'āmma* has, in these works, become part of *al-fiqh al-siyāsī* (jurisprudence dealing with the creation of Islamic polity). This treatment of *al-wilāya* was a necessary development of the implications we have observed in the discussions of *wilāya* in jurisprudence. It is plausible that the topic was not only academically important for the *Makāsib* section of jurisprudence and the discussion of the discretionary control of the *walī* in the legal affairs of the community, but more essentially, as the institutionalization of *al-wilāyat al-'āmma* has shown, it appears to reflect the urgent need to clarify the position of an Imamite jurist as a *walī al-'amr* in the Imamite state.

The development of political jurisprudence must be seen in the light of the failure of those invested with power (i.e., the monarchs) to promote the general well-being of Imamites. Thus, when Bihbahānī declares that if a *faqīh*, appointed by the Imam on the basis of his "general permission," were to be appointed as a *sultān* or *hākim* for the people of Islam, there would be no unjust rulers,[103] this is not merely wishful thinking on the part of jurists. On the contrary, it demonstrates dissatisfaction with temporal authority. The alternative—to redress the wrong in the interests of society—is seen in the establishment of order under *al-wilāyat al-'āmma*.

Before the establishment of Imamite temporal authority, such as that of the Safavids and the Qājārs, suggestions were made to the effect that it would be permissible for the *faqīh* to accept office under the *sultān*, though *sultān*s were "unjust." Then, once Imamite temporal authority was established, the permission was taken for granted because it was unthinkable to describe a protector of the Imamite religious system as *al-jā'ir* (unjust). At this stage, there arose a logical question in the mind of Bihbahānī, Muhaqqiq a-Karakī, and all those who supported *al-wilāyat al-'āmma*, including the Āyatullāh Khumaynī: If a *sultān* can assume and exercise temporal authority in the name of the twelfth Imam, pending his return, then is not a well-qualified *faqīh* a more fitting candidate for this position, especially when he has exercised most of the Imam's authority as a legitimate functional Imam effectively during the occultation?

It is important to bear in mind that an Imamite state in which the *wilāya* and the *saltana* of the Hidden Imam could be invested, however temporarily, in the jurist is, from these scholars' viewpoint, a desirable end only inasmuch as the community requires it for the actualization of the ideals of Islamic law. Accordingly, the Imamite state, indispensable for the religious welfare of the community, may be justifiably regarded as the sign of divine grace (*lutf*) that enables believers to draw close to obedience to God and away from disobedience. However, the eventual unity of the *wilāya* and the *saltana,* separated at the beginning of the Imamite political experience following the assassination of the Imam 'Alī, had to await the return of the messianic Imam, the awaited Just Ruler.

It was in response to the question regarding the candicacy of the *faqīh* as a legitimate functional imam that *al-wilāyat al-'āmma* was taken up by prominent jurists in their political jurisprudence in order to establish the permission of the Imam for the jurist to assume that authority. From the early days of Imamite jurisprudence it was acknowledged that the permission of the Imam (*idhn al-imām*) is obligatory in all those rulings that affect the general membership of the Imamite community, because it was only the Imam who, in his position as the absolute *walī,* could exercise discretionary control (*wilāyat al-tasarruf*) over Muslims, and make binding decisions affecting them. Accordingly, in jurisprudence, as we have seen in this study, the matter of deputyship for the Imam was raised whenever someone had to exercise the Imam's authority over his followers.

In all those matters where a judicial decision affected a social aspect in the life of Imamites, the permission of the Imam or his deputy was considered a precondition. Thus, for instance, in the discussion on *zakāt,* which affects the general membership of the community, it is only with the permission of the Imam that one can undertake to distribute it, and it is only the Imam who can appoint a tax collector. In the section dealing with *al-khums,* it becomes evident once more that the Imam, in his position as the successor of the Prophet, is the only person authorized to collect *khums.*[104] During the occultation, when the actual authority is absent, the jurists discuss the "general permission" given to those who pay these taxes, either to distribute the Imam's share to deserving recipients themselves, or (as consensus has emerged among present-day jurists) to hand it to the "general deputy" of the Imam—namely, the jurist who has permission to collect and distribute the *zakāt* and *khums,* and who can ensure the proper use of these funds in accordance with the wish of the Hidden Imam.

When a question arose regarding the administration of justice, holding a political office, and so on, the permission of the Imam was of a paramount consideration, as we have seen, and it was by rational as well as revelational proofs recognized as authoritative in making a judicial decision that the *wilāya* of the jurists was established in those and related areas. What still remained problematic was *wilāyat al-tasarruf,* the total discretion of a jurist in all those matters affecting the lives and property of Muslims. As treated by Ansārī and others, it was the question of assumption of *saltana* by a jurist under the rubric of *al-wilāyat al-'āmma,* similar to the authority of the Imam *al-ma'sūm* (the infallible), that prompted the jurists to examine the underlying source of authority in Imāmī Shī'ism. The theory of political authority in Imāmī Shī'ism has underscored the importance of designation (*nass*) by the Imam, and it was this question that came to be raised in the case of *wilāyat al-faqīh:* Does he have the "designation" or—to use the juristic term—"permission" of the Imam to assume *al-wilāyat al-'āmma?*

In the process of answering this question, the jurists, in the first place, under the influence of the long history of their leadership, began to treat *fuqahā'* as a special class of Imamites,[105] who, because of their "sound belief," "sound knowledge," and "sound character," were regarded as the most qualified Shī'ites to assume any public responsibility in the Imamite community. In the second place, they disregarded the whole question of the necessity for public recognition of that qualified individual among themselves who could exercise the Hidden Imam's authority during the occultation. The jurists had regarded *al-wilāyat al-'āmma* as dependent upon special designation of the Imam. Because the *fuqahā'* did not possess that designation, their *wilāya* in other areas could not be taken to include the exercise of the Imam's constitutional authority.

As a corollary of this argument against the illegality of *al-wilāyat al-'āmma* for the jurist, one would have expected to find a statement in these writings condemning Imamite *sultāns* when they assumed constitutional authority—condemning them as usurpers in the same vein as were the jurists condemned who supported *al-wilāyat al-'āmma* for a well-qualified jurist criticized for loving "leadership (*ri'āsa*) and power (*saltana*) in Iran."[106] The absence of such condemnation of the Shī'ī rulers corroborates my observation that the problem that Ansārī and others who followed him encountered did not center on the lack of special designation of the *fuqahā'* to assume *al-wilāyat al-'āmma*, especially in view of the fact that they did not object to the *sultāns'* assuming it. The problem was their unwillingness to accept the *faqīh* in the political position in which they had seen the monarchs and the caliphs, who all fell short in fulfilling the Islamic ideal. If they had failed because of the injustice inherent in power except when the wielder of authority is divinely protected, would the *faqīh* succeed?

It was in this vein that Narāqī ruled political education a precondition for a *mujtahid* to assume political authority under *al-wilāyat al-'āmma*.[107] Whether the jurist was invested with *al-wilāyat al-'āmma* through special designation or not, and whether there was an absolute proof to support the legality of his assuming this authority during the occultation or not, the *faqīh* still enjoyed the confidence of believers as their *marja' al-taqlīd*. The Shī'a regarded him as the deputy of the Imam, who, they believed, had designated him as the *walī al-'amr* in their affairs. Undoubtedly, it was this acknowledgment on the part of the Shī'a, from the early days of Imamite history, which was prudently recognized by many a *marja' al-taqlīd*, that ultimately helped to institutionalize *al-wilāyat al-'āmma* of a well-qualified jurist in the constitution of a Shī'ī nation in modern times. Furthermore, without the rational procedures outlined in works on the theoretical basis of Islamic law, the *usūl al-fiqh* (which attained perfection under Ansārī and his eminent disciples who provided the Practical Principles and the necessary terminology to derive fresh rulings in the area of political jurisprudence), it would have been impossible for the proponents of *al-wilāyat al-'āmma* to legalize as well as institutionalize the all-comprehensive authority of the *mujtahid* in his role as the *marja' al-taqlīd* and *al-walī al-faqīh*.

6

Conclusion

The earth shall not remain but there will always be a learned authority
(*'ālim*) from among us, who will distinguish the truth from falsehood.

The prophecy of Imam al-Sādiq, as the present study has demonstrated, has been
fulfilled for the Shī'ī community throughout their history following the occultation
of their last Imam, al-Mahdī, during which there has always been an *'ālim* to direct
the community to its goal. The presence of a learned authority has, moreover,
generated confidence in the Shī'a community to manage its socio-political affairs
through the contingencies that have occurred since the occultation of the Imam.
There is no doubt that this confidence was generated, in large measure, by the
personal character of an *'ālim* who had taken upon himself to "distinguish the truth
from falsehood," preserve it in his oral and written pronouncements, and disseminate it among the populace by training students who were made responsible to
teach the Shī'a. This contribution of the Shī'ī scholar to the preservation of the
Imamite worldview strengthened belief of the Shī'a in the continuation of religious
leadership, and this gradually prepared the way for the Shī'ī learned authority to
assume the much wider socio-political leadership of the Imamite community as the
functional Imam. More significantly, the acknowledgment of the faithful made it
possible for the Shī'ī jurist to assume the all-comprehensive authority (*al-wilāyat al-
'āmma*) of the Imam as a just ruler, when historical circumstances made it necessary
for him to do so.

This development in the authority of the Shī'ī jurist was not something that
awaited the period of the Complete Occultation of the last Imam; rather, it had
begun during the time when the Shī'ī Imams, living under difficult political circumstances (eighth-ninth centuries A.D.), could not assume full responsibility for directing the affairs of their Shī'a. Hence they delegated their partial authority to those
among their close associates who proved their loyalty to them by believing in them
("sound belief") and learning from them ("sound knowledge"). This was the beginning of the deputyship (*niyāba*) of the Imams, on whose authority those trusted
disciples managed the religious and socio-political affairs of the Shī'a in widely
separated areas like Kūfa, Qumm, Rayy, and so on.

With the occultation of the Imam, his deputies assumed wider authority among
the Imamite community, responding to the urgent need of preserving the Imamite
ideology. Historical vicissitudes required the community to respond without com-

promising the central doctrine of the Imamate. This response was worked out by the Imamite jurists, as the spokesmen of the community and the deputies of the Imam, with due intellectual caution about the nature of the authority of the Imamite jurist during the occultation.

It is methodologically convenient to divide the responsa of the Imamite jurists in this regard into four historical periods when major works dealing with political jurisprudence were written.

The first was the Būyid era (945–1055), when the eminent personages of the Baghdad school of Imamite jurisprudence—Mufīd, Sharīf al-Murtadā, and Tūsī—discussed the wilāya (authority) of the faqīh and delineated this authority in their jurisprudential works wherever necessary.

Then came the post-Seljūq and the Ilkhānid era (twelfth–fourteenth centuries), when prominent jurists of the Hilla school of Imamite jurisprudence, like Ibn Idrīs al-Hillī, Muhaqqiq al-Hillī, ʿAllāma al-Hillī, and Shahīd I, wrote their works, expounding further on the discussions of the Baghdad school in the matter of wilāyat al-faqīh.

The third was the Safavid era (1501–1786), when jurists like Muhaqqiq al-Karakī, Shahīd II, Muhaqqiq al-Sabzawārī, and Fādil al-Hindī wrote their commentaries on the works of the Hilla jurists and discussed the authority of the jurists, explaining its wider implications for Imamite temporal authority.

The fourth was the Qājār and the post-Qājār era (late eighteenth–early twentieth centuries), when the monumental works on jurisprudence produced by great scholars like Najafī, Ansārī, Muhammad Kāzim Yazdī, Muhsin al-Hakīm, Rūh Allāh Khumaynī, and other modern jurists defined and set the authority of the jurists, as adumbrated in the works of the previous generations of Imamite jurists, on a firm theological basis of the "general" deputyship of the Hidden Imam.

During the first period (945–1055), the juridical works produced under the guidance of Mufīd and his Baghdad school indicate that assuming the leadership of the Imamite community was not only a logical extension of the Imamate of the Hidden Imam, but also that the necessity to do so was not problematic legally. The Imamite jurists had to preserve the Imamite theory of leadership, which included the spiritual and temporal authority of the Imam. They witnessed the establishment of an Imamite temporal authority under the Būyids, which had to be explained without compromising an ideal threatened by political contingencies. According to this ideal, the Imam, as an infallible leader and authoritative interpreter of Islamic revelation, was the sole legitimate authority (al-sultān al-ʿādil—i.e., the Just Ruler) in whom temporal authority could be invested. This ideal was the classical theory of the Imamate, which assumed the unity of religious and political authority under the explicitly designated Imams. However, reality mirrored the absence of the temporal authority of the Imams. The Imams, troubled by the role of the sinful occupants of political power and sensing their inability to replace them, worked toward freeing the Imamate from being conditional upon the Imam's actual investiture as a political leader, as required, for instance, by the Zaydī Imamate. Accordingly, historical realities necessitated a division of the Imamate into religious and political spheres. The political authority of the Imam was considered usurped by "unjust" rulers, but his religious authority remained intact to enable him to provide the necessary socio-religious prescriptions for his Shīʿa. Nevertheless, in theory, he was the Just Ruler (al-sultān al-ʿādil), divinely designated to exercise religious as well as political authority. In this way the Imamite ideal continued through all political circumstances until the twelfth Imam went into occultation (A.D. 874).

Following the occultation, the Imamite theologians disregarded the cleavage in their theoretical pronouncements on the Imamate and, following the examples of the Imams, kept themselves unsullied by the injustice inherent in political as such. Under these concrete and contrived circumstances, the power of the Imam came to be measured in terms of his knowledge (al-'ilm), which he had inherited from the Prophetic source, through the proper designation (nass). Consequently, in Imamite theology "power" in the sense of "authority" having moral and legal supremacy with the right of enforcing obedience came to be recognized in the 'ilm, especially when the power to exercise control or command over others remained in the hands of those who are frequently mentioned in our sources as instances of al-sultān al-jā'ir (tyrannical or unjust ruler).

When the Būyids came to power some seventy years later, the question of Imamite political authority began to be addressed in jurisprudence. However, the jurists realized all too soon that the continuation of the socio-religious structure of the Imamite community depended upon consolidation of the institution of the deputyship of the Imam, not on the Būyid sultāns, who had allowed the Sunnī caliphate to continue as symbolic of the unity of the majority of Muslims. The deputyship, for these jurists, became a sort of trust in the Hidden Imam, one that could assume functions that the Imam, had he been present, would have undertaken himself or would have delegated to someone qualified to represent him, such as the khalīfat al-imām. Thus, deputies were authorized to undertake functions with theologico-political implications, as functional Imams, with the potential of becoming sultān al-zamān (the ruling authority of the time), in the general interest of the Shī'a.

During the second period (twelfth–fourteenth centuries), the rulings of Mufīd and Tūsī regarding the authority of the deputyship of the Imam came to be identified with the well-qualified jurist who could exercise wilāya in the administration of justice and whatever was related to it. Their rulings went through further clarification, with the result that in subsequent works on jurisprudence the possibility of the existence of al-sultān al-'ādil (the Just Ruler), along with the Imam proper, was considered to be grounded. Jurists during this period were witnesses to the political turmoil following the breakdown of Sunnī Seljūq authority and the destruction of the 'Abbāsid caliphate by the Mongols. This unfavorable situation more than anything else convinced them that the existence of Shī'ī political authority (other than that of the Imam) willing to consider the implementation of the Sharī'a was not only expedient but necessary, because it fulfilled the obligation of "enjoining the good and forbidding the evil." The fulfillment of this obligation also provided legitimate grounds to apply the phrase al-sultān al-'ādil to any Shī'ī authority committed to the promulgation of the Sharī'a.

In this regard, Shī'ī jurisprudence reflected a similar development in the juridical corpus of Sunnī scholars, who, following the termination of the "rightly guided" caliphate, had rationalized the existence of the sultān (theoretically, the Just Ruler) and the public order he protected, not so much as identical with the faith, but indispensable for safeguarding and widening the application of the principles provided in the normative Sharī'a. In other words, the Islamic state under the sultān was an end in itself only inasmuch as the community required it for "drawing close to obedience to God and away from disobedience" (the purpose of *lutf*). Significantly, the legitimacy of a public order under the *sultān* was measured in terms of the glory of Islam under his authority. This glory was regarded an explicit sign of divine approbation of a rule that otherwise lacked theological validity.

Among the administrative institutions that had grown up in the Islamic empire, *al-qadā'*, administration of justice, became one of the most important in preserving the popular sense of universal justice. As such, at times when the central power of Muslim rulers had disintegrated, the prestige of *al-qādī* in general became immeasurable. In Shī'īsm, *al-qadā'* became the most fundamental aspect of the growing political power of the jurists, who, in their position as the lawful administrators of justice, were regarded as the protectors of the people against the unjust behavior of those in power. The license and tyranny of those in power, in addition to the expectations of the Imamite community, required jurists to undertake the wider role of the functional Imam (beyond their already well-established role as the interpreters of the Islamic revelation), in their capacity as the "general" deputy of the Hidden Imam. The new role was carefully worked out in all its details under the rubric of *wilāyat al-faqīh* in subsequent eras.

During the third period (1501–1736), when the Shī'a state was established in Iran under the Safavids, the works of jurisprudence took note of the development of the temporal authority of the professing Imamite *sultān* by further interpretation and even legitimation of *sultān al-zamān* for management of the affairs of the Imamite community. The most significant argument justifying the authority of the jurist was that it made it possible to implement Islamic laws before the return of the Imam from occultation by asking jurists to undertake the responsibility of their execution, because they had been considered as the *ulū al-'amr* (those who possess authority) through the exegetical extrapolation of the rescript written by the twelfth Imam. In other words, the jurist among the Imamites could exercise authority similar to that of the *sultān* among the Sunnīs inasmuch as the *sultān*s likewise held the position of *ulū al-'amr* (those who were invested with authority) in the Sunnī community. The *wilāya* of the jurist, more specifically his wielding of *al-wilāyat al-'āmma* (the comprehensive authority), had the same legal validity as the authority of *ulū al-'amr* in the Muslim public order. Moreover, just as the investiture of *wilāya* was necessary to carry out any official political function, so the investiture of *al-wilāyat al-'āmma* was necessary to carry out the obligation of "enjoining the good and forbidding the evil." This obligation was the main revelational justification for the existence of any government during the occultation. Because the Imam had invested the "general" deputyship in the jurist, he was the most qualified among the Imamites to undertake the implementation of the obligation in his position as one of the category of *ulū al-'amr*.

The implications of the comprehensive authority of the Imamite jurist became obvious in the fourth period of the political jurisprudence of the Imamites, from the eighteenth century onward.

During this period, the position of the jurist became centralized and institutionalized in the position of *marja' al-taqlīd* (the authority who was followed by the general membership of the Shī'a as its religious leader). With the failure of those invested with power to uphold justice, the comprehensive authority of the jurist began to be seen as an alternative authority that could replace the corrupt ruler and fulfil the function of *al-sultān al-'ādil*. In the eyes of the Shī'a the jurist, who was appointed by the Imam on the basis of his general permission, had more legitimate claim than did the monarch to exercise *al-wilāyat al-ʿāmma* in the name of the Hidden Imam pending his return. Moreover, he enjoyed the confidence of the faithful as their religious guide, for they regarded him as the deputy of the Imam, sharing in the Imam's charisma. In a distinct way, then, the jurist had a popular mandate to function as the *walī al-ʾamr* of the Shīʿa, in a way the monarch lacked. Thus, *al-wilāyat al-ʿāmma* of a jurist who possessed "sound belief," "knowledge," and "character" substituted for the *wilāya* of the Imam, and generated loyalty to Imāmī Shīʿīsm through the process of pledging obedience to the authority of a learned jurist (*taqlīd*).

The eighteenth–twentieth centuries marked the introduction of a modern system of administration, modern education, modern values, and so on. More essentially, this period affected both the interpretation of *al-wilāyat al-ʿāmma* of a jurist and the expectations of Shīʿites from the *nāʾib al-imām* (deputy of the Imam). The most important aspect of the *wilāya*, which had been neglected in all the discussions on political jurisprudence, was the right of the Shīʿa to determine the jurist's qualification to assume the comprehensive authority of the Imam. This had been a well-established point in the earlier works, where the difference between the *wilāya* of the Imam and the jurist was explicitly seen in the acknowledgment afforded by believers. Whereas the former did not depend on the people, because it was a divinely ordained office, the latter clearly depended on the acknowledgment by the Shīʿa of the characteristics that would qualify a particular scholar to assume the *wilāya* invested in him through a general designation of the Imam.

Although this aspect of the right of the Shīʿa has received careful elucidation in some works of jurisprudence, *al-wilāyat al-ʿāmma* has now become entangled with the national identity of a particular group of the Shīʿa living in a particular modern state. This development, from which the Shīʿī vision of *al-wilāyat al-ʿāmma* was spared until now, has given rise to a tension similar to that experienced by Sunnī Muslims, who have become divided under different national identities to which they must adhere while maintaining the universalistic vision of Islamic revelation. The nationalistic and particularistic vision has created obstacles to the revival of the universal Sunnī caliphate. Similarly, it is doubtful that the particularized *al-wilāyat al-ʿāmma* of the Imamite jurist in the constitution of modern Iran will allow the recognition of the *walī al-faqīh* as the just ruler of *all* the Shīʿa, pending the return of the Mahdī.

APPENDIX

The Imam's Share in the Fifth (*al-khums*) during the Occultation

I have dealt with the subject of *al-khums* ("the fifth") in an article entitled "*Al-khums:* the Fifth in the Imāmī Legal System," where I have examined in detail the concept of *al-khums* in Imāmī Shī'īsm and have discussed it in the context of Islamic jurisprudence in general.[1] For the present study, I need to elaborate on the Imam's share in this tax during the occultation, for it has relevance to the development of *al-wilāyat al-'āmma* of the jurist.

In the Imamite legal system *al-khums* applies to a variety of items and is used in a much wider sense than the booty taken from *dār al-harb* (abode of [the enemies of Islam defeated in] battle). This is one of the basic differences between the Imāmī and other schools of legal thought in Islam. According to Tūsī, in addition to the spoils of war, *al-khums* applies to all that is acquired from hoarded treasures, from mines, or from the sea, regular earnings, and so on.[2] Moreover, the allocation of the *khums* payable on all these items forms another major difference between Imamite and non-Imamite legal systems. In Imāmī jurisprudence the *khums* was apportioned according to a sixfold division: so much each for God, the Prophet, his family, orphans, the needy, and wayfarers. According to the majority of Imamite scholars, the shares of God and the Prophet belonged to the latter's successor, his *walī al-'amr* (the executor of his testamentary injunction). Thus the Imam received three shares, two as the rightful heir of the Prophet and one allotted to him on God's behalf; the remaining three shares belong to those among the Banū Hāshim, the Prophet's clan, who are orphans, poor, or wayfarers.[3]

The division of the *khums* into six portions conformed with the literal application of the injunction in the Qur'an. They were to be administered by the Imam himself, because *al-khums,* in the jurisprudence of the Imamites, constitutes the state's share in the Imamite political system, with the Imam as rightful head of the government invested with *wilāyat al-tasarruf* (total discretion in administering the

237

affairs of the Muslim community).[4] As a consequence, administration of the *khums* requires the perpetual presence of the Imam or his specially appointed deputy. As in other areas of Imamite jurisprudence, a question arose not only about the administration of al-khums during the occultation, but also whether or not the *khums* is payable at all during the absence of the *walī al-mutlaq* (the absolute authority—i.e., the Imam). The latter question leads to the highly debated issue of Imamite authority during the period of the occultation: Who is authorized to act as the leader of the Imamite community pending the return of the twelfth Imam?

The *khums* during the Occultation of the Imam

One of the fundamental problems in rulings about the *khums* during the occultation has been the lack of unambiguous documentation to support its incumbency during the absence of the administrator of "the fifth"—namely, the Imam. Mufīd has acknowledged this situation in the matter of *khums* by confessing that it is one of the truly testing and perplexing questions in jurisprudence. Major Imamite figures, as we shall see presently, have experienced confusion in the management of the entire *khums*, not just the Imam's share of it.[5] Unlike *zakāt* which was, at least in its outlines, laid down explicitly in the Qur'ān, reference to *khums* in the Qur'ān is limited to one passage (8:41). Moreover, there does not seem to be a consensus of opinion even among the major Sunnī jurists as to the manner of distribution of the *khums*, and the sources reveal that the whole controversy arose shortly after the Prophet's death. For political reasons, which cannot be elaborated here, the early caliphs had to deal with the shares of the Prophet and his family in the *khums*, especially when the nearest of kin demanded that their entire share be handed over to them.[6]

Differences of opinion among Sunnī jurists, sometimes among the jurists of a single school, in solving this question suggest that there were two important considerations that influenced their final rulings with regard to the Prophet's and his family's share. First, if these shares were acknowledged in the *khums* even after the death of the Prophet and handed over to the Imams from among the *ahl al-bayt*, then it would have meant, however indirectly, recognition of their right to leadership in the community. This is by virtue of the fact that the Prophet was entitled to the *khums* because of his being the leader of the *umma*. Secondly, it was important to uphold the *wilāyat al-tasarruf* of the caliph and give him complete discretion in the disposal of the *khums*, thus limiting or even declaring null and void any claims by the *ahl al-bayt* to the *khums*.

But for Imamite jurists, *al-khums* was the right of the Imam and his nearest of kin, and, accordingly, its management was equivalent to the management of the property belonging to an absent person, a transaction that called for extreme care. Its management, moreover, required proper investiture because it was part of the *wilāyat al-tasarruf* in which, as a *walī*, the "guardian" could exercise his free discretion in disposing of the goods belonging to the Hidden Imam. In other words, administration of the *khums* was part of *al-wilāyat al-'āmma*, which, if proven valid in the case of a jurist, could also empower him to exercise his discretion in the distribution of the *khums*. However, as I have indicated in this study, there is no consensus about *al-wilāyat al-'āmma* of the jurist, and consequently there is no agreement in the jurist's right of administration of the *khums*.

All major works of Imamite jurisprudence have a long section discussing *al-khums*, more specifically the three shares that belong to the Imam in occultation, citing the opinions of all major Imamite jurists like Mufīd, Sharīf al-Murtadā, Tūsī, and others. One of the most comprehensive discussions on the subject is recorded by 'Allāma in his work on comparative jurisprudence, *Mukhtalaf*,[7] where he chronologically takes up the opinions of major jurists from the classical period. This indicates the significance attached to the problem of the right of the Imam in the *khums* and the way it should be administered during the occultation, in the absence of any explicit text regarding it. This latter fact, according to Ibn Idrīs, led to different judicial decisions about the disposal of the Imam's share during the occultation.[8]

The Imamites, according to Mufīd, have differed markedly in the matter of the Imam's share during the occultation. One group maintains, without citing any documentary evidence, that the giving of this tax is void during the occultation. Some say it is obligatory to bury it. To support their view, they cite the statement about al-Qā'im's (i.e., the twelfth Imam's) reappearance when the earth will bring forth its treasures. Others maintain that it is a recommended duty to distribute it among the poor of the Shī'a. Still others say that it is necessary to preserve it (i.e., put it aside) for the Imam as long as he is alive, and when death approaches him he should make a will (*wasiyya*) enjoining one of his trusted followers to hand it over to the Imam when he reappears; and if this trusted follower dies without seeing the Imam, then he too should make a similar will enjoining some other reliable person until the funds reach the Imam. This is the opinion accepted by Mufīd because the *khums* is the right of the Hidden Imam. Inasmuch as there is no prescribed rule as to its disposal before his occultation ends, this is the only way that the *khums* can ultimately reach him. Thus, rules Mufīd, it is incumbent to save it for the Imam through perpetual *wasiyya* until the time when some appointed person reaches the Imam or is able to deliver the *khums* to him, or at least there is someone who could deliver it to him in due form. If someone prevents the spending of the Imam's share, while distributing the other shares of the *khums* among their due recipients, as specified in the Qur'ān, then (asserts Mufīd) he has acted in perfect accordance with the injunction.

Next, Mufīd proceeds to give reasons for the existence of different opinions in this matter. He also acknowledges, as other jurists have done, that there is no textual proof to support his opinion, although there is a supporting rational proof in the form of a moral axiom that regards it improper to dispose of any property without the owner's permission.[9]

However, Mufīd seems to have revised his opinion regarding the Imam's share. In his *al-Risālat al-gharawiyya*, he says that when the rightful Imam is absent, and the Shī'ites cannot reach the Imam or do not know his whereabouts (due to the severity of the *taqiyya* and the necessity for the Imam to conceal himself), when they have collected the *khums*, they should give it to those among the descendants of the Prophet who are orphans, needy, and wayfarers, and they should even increase their share "because of the severity of their need and the obstruction of the Sunnites to their right, and their [suffering under] oppression and suppression." Their share, says Mufīd, is not the same when the Imam is present, because of the

changed circumstances. He then cites a *hadīth* report that supports the consideration of an increase of shares when there is need of it.[10]

Tūsī has followed Mufid's opinion in general, but he provides a further development in the ruling when he says that the Imams have given permission, to their
Shī'ites during the occultation, to dispose freely of their goods that would normally
go to the *khums*, for what they need in order to cover the expenses of a marriage,
the transaction of business, or buying or building a residence. But for other than
these stated purposes, Tūsī cautions, it is not permissible to freely dispose of the
khums at all. As for the Imam's share from treasures and other sources during the
occultation, Imamites have held differing views, but there is no textual evidence to
support their views, and every jurist, says Tūsī, has ruled in conformity with the
principle of precautionary measure (*ihtiyāt*). Some have maintained that it is permissible that even the Imam's share be used for marriage, business, and residence
purposes; others have ruled it necessary to hold it in quasi-perpetual escrow, as
described above.[11]

Ibn al-Barrāj, who had studied with Mufīd and Sharīf al-Murtadā, and who was
the *qādī* in Tripoli as well as Tūsī's representative in religious matters in Syria, was
apparently the first person to explicitly maintain that the Imam's share should be
taken to the Imam. If a person cannot meet the Imam, then he should hand it over
to someone whom he trusts, as to religion and reliability, among the *fuqahā'* of the
Imāmiyya. He should also lay a testamentary injunction upon him to hand it over to
the Imam if he meets him; but if he does not meet him, then he should lay a similar
injunction upon another person, and so on.[12]

Sallār al-Daylamī, another student of Mufīd and Sharīf al-Murtadā, and Tūsī's
contemporary, ruled that the Imam's share cannot be disposed of freely without his
permission. However, he says, during the occultation, the *khums* has been made
lawful for Shī'ites to dispose of as a special gift from the Imam.[13]

A similar opinion was expressed in the work *Wasīla* by Ibn Hamza al-Tūsī, who
had studied with Tūsī or Tūsī's son Abū 'Alī. According to Ibn Hamza, if the Imam
is not present, then the Imam's share should be distributed among his Shī'a
(*mawālīhi*), who acknowledge his right in the *khums*, and who are needy, honest,
and pious. As to its distribution, the Imam must supervise it if he is present;
otherwise it is the duty of the person on whom it is incumbent to distribute it
fittingly when the Imam is in occultation. He should be informed of the identity of
rightful recipients and should distribute it to the best of his ability. It is, however,
better to hand it over to some pious *fuqahā'* to undertake its distribution. Indeed, if
he thinks that he cannot perform this duty well, then it is *incumbent* on him to hand
it over for distribution to scholars learned in *fiqh*. Ibn Hamza even maintains that
all six shares of the *khums* (i.e., the Imam's and the Hāshimites' shares) must be
handed over to an Imamite jurist because, in his opinion, the three shares of the
Imam should be distributed among the Shī'ites in general, as specified above;
whereas the other shares must go to the Hāshimites, in accordance with the
Qur'ānic injunction.[14]

Ibn Idrīs, the leader of the Imamite jurists in Hilla, also discussed the Imam's
share in great detail. After citing and commenting on opinions of all the major
Imamite jurists, he gives his own decision—namely, that one should make a will
regarding the Imam's share and leave it as a trust. It is not permissible, according to
him, to bury the Imam's share, as maintained by some on the basis of the *hadīth*

that says "the earth will bring forth its treasures when al-Qā'im [i.e., al-Mahdī] rises," for there is no well-argued reason for doing so. It is significant that he does not deal with the right of the Imamite jurist to collect the Imam's share and administer its distribution.[15]

Muhaqqiq al-Hillī is perhaps the first Imamite jurist to rule that the Imam's share must be administered by the one who possesses *wilāyat al-hukm* (the authority to administer justice) by virtue of the right of deputyship. Commenting on this opinion, Najafī says that it is conditioned by the circumstance of the occultation, and is intended to ensure that the Imam's share be expended in the way the Imam himself would have determined.[16]

The culmination of all the opinions in the matter of the Imam's share in the *khums* was marked by Muhaqqiq's nephew, and his most erudite disciple, 'Allāma, whose various works on jurisprudence have dealt with *al-khums* in different contexts. In his *Tadhkira*, 'Allāma upholds the lawfulness of the Shī'a's applying the Imam's share in the expenses involved in marriage, buying a home, and engaging in trade, which, according to him, was permissible both during the lifetime and in the occultation of the Imam. The reason for this permissibility was to protect the Shī'ites from sinning, given the absence of any other way for them to receive the *khums*. This, says 'Allāma, is the greatest category of need for the Shī'a during the occultation.[17]

In his *Qawā'id*, 'Allāma takes up the matter of the Imam's share in the *khums* in the section dealing with *zakāt*, and rules that during the occultation a believer has a choice between preserving the Imam's share through ongoing testamentary provisions or distributing it among the three specified categories of recipients among the Prophet's descendants.[18] This ruling is elaborated in his *Mukhtalaf* (in the section called *Kitāb al-khums*) at the end of his lengthy discussion of the general issue. He begins by stating that the most preferable among the opinions is that the *khums* must be divided into two parts. What belongs to orphans, the poor, and wayfarers among the descendants of the Prophet should be distributed among them according to their needs. What belongs to the Imam must be preserved for him until he reappears. It should be handed to him either in person or through ongoing testamentary provisions to thoroughly reliable persons until it reaches the Imam. As for the distribution of his share among the destitute of the Hāshimites, some *hadīth* reports show that this was licit when the Imams were present. All the more so, then, can it be presumed permissible for their descendants when they are in need during the occultation.

Furthermore, it is established in tradition that if the share alloted to the destitute is less than their need, then it is for the Imam to compensate them from his own share so long as he is present. If so, it ought also to be in effect during the occultation, because that which is obligatory in *khums* does not become void during the occultation, especially when God's own right is involved. If this is granted as valid, then to administer the distribution of the Imam's share to the destitute among the Hāshimites is the duty of the one who possesses *wilāyat al-hukm* on behalf of the Hidden Imam. The *walī al-hukm* has to fulfil the responsibility of administering this share, exactly as he would in the case of an absent person. This *walī al-hukm* is the *faqīh*, the reliable person, qualified to issue legal decisions and to administer justice. Anyone besides him who undertakes to distribute the Imam's share must account for his action.[19]

It should be stated that from the time of Muhaqqiq the *faqīh* was clearly seen as the deputy of the Imam fully invested with *wilāyat al-hukm,* and authorized to manage the Imam's share as the most qualified individual to ensure the proper use of the *khums* in accordance with the wish of the Imam in occultation. In the works of the Shahīd I and Shahīd II the opinion that the Imam's share must be given to the *nāʾib* (deputy) of the Hidden Imam—that is, a well-qualified jurist—was well established and was regarded as the preferable one, even when other opinions were mentioned alongside it for academic reasons. Also, it appears from Shahīd I's ruling that no part of the Imam's share can be disbursed among other legal recipients of the *khums.* An exception, however, was made regarding expenses for marriage, trade, and residence, which could be provided from the Imam's share unrestrictedly.[20]

The case for the administration of the Imam's share by the one who exercised *wilāyat al-hukm* was further strengthened during the Safavid era, as may be seen in the juridical works of the period. It was argued that the jurist has the *wilāya* over an absent person's property, and this included the Hidden Imam's *amwāl* ("goods," in the sense of the fifth). According to Najafī, Majlisī, in his *Zād al-maʿād,* indicated that most scholars have taken the position that if a person has collected the *khums* and then decided to dispose of the Imam's share himself, this still does not relieve him of his responsibility. Rather, says Majlisī, it is incumbent upon him to hand over not only the Imam's share to *al-hākim,* but "it is my supposition that the ruling affects the entire *khums.*"[21]

Sabzawārī, who studied with Majlisī's father, Muhammad Taqī, after discussing various items on which the *khums* must be paid (and which have been made lawful to the Shīʿites), comments that the opinion to the effect that the Imam's share is licit for the Shīʿa, unrestrictedly, is not without strength. However, for safety's sake, it must be spent among the existing categories of the rightful recipients, under the supervision of the righteous, well-qualified jurist who can issue legal decisions. The jurist should supervise it carefully, as far as possible, and distribute it adequately to the needy, not exceeding one year's provision. Then he cites an *ijmāʿ,* on the authority of Shahīd II, stating that the Imam's share must be distributed among the three categories specified above. Moreover, some scholars have required that the share be distributed by *al-hākim;* otherwise the person on whom it was incumbent to pay the *khums* would be held responsible to pay it again; whereas Mufīd, says Sabzawārī, in his *al-Risālat al-gharawiyya* indicates permission for the owner to undertake its distribution directly. This is also Sabzawārī's conclusions in his *Kifāyat al-ahkām.*[22] In his *Dhakhīra,* however, his ruling mentions the option of either preserving the Imam's share until the Imam reappears, or giving it to the jurist who has *wilāyat al-hukm.* The *faqīh,* says Sabzawārī, can distribute the *khums* to needy recipients on the basis of his *niyāba* (deputyship) from the Imam in general matters affecting the Imamite community.[23]

During the Qājār period, Imamite jurists had almost reached consensus regarding the disposal of the Imam's share in *khums,* which clearly reflects the consolidation of the jurist's position in the Imamite community. One should also note the frequent use of the title *al-mujtahid* for the *faqīh* in juridical writings during the late Safavid and Qājār era. Thus, Shaykh Jaʿfar Kāshif al-Ghitāʾ (d. 1228/1813), celebrated for his opinion on *jihād* during Fath ʿAlī Shāh Qājār's rule, takes up the question of the Imam's share in the *khums* in his *Kashf al-ghitāʾ.* He believes that it should be given to the Imam when he is present and when it is possible to reach him. During the period of *taqiyya* (precautionary dissimulation) and *ghayba* (occultation) it should be distributed among the three categories of the Hāshimites (as specified above). This, says Shaykh Ja'far, is more appropriate than any other procedure.

This distribution should be undertaken by the *mujtahid*, as a precautionary measure, and he should be the best one among them. The *mujtahid* should undertake to spend it in the most fitting way. However, when it is difficult to deliver it to the *mujtahid*, or there is a valid excuse for not doing so, or when it is impossible to guard the items belonging to the Imam until permission regarding its disposal from the *mujtahid* reaches a suitable person, then it is proper for any righteous believer to undertake its distribution. If someone gives the Imam's share to a person other than the *mujtahid* or his appointed agent (*wakīl*) or the one who has his permission to collect, when it is in fact possible to deliver it to one of them, then he should pay it again. In any case, it is for the *mujtahid* to give permission, although precautionary measure (*ihtiyāt*) requires that the donor pay the *khums* again.

Also, someone could give it to a person whom he believes to be a *mujtahid*, and it later transpires that this was not the case; if the *khums* is still with the recipient, then the donor should take it back from him. If it has been spent, and if he is certain that the Imam's share is still undistributed among the other three categories of the Hāshimites, and if he is unable to repay it himself and the one who received it in the first place is also unable to do so, then the donor is not responsible.[24]

This is the first time that we read a detailed ruling about the administration of the Imam's share in the *khums* where the *mujtahid's* role is clearly defined. Najafī, who is not only the contemporary of Shaykh Ja'far but also his colleague, and who studied under the *usūlī mujtahid* Wahīd al-Bihbahānī, consolidates Shaykh Ja'far's position regarding his ruling on the permission given to a righteous believer to undertake the distribution of *al-khums* when the *mujtahid* is inaccessible. However, Najafī adds, a pious individual would not dare to undertake such a great responsibility, "because of lack of clarity in the rulings connected with it, [a situation] where both reason and revelation require a person to adopt precautionary measure."[25]

As for its being licit for the Shī'a to receive the Imam's share of the *khums*, Najafī cites several jurists who believed that the Imam's share was licit for the Shī'ites. Their opinion (as cited by Muhaqqiq), says Najafī, implies that the entire *khums* is licit for the Shī'ites and they do not have to transmit it to some duly constituted authority. In support of this opinion these jurists cite several traditions that refer to the lawful authority granted to a particular individual at a particular time, not signifying general application of the permission. Moreover, what the Imam can make licit is only that which belongs to him, and as such the half of the *khums* that belongs to the Hāshimites cannot be made licit, as documented by Ibn Junayd al-Iskāfī (d. 381/991), Mufīd's teacher. The evidence that has been gathered in the communications of the Imams supports the licitness of the Imam's share only, not that of the entire *khums*. Furthermore, even in the case of the Imam's share, only part of it was made licit for certain of the Shī'ites when they sought the Imam's permission for stated purposes.

It is evident, says Najafī, that the Imams had agents who were collecting the *khums* on their behalf, especially for the nearest of kin of the Imam who could not be helped from the *zakāt*, as ruled in the laws governing *zakāt*. Even during the

Short Occultation, says Najafī, which lasted for some seventy years (A.D. 874–941), the four deputies of the Imam used to collect the *khums* and were fully aware that it was the will of the Imam that they should do so. Also, the traditions regarding the distribution of the *khums* confirm that it was not made licit in its entirety, nor was there permission for the Shī'ites to make disbursements from it. It is, furthermore, evident from the principles governing juristic practice, and from the body of textual proof, that no one has a right to freely dispose of that which does not belong to him.[26]

The Akhbārī jurists, who, in general, were averse to increase in the *mujtahid*'s authority, which according to them had no textual basis in the *akhbār* (traditions), were supporters of the opinion favoring the licitness of giving the Imam's share of the *khums* to the Shī'a. Shaykh Yūsuf al-Bahrānī, the Akhbārī contemporary of Najafī, has a detailed study of the traditions on *al-khums* in his *Hadā'iq,* along with his discussion of various opinions on the handling of the Imam's share during the occultation. The Akhbārī opinion is summed up at the end of the section on *al-khums,* where Bahrānī says that the entire *khums* should be transferred to the Imams when they are present and in authority, or to their designated agents. It is not permissible to dispose of it without their permission, whether their permission is considered either obligatory or recommendable (the two possibilities in the injunction, with the former being more probable). It is not incumbent on believers to ask the Imams what they should do with it after it reaches them, except that it is understood from some reports that the Imams made it licit for the one who brought it to them. It is the believers' right to spend it on rightful recipients in whatever amounts they see fit.

During the occultation, however, says Bahrānī, the more plausible view is that the shares of the Hāshimites should be given to them, as maintained by the majority of Imamite jurists. It is obligatory to deliver those shares to them, if there is no obstacle to doing so. As for the Imam's share, the most plausible view is that it is licit for the Shī'ites to accept it, on the basis of the rescript from the twelfth Imam as recorded in Kulaynī's *Kāfī.* Nevertheless, regard for *taqiyya* precautionary measure (*ihtiyāt*) requires that it be spent on deserving Hāshimites. But as for the licitness of the entire *khums* for the Shī'ites until the day of judgment, this idea, agrees Bahrānī, is difficult to accept, and it is contrary to the obvious sense of the verse of the Qur'ān and the traditions reported in support of the obligation of paying *al-khums.*[27]

Opinions on the Imam's share formulated during the Qājār period remained the standard view for subsequent Imamite jurists who refined the insistence of the Usūlī jurists on the *mujtahid*'s role in the administration of the entire *khums.* Thus, in his commentary on Muhammad Kāzim al-Yazdī's opinion on the Imam's share in the *khums* (in his work *al-'Urwat al-wuthqā,*[28] which states that it must be handed to the Imam's deputy [the well-qualified *mujtahid*]), the prominent jurist of the present century, Muhsin al-Hakīm (d. 1975), says that a *mujtahid* well-qualified in religious sciences could act as the representative (*wakīl, nā'ib*) of the Imam and collect the *khums* on his behalf. The *nā'ib* had to be the righteous (*'ādil*) and the most learned (*a'lam*) jurist. Both these conditions were desirable in a *mujtahid* to ensure the proper use of the *khums* in accordance with the wish of the Imam in occultation.

The *mujtahid,* having received the *khums,* was to divide it into two parts: the *sahm al-imām* and the *sahm al-sādāt* (for the three categories of the Hāshimites).

The first portion, which includes the three shares of God, the Prophet, and the Imam, was to be used for the propagation of the faith and related activities. By this specific use of the Imam's share, Hakīm contended that the administration of this part of the *khums* truly belonged to a person who was well-informed in matters pertaining to the faith. Thus, it was the *mujtahid* who should look after the implementation of the purpose of the Imam. The other portion, *sahm al-sādāt*, was to be distributed strictly among the Banū Hāshim only.[29] What Hakīm ruled was based on the precedent provided during the Short Occultation of the twelfth Imam, when the special deputies of the Imam undertook the duty of collecting and administering the *khums* under the "indirect supervision" of the Hidden Imam.

There is no doubt that the revenue from the *khums* at that time, as today, remained an important source of the *mujtahids*' power and prestige as those invested with *wilāyat al-hukm*. Moreover, it was the revenue from the *khums* that also made it possible for the *mujtahids* to remain independent of any direct control by the contemporary de facto ruler, the *sultān al-waqt*. Indeed, it was the financial independence of the *fuqahā'* that made it possible for the Shī'ī religious infrastructure to escape penetration by the ruling authority, whether in Iran or Iraq, where the Shī'ī centers of learning and an authoritative reference, *marja'iyya*, were located.

The ruling that the *mujtahid* can ensure the proper use of the *khums* in accordance with the wish of the Imam in occultation has given much wider authority to the jurist to interpret the will of the Imam. This has led to the opinions in modern times authorizing the spending of the Imam's share on troops and weapons engaged in the overthrow of unjust governments. This ruling carries with it the corollary that the Imam's share can also be used in defensive *jihād* against the enemies of Islamic public order.

It is reasonable to maintain that the administration of *al-khums*, especially of the Imam's share, went toward further consolidation of the leadership of the functional Imams of the Shī'a during the occultation of the twelfth Imam. The Shī'a who handed their *khums* to the *mujtahids* did so with the conviction that the *mujtahid* was truly the representative of the Imam (*nā'ib al-imām*), with full authority to dispose of the funds in accordance with the will of the Imam and under his "indirect supervision."

The practice of giving the *khums* to the well-qualified jurist also led to centralization of Imamite religious authority in the institution of authoritative reference, *marja' al-taqlīd*. This is the point that Hakīm's opinion is furthering when he lays down the qualification of *a'lamiyya* (being most learned) in addition to *'adāla* (righteousness) for the jurist who is permitted to administer the *khums*. This was a prerequisite in the matter of *taqlīd* (following the rulings of the *marja' al-taqlīd*), as we have seen in the treatment of Ansārī's discussion on the *wilāyat al-faqīh* in chapter 5, where the principle was to accept and follow the authority of the "most learned" *mujtahid*, who became the point of reference by the very fact of his being righteous and most learned.

One may even suggest that the gradual development of the present-day opinion regarding the *khums* and its administration by a *mujtahid*, as traced in this appendix, helps us to understand the overall evolution of *al-wilāyat al-'āmma* (the comprehensive authority) of a jurist during the occultation of the Imam. It is the chapters on *al-khums*, *al-jihād*, and *al-'amr bi al-ma'rūf wa nahy 'an al-munkar* in Imamite jurisprudence that set forth the role of the jurists in the Imamite community in such a way as to culminate in the "government of the jurist" in modern times.

GLOSSARY

'adāla   "righteousness," "moral probity." In Imamite jurisprudence 'adāla is a quality required of the transmitters (ruwāt, rijāl, faqīh, mujtahid [q.v.]) of the Imams' teachings.

al-'adl   the doctrine of divine Justice. God is infinitely removed from every evil act and from being remiss in anything considered obligatory.

ahl al-bayt   "the People of the House." In Shī'ī writings it includes the Prophet, Imam 'Alī, Fātima, al-Hasan, al-Husayn, and other Imams from among the descendants of al-Husayn.

ahl al-ijmā or ashāb al-ijmā'   "those who participated in reaching a consensus." In Shī'ī writings the phrase refers to some eighteen early associates of the Imams who took part in the ijmā' with the infallible Imam in reaching a legally authoritative consensus.

al-akhbār, al-āthār (pl. of khabar and athar)   traditions reported on the authority of the Prophet and the Imams. Sometimes hadīth is used as synonymous with khabar. But technically hadīth is a tradition related by the Prophet and the Imams, whereas khabar is a report related by an associate of the Prophet or the Imam. Athar is applied to the saying of the Prophet and his associates.

akhbārī   Imamite jurists who depended on the traditional methodology of understanding the text of the akhbār (q.v.) and discussed its implications within the limited bounds of the apparent sense of the text of the Prophetic communication.

'ālim   See faqīh.

al-'amr wa al-nahy   referring to the obligation of "enjoining [the good] and forbidding [the evil]" in applied jurisprudence.

asl (pl. usūl)   in the context of hadīth, referring to traditions that had been directly "heard" from the Imams.

al-dalīl al-'aqlī   "intellectual reasoning" and "demonstrative proof" used in support of the fundamental principles and their derivatives, supplementing al-dalīl al-sam'ī (q.v.), and independent of it. Usually the inference is from effect to cause.

al-dalīl al-sam'ī   "scriptural" and "traditional" proof used in explaining the fundamental principles of the faith (usūl al-dīn) and their derivatives (furū' al-dīn).

dalīl qat'ī   "absolute proof." In Imamite jurisprudence it refers to the proof that becomes authoritatively binding because it generates absolute certainty ('ilm) in the mind of a jurist who determines this certainty through rational enquiry.

dalīl zannī   "presumptive proof." In Imamite jurisprudence it refers to a proof that cannot become binding, because it lacks absolute certainty. As such, in itself it is invalid for use as a proof to deduce a legal injunction.

fatwā (pl. fatāwā)   legal, judicial decision deduced by a mujtahid (faqīh, jurist) after re-

246

searching the sources of jurisprudence—the Qur'ān, the Sunna, and the *ijmā'*— exerting one's mental faculties to the utmost.

fuqahā' (pl. of *faqīh*) = *rijāl, ruwāt* "jurisconsult," "jurist."

fiqh "knowledge of religious law" in Islam. In the earlier period the term also included the speculative side of the faith in addition to jurisprudence, and was used in the sense of comprehending religious sciences through independent exercise of one's well-informed judgment.

ghayba "occultation," "disappearance" of the twelfth Imam, with the understanding that there would be no Imam after him and that he would appear to establish the rule of justice and equity as the Just Ruler (*al-sultān al-'ādil*) (q.v.) of Islam, the Mahdi.

al-hākim "administrator of justice." In the context of Imamite jurisprudence, he is the one in whom *wilāyat al-hukm* (q.v.) has been invested. Accordingly, the title is given a much broader political signification of "guardianship" of the Imamite community during the occultation.

hākim al-shar' in Imamite jurisprudence, a title referring to the well-qualified Imamite jurist who can decide on legal matters and supervise the affairs of Muslims in the area of the application of the Sharī'a.

al-hawādith al-wāqi'a or simply *al-hawādith* "future contingencies," a phrase derived from the rescript of the twelfth Imam and used as a technical term in jurisprudence to refer to future difficulties that might be encountered by the Shī'a, for which jurists (*ruwāt, rijāl*) must be consulted. It is also known as one of the three fundamental "investiture traditions" that were extrapolated for the all-comprehensive authority of the jurists.

hudūd (sing. *hadd*) "legal" or "statutory punishments."

al-hukūma "administration of justice."

'ibādat the section of applied jurisprudence that deals with "religious observances"—that is, actions done by virtue of human-divine relationship, with the intention of drawing close to God (*al-qurbā*).

ijāza "license to teach" or "permission to transmit" Imamite religious sciences given by well-established jurists to their pupils. It also refers to the certification given to a qualified jurist to employ *ijtihād* (q.v.) for the purpose of deriving judicial decisions.

ijtihād "a lawyer's exerting the faculties (of the mind) to the utmost for the purpose of forming an opinion in a case of law (respecting a doubtful and different point)." It is an independent estimation of the jurist in a legal or theological question based on interpretation and application of the authoritative sources of Islamic law: the Qur'ān, the Sunna, and the *ijmā'*.

al-ijtihād al-shar'ī in Imamite jurisprudence, the independent estimation of a jurist to obtain "absolute proof" based on Islamic revelation (the Qur'ān and the Sunna of the Prophet and the Imams) for the purpose of deriving a legal injunction. As such, it rejects the *ijtihād* based on "presumptive proof" (*al-zann*), which is sometimes derived from inductive reasoning (*al-qiyās*).

'ilm al-hadīth the religious science employed to determine the authenticity of the *akhbār* reported on the authority of the Prophet and the Imams.

'ilm al-rijāl critical study of the transmitters (*al-ruwāt*) who figure in the chains of transmission (*isnād*) that are appended to *hadīth* reports to determine the level of authenticity of an informant, which directly affected the authenticity of a tradition reported by that person.

'isma "infallibility." It is a faculty of avoiding or shunning acts of disobedience (or of self-preservation therefrom), with (i.e., despite) the power to commit them. It also means infallibility in the total knowledge of the meaning of revelation and its perception.

khabar al-wāhid "Single-individual tradition" or "virtually unique tradition" not reported with frequency through various "ways" of transmission, regardless of the number of transmitters who have reported it. Such a tradition is regarded as *dalīl zannī* (presumptive proof) even when the report is related by a reliable authority, unless the report is elevated to the level of *dalīl qaṭʿī* (absolute proof).

al-khums   "the fifth." In the Imamite legal system, it is a form of taxation incumbent upon all Imamites—in addition to almsgiving (*al-zakāt*).

khulafāʾ al-jawr or *al-ẓalama*   "tyrannical or unjust caliphs." In Imāmī Shīʿism the phrase refers to those who prevented the Imams among the *ahl al-bayt* from assuming leadership of the Muslim community.

majlis   in Imamite piety, the "gathering" in which all believers participate to mourn the martyrdom of al-Husayn, the third Imamite Imam, who was killed in Karbalāʾ in the year 61/680.

makāsib   lit "earnings," a section in Imamite applied jurisprudence dealing with lawful and unlawful means of livelihood and gain. Also known as *matājir* ("trade").

maqbūla   "Approbated," "accepted" *hadīth* that has attained approbation through practice of the terms of the text and has been placed among the "sound" traditions.

marjaʿ al-taqlīd   the most learned juridical authority in the Imamite community. His rulings on the Sharīʿa are followed by those who acknowledge him as such, and commit themselves to base their religious practice in accordance with his *fatwā*s (*taqlīd*).

al-masālih al-ʿāmma   the "general welfare" of the Muslim community. It has served as a source for legal injunctions in jurisprudence not based on purely textual evidence but on rational inference.

muʿāmalāt   the section of applied jurisprudence that deals with the affairs of this world—that is, person-to-person relationships. As such it includes *al-ʿuqūd* (contracts), *al-ʾīqāʿāt* (implementation of judgments), and *al-ahkām* (giving of judgments).

mujtahid (= *faqīh*)   a "jurist" who applies *ijtihād* (q.v.) to deduce laws.

muqallid   in Imamite jurisprudence, a believer who has committed himself to act in accordance with the rulings of a *mujtahid* in whom he has put his trust, even if he does not act in accordance with those rulings subsequently. This is the meaning of *taqlīd*.

mursal   a kind of *hadīth* in which proper *isnād* (chain of transmission) is not considered an absolute necessity.

musnad   a kind of *hadīth* in which *isnād* (chain of transmission) is incorporated.

al-mutaʾakhkhirūn   the "modern" Imamite jurists, from ʿAllāma Hillī (d. 726/1325) to the modern period. Sometimes Imamite jurists speak of *mutaʾakhkhiru al-mutaʾakhkhirīn*, meaning "modern of the modern"—that is, the Qājār and post-Qājār jurists.

al-mutaqaddimūn   the "ancient" Imamite jurists, from the beginning of Imamite jurisprudence until the death of Muhaqqiq al-Hillī in 676/1277.

al-naṣṣ al-sarīh   "explicit" or "clear designation" of the Imam to the position of the Imamate, successorship from the Prophet.

al-niyāba (= *al-wikāla*)   "deputyship."

al-niyābat al-ʿāmma   "general deputyship" of the twelfth Imam during the Complete Occultation (from 941 A.D. onward), on the basis of the rescript received from him regarding *al-hawādith al-wāqiʿa* (q.v.).

al-niyābat al-khāssa   "special deputyship" of the twelfth Imam during the Short Occultation, which terminated in the year 941.

rijāl (pl. of *rajul*) (= *fuqahāʾ*)   literally, "a man"; technically, "prominent personage" who transmitted teachings of the Imams.

risāla   in the Imamite context, a book of reference issued by *marjaʿ al-taqlīd* for the religious guidance of his *muqallidūn*. It consists of his legal rulings in all the sections of applied *fiqh*.

al-ruwāt   "transmitters of the traditions." In Imamite usage transmitters of the Imamite teachings who were jurists (*fuqahāʾ*) of the Shīʿa.

saltana or *qudra*   "power" in the exercise of moral and legal authority by the demand of obedience.

shahāda "martyrdom" in the context of the Shī'ī Imams' struggle for justice and truth against oppression and falsehood.

sultān the ruling authority of the community in whom power to exact or enforce obedience is invested.

sultān al-zamān "the ruler of the age," used in reference to both "just" and "unjust" rulers. He is also known as *sultān al-waqt li al-ri'āsa* (the ruling authority of the time to head the community).

sultān al-islām "the ruler of Islam." The title refers to the twelfth Imam, especially when followed by the phrase *al-mansūs min qibal allāh* (designated by God).

al-sultān al-'ādil or *al-sultān al-haqq* "the just authority" designated by God. it could also refer to a just Imamite authority exercising political power during the occultation.

al-sultān al-jā'ir "unjust ruler," usurper of the rightful Imam's authority.

taklīf "imposition of a task, obligation." It can also mean "religious obligations" imposed on a believer, the fulfillment of which carries great reward in the future life.

al-takālīf al-shar'iyya (pl. of *taklīf*) obligations imposed on believers by the Sharī'a.

taqiyya technically, "precautionary dissimulation"; "dispensation from the requirements of religion under compulsion or threat of injury." Imamites also use it in the meaning of "protection of the true religion from enemies by hiding it, in circumstances where there is fear of being killed or captured or insulted."

wājib bi'l-kifāya "representatively incumbent obligation." It is fulfilled by one or more in the community on behalf of everyone else.

wājib al-'aynī "personally and individually incumbent obligation." It must be fulfilled by everyone legally competent.

al-walī the person who possesses authority (*wilāya*).

al-walī al-mutlaq "the absolute authority." It refers to God and, in some particular contexts, to the Prophet and the Imams.

al-walī bi al-niyāba the person who possesses authority through "deputization."

walī al-'amr the person in whom authority to manage affairs has been invested. As such, he is part of the *ulū al-'amr,* with the right to dispose of property and make decisions in the interests of the community.

walī al-hukm the jurist in whom the authority to administer justice (*al-hukm*) is invested.

wilāya "Authority." Its possession enables a person to assume a position of responsibility and confers on him the right to demand obedience depending on legal-rational circumstances.

al-wilāyat al-'āmma the all-comprehensive authority of the Imamite jurist, which includes *wilāyat al-tasarruf* (q.v.).

wilāyat al-hukm posssession of the authority to judge, on the basis of divine revelational sources, matters pertaining to public welfare.

al-wilāyat al-ilāhiyya "the divine sovereignty."

al-wilāyat al-i'tibāriyya "relative authority," also known as *al-wilāyat al-tashrī'iyya,* "religious-legal authority," invested in the Prophet to undertake the legislation and execution of the divine plan on earth.

wilāyat al-nizām the authority to manage public order.

wilāyat al-nusra the authority of "backing" the religion of God by promulgating God's laws.

wilāyat al-qadā', wilāyat al-hukm the authority to administer justice.

wilāyat al-siyasa the authority to administer government and hold political office.

wilāyat al-takwīnī the authority of God, which "originates" in the godhead. It is absolute, unconditional, and all-encompassing authority.

wilāyat al-tasarruf possession of authority that entitles the *walī* to act in whatever way he judges best, according to his own discretion, as a free agent in the management of the affairs of the community.

al-'urf "ordinary language"—the conventional and customary sense of terms used in social transactions. It is regarded as one of the important sources in deriving judicial injunctions in the Imamite understanding of *al-ijtihād al-sharʿī* (q.v.).

ulū al-ʾamr those in possession of authority to manage the affiars of the people.

al-ʿulūm al-naqliyya "transmitted religious sciences," also known as *al-manqūlāt.*

usūl al-fiqh "principles of jurisprudence" or "theoretical basis of Islamic law." These principles include "traditional" sources, such as the Qurʾān, the Sunna, and *ijmāʿ,* as well as rational sources based on *al-ʿaql* ("reason, intellect, intelligence").

usūlī those Imamite jurists who depend on the *usūl al-fiqh.* Their work includes the utmost use of faculties of the mind to derive a legal decision inferrentially.

ziyāra "visitation" to the shrines of the Shīʿī Imams (*al-mashhad*), in Medina, Karbalāʾ, Najaf, Sāmarra, Kāzimiyya, Mashhad, and elsewhere.

NOTES

Introduction

1. Joseph Schacht, "Islamic Religious Law," in *The Legacy of Islam*, ed. Joseph Schacht with C. E. Bosworth, 2nd ed. (Oxford, 1974), has pointed out this characteristic of Islamic law, which did not grow out of preexisting law, nor did its formation take place under the impetus of the needs of practice or of juridical technique. It came into being and developed by private "pious specialists" against a varied political and administrative background. It is for this reason that Schacht identifies Islamic law as "jurists' law." In the first decades of the 2nd/8th century it developed into ancient schools of law, and derives from private individual jurists who in their rulings reflect the social realities of the various parts of Islamic empire. See also his essay "Law and Justice," in *The Cambridge History of Islam*, ed. P. M. Holt, A. K. S. Lambton, and B. Lewis (Cambridge, 1970), vol. 2B, pp. 539–68.

2. In this regard see the valuable contribution of Ignaz Goldziher, *Muslim Studies*, vol. 2, ed. S. M. Stern (London, 1971). In his discussion of the early development of the *hadīth* and the ancient schools of law, Goldziher has shown the individualism of the early schools of Islamic law and the way early scholars individually contributed to the legitimation of the authority in power.

3. In his study entitled *The Sectarian Milieu: Content and Composition of Islamic Salvation History* (Oxford, 1978), John Wansborough has discussed the development of the practical application of the Qur'ān during the process of community formation. In the chapter dealing with authority (which he defines as "the immediate and tangible instruments of legitimation: those means by which the sanctions of a transcendent diety are realized in practice, those terms within which a theodicy becomes credible and workable"), he takes up the role of the Qur'ān in the process of regulating the Muslim community, and concludes with due intellectual caution and evidence from classical sources that in the development of the early Sharī'a, especially in the derivation of juridical decisions, material was drawn from an equally authoritative source outside the Qur'ān—namely, the Prophetic Sunna.

4. Historians, sociologists, and political scientists have undertaken to explain the development of religious and political factors in the history of Iran, which culminated in the political dominance of the Shī'ī jurists in 1978–79. See, for instance, Nikki R. Keddie, *Religion and Rebellion in Iran: The Tobacco Protest of 1891–1892* (London, 1966); Hamid Algar, *Religion and State in Iran 1785–1906: The Role of the Ulama in the Qajar Period* (Berkeley, 1969); Shahrokh Akhavi, *Religion and Politics in Contemporary Iran* (Albany, 1980); Ervand Abrahamian, *Iran Between Two Revolutions* (Princeton, 1982); Shaul Bakhash, *The Reign of the Ayatollahs* (New York: 1984); Said Amir Arjomand, *The Shadow*

of God and the Hidden Imam (Chicago, 1984); Roy Mottahedeh, *The Mantle of the Prophet: Religion and Politics in Iran* (New York, 1985). These are a few among many works that provide a wealth of information on the socio-political and to some extent religious factors that led to the 1978–79 events in Iran. Arjomand's *The Shadow of God* presents us with a substantial sociological study of religion and politics in Shīʿite Iran. He has read his sources, including some theological and juridical ones, through the conceptualizations created by Weber's sociological study of world religion and the legitimation of traditional authority. But in the present work I have sought to explain the development of Shīʿī authority during the occultation of the twelfth Imam by allowing the juridical sources to explain the development of Shīʿī Muslim authority. I believe that Weber's basic model in the interpretation of a traditional authority is far too inflexible to permit analysis of the complex emergence of authority in medieval theologico-juridical Muslim sources. Ideally, as some sociological studies have shown, Weber's analyses are suited to the empirical study of contemporary Islam. My approach is inspired by observations made by Joseph Schacht regarding Islamic law in general—namely, that juridical sources are our most important sources for the investigation of Islamic society. See his aforementioned essay in *The Cambridge History of Islam*, vol. 2B.

5. Tūsī, *Fihrist*, p. 160; Bihbahānī, *Sharh*, vol. 1, fol. 62–63.

6. Najāshī, *Rijāl*, p. 273.

7. Cited by Qummī, *Kunā*, 1/26–27.

8. Bihbahānī, *Sharh*, vol. 1, fol. 62–63. It was also this kind of *ijtihād* based on *qiyās* that was criticized by the Akhbārī jurist Muhammad Amīn al-Astarābādī in his *Fawāʾid al-madaniyya fī radd man qāla bi al-ijtihād*. See Khwānsārī, *Rawdāt*, 1/197ff., where he discusses the contents of *Fawāʾid*.

9. *Rijāl al-sayyid*, 2/211ff.

10. Ibid. Even he is criticized in the Akhbārī work *Fawāʾid al-madaniyya* as practicing *qiyās*. See *Rawdāt*, 1/178.

11. ʿAllāma, *Muntahā*, pp. 3–4.

12. Ibid.

13. See below, chap. 3.

14. ʿAllāma, *Rijāl*, p. 28; *Rijāl al-sayyid*, 2/131–34.

15. Halabī, *Kāfī*, fol. 1–2.

16. Al-Tabaristānī is mentioned by Suyūtī, *Bughyat al-wiʿā*, p. 259. In Imamite works he has been known simply as "Sallār" or "al-Daylamī."

17. See below, chap. 5. For biographical information on Sallār, see Najāshī, *Rijāl*, p. 206; Qummī, *Kunā*, 2/228; ʿAllāma, *Rijāl*, 86; *Rijāl al-sayyid*, 3/6–17.

18. This is mentioned by Khwānsārī, *Rawdāt*, 3/191.

19. *Maʿālim*, p. 80, item #545.

20. *Rijāl al-sayyid*, 3/60–63; Qummī, *Kunā*, 1/224

21. Qummī, *Kunā*, 1/267.

22. Khwānsārī, *Rawdāt*, 7/190.

23. Qummī, *Kunā*, 1/267.

24. For a brief biographical note on Ibn Idrīs, see Qummī, *Kūnā* 1/210; Ibn Hajar, *Lisān al-mīzān*, 5/65.

25. Khwānsārī, *Rawdāt*, 2/443.

26. Tūstarī, *Qāmūs al-rijāl*, 2/378.

27. ʿĀmilī, *Aʿyān al-shīʿa*, 16/228.

28. Tihrānī, *Dharīʿa*, 13/47–48, 316–32; he also lists the addenda to *Sharāʾiʿ* (marginal glosses) in 4/108 and 6/106–9, respectively.

29. Article "Ilkhans," in *Encyclopedia of Islam*, 2nd ed. (EI²), 5/1122.

30. Ibn Hajar, *Durar*, 2/49; *Lisān al-mīzān*, 2/317. Also *Rijāl al-sayyid*, 2/257–94.

31. Khwānsārī, *Rawdāt*, 7/258. A historical sketch of Shahīd I (and Shahīd II) appears in the Najaf edition of *Lum'a* (ed. Madhī al-'Āsifī). See also Qummī, *Kunā*, 3/61.

32. 'Āsifī's introduction to *Lum'a*, 1/131–32

33. Majlisī, *Bihār, Kitāb al-ijāzāt*, 24/45. Ibn Fahd had also "permission" from 'Allāma's son, Fakhr al-Muhaqqiqīn, to transmit the works of his father.

34. *Rijāl al-sayyid*, 2/107–11. Tihrānī, *Dharī'a*, 2/21, mentions the *risāla* as *Istikhrāj al-hawādith* among Ibn Fahd's writings, which indicates Ibn Fahd's leanings toward Sufism and asceticism.

35. Rawdatī, *Jāmi' al-ansāb*, 1/123.

36. The manuscript at the Astane Library in Mashhad (#8033) mentions a variant title: *al-Muhadhdhab al-bāri' fī shar' mukhtasar al-sharā'i'*. This seems to be erroneous, for Muhaqqiq has not left us an abridgment of his *Sharā'i';* but he has a work entitled *al-Mukhtasar al-nāfi'*.

37. In his analysis of the Safavid administrative system, R. M. Savory makes this observation in his article "The Principal Offices of the Safavid State During the Reign of Tahmāsp I (930–84/1524–76)," *BSOAS*, 24 (1961) p. 81.

38. Khwānsārī, *Rawdāt*, 5/170.

39. 'Āsifī's introduction to *Lum'a*, 1/166–67

40. See W. Madelung, *"Akhbāriyya,"* in *EI²*, 5. For critique of the traditions transmitted by Ibn Bābūya, see Mufīd, *al-Masā'il al-sarawiyya*, pp. 55–56. For other socio-political factors leading to the Akhbārī-Usūlī controversy in Imāmī Shī'īsm, see Algar, *Religion and State*, pp. 33–36. See also D. M. MacEoin, "From Shaykhism to Bābism: A Study in Charismatic Renewal in Shī'ī Islam," Ph.D. dissertation submitted to Cambridge University, 1979, Introduction, pp. 25ff.

41. H. Algar, "Bihbahānī," in *EI²* , 5:134–35.

42. Algar, *Religion and State*, p. 34.

43. Majlisī, *'Ayn al-hayāt*, 560–7. For the Sunnī attitude to political authority in the medieval period, see A. K. S. Lambton, *State and Government in Medieval Islam* (Oxford, 1981), pp. 108–29, 138–51.

44. Algar, *Religion and State*, p. 34.

45. Ibid., pp. 53–54. See the article "Kāshif al-Ghitā'," in *EI²* , 4/703.

46. See below, chap. 3.

47. Narāqī's works *'Awā'id al-ayyām* and *Mustanad al-shī'a fī ahkām al-sharī'a*, and Shaykh 'Alī's *al-Nūr al-sātī' fī fiqh al-nāfi'* (which has one of the most detailed studies on the question of *taqlīd* in Imāmī Shī'īsm), have not enjoyed the popularity of *Jawāhir* among scholars. *Jawāhir* has become a sort of reference work for jurists in modern times.

48. Muhammad Taqī al-Hakīm, *al-Usūl al-'āmma li al-fiqh al-muqārin* (Beirut, 1983), pp. 419–26, discusses the historical development of the concept of *al-'urf*, identifying different kinds of *'urf* through various definitions in the works on *usūl al-fiqh*.

49. Ansārī, *Makāsib*, p. 154.

50. Muhammad Hasan Shīrāzī's role in the repeal of the Tobacco Concession to the British in 1890 and Ākhund's role in the constitutionalism of 1906 are well known to the student of modern Iran. See, for instance, Algar, *Religion and State*, pp. 205–21.

51. See Algar, *Religion and State*, pp. 251–54. See also Said Amir Arjomand, "Religion and Ideology in the Constitutional Revolution," review article in *Iranian Studies*, vol. 12, nos. 3–4 (1979) pp. 283–91.

52. See Abdul-hadi Hairi, *Shi'ism and Constitutionalism in Iran* (Leiden, 1977), for a discussion on the problems of legitimizing constitutionalism, the dispute among the jurists, and the question of legitimacy (especially chap. 6).

53. Their opinions on modern issues need to be compiled and commented upon. Some work in this direction has been undertaken in recent years as evidenced by the following

publications of the responsa of the major figures of Imamite jurisprudence in modern times:

Abū al-Qāsim al-Khū'i, *al-Masā'il al-mustahdatha.*
Rūh Allāh al-Mūsawī al-Khumaynī, *Tahrīr al-wasīla.*
Muhammad Sādiq al-Rūhanī, *al-Masā'il al-mustahdatha.*

These works are not usually part of their *risāla*s, written for the religious guidance of their followers, or in their commentaries upon classical works of *fiqh.* The *risāla*s of these scholars still retain the classical opinions in simplified form, without any reference to the change of circumstance in the sections that need revision. On the other hand, in their responsa to the questions about modern issues, one can discern their application of the methodology of the *usūl* to deduce new judicial decisions. Such is the case with women's rights in the modern world within the limits recognized by Islam. These rulings are not part of the *risāla*s. Accordingly, studies by some social scientists on these issues should not be carried on without understanding the way the responsa are given outside the usual published works of the jurists.

Among the published works of modern jurists who have followed the methodology of Ansārī and Ākhund al-Khurāsānī, the following are in the tradition of the classical works in jurisprudence:

1. Abū al-Hasan al-Isfahānī's *Nihāyat al-dirāya* and *al-Usūl 'ala nahj al-hadīth* deal with the *'ilm al-hadīth*, whereas *Hāshiya 'alā al-makāsib* of Ansārī deals with applied *fiqh.*

2. Husayn al-Burūjirdī's *al-Badr al-zāhir fī salat al-jum'a wa al-musāfir* deals with his lectures on the Friday and traveler's prayer, including the question of juridical authority in the Imamite community, annotated and documented by the Āyatullāh al-Muntazarī.

3. Muhsin al-Hakīm's *Mustamsak al-'urwat al-wuthqā* is a detailed, demonstrative commentary on Muhammad Kāzim al-Yazdī's *'Urwat al-wuthqā.*

4. Muhammad Hādī al-Mīlānī's *Muhādarāt fī fiqh al-imāmiyya* consists of lectures on all chapters of *fiqh,* annotated and thoroughly documented by his grandson Sayyid Fādil al-Mīlānī.

5. Abū al-Qāsim al-Khū'ī's *Minhāj al-sālihīn* and *Tanqīh* deal with jurisprudence; *Mu'jam al-rijāl al-hadīth* has to do with eminent personages who have appeared in the chains of transmission in the *hadīth.*

6. Rūh Allāh al-Mūsawī al-Khumaynī's *al-Makāsib al-muharrama* and *Kitāb al-bay'* treat of the *mu'āmalāt* section of jurisprudence.

Chapter 1

1. 'Āmilī, *Wasā'il,* 11/483, *Hadīth* #2. However, the text has been interpreted to exclude any other "learned authority" than the Imams themselves.

2. Kashshī, *Rijāl,* 4/5 (the first number refers to the page and the second to the item in the 1348/1964 edition of *Rijāl* edited by Mustafawī). See also *Wasā'il* 11/510, Hadīth #1.

3. 'Āmilī, *Wasā'il,* 11/510, *Hadīth* #1. A variant of the tradition reads: "If he does not do so, the light of the faith will be taken away [from him]." *'Ilm* (knowledge) during the early period of Imamite religious history was confined to the study of Islamic religious law and it was expected that a person claiming any religious authority had to demonstrate competence in discussing its derivation from Islamic revelation. Even the Imams among the *ahl al-bayt* were expected to demonstrate their thorough comprehension of the *shar'ī* laws. The founders of the *shar'ī* schools were tested in their legal expertise, as in the case of Imam al-Sādiq who was invited by the 'Abbāsid al-Mansūr from Medina and was confronted by Abū Hanīfa on *fiqh*-related questions. See Qummī, *Muntahā,* 2/124–25, who cites the event on the authority

of Ibn Shahrāshūb. Kashshī, *Rijāl*, 11/23, mentions the knowledge (ʿ*ilm*) given to the early disciples of the Imams. For relevant discussion on the role of *sunna* in the formation of Islamic religious-legal life, see Joseph Schacht, *Origins of Muhammadan Jurisprudence* (Oxford, 1950), pp. 58–81; and, for correction of some of Schacht's views and new interpretation of the concept of *sunna*, Fazlur Rahman, *Islamic Methodology in History* (Karachi, 1965), pp. 1–84.

4. "Apostolic succession" is the term used by Wansborough, *Sectarian Milieu*, p. 81, in his structural analysis of Islamic authority and the particular role of precedent, which constituted the only valid basis for prescription in Islam.

5. See n. 3, above.

6. Kashshī, *Rijāl*, 209/369.

7. Ibid., 4, *Hadīth*, #6.

8. Ibid., 12/25; 16/37. The *hadīth* that "Salmān belongs to the *ahl al-bayt*" is one of the well-known traditions among Muslim traditionists. But it is in the Shīʿī biographical dictionaries that esoteric knowledge is attributed to Salmān, as the two reports in Kashshī indicate. Kashshī also reports the *hadīth* related by ʿAlī in which the Imam informs Abū Dharr that Salmān is the gate of God on earth. "The one who acknowledges him is a believer; and the one who rejects him is a nonbeliever. Indeed, Salmān belongs to us, the *ahl al-bayt*" (15/33). Imam Muhammad al-Bāqir instructed his followers to address Salmān al-Fārisī as Salmān al-Muhammadī—i.e., "belonging to the Prophet" (Kashshī, *Rijāl*, 12/26). Imam al-Sādiq was heard by one of his prominent associates, Abū Basīr, as saying: "Salmān was taught *al-ism al-aʿzam*" (the Greatest Name [of God]), that was supposed to empower its possessor with extraordinary power in conceiving hidden matters (Kashshī, *Rijāl*, 13/29). In another place, al-Sādiq says that Salmān possessed both "ancient" and "modern" learning. Kashshī also mentions ʿUmar b. Yazīd Bayyāʿ al-Sābirī and ʿIsā b. ʿAbd Allāh, who were declared by al-Sādiq as belonging to the *ahl al-bayt*, with similar implications (ibid., 331/605 and 607).

9. Kashshī, *Rijāl*, 6/12 and 15.

10. Ibid., 123/194; 115/184. In the forthcoming work, *Human Rights and the Conflict of Cultures: Western and Islamic Perspectives on Religious Liberty*, (Columbia, S.C., 1988) co-authored by David Little and John Kelsay, I have argued that "apostasy" in Islam is not purely a religious violation; rather, it is a moral violation, which then requires the Islamic political authority to punish it.

11. This difference between al-Bāqir and al-Sādiq's time and the period of the previous Imams is evident in the topics discussed by their disciples. See Kashshī, *Rijāl*, 138–40/221.

12. Ibid., 148/234, mentions Zayd b. Abī Hilāl's discussion with Imam al-Sādiq on the question of *istitāʿa* and the dispute that had occurred among the disciples of the Imam in Kūfa. Also, at 144/226, Kashshī reports *fiqh* questions pertaining to ritual performances being asked of the Imam.

13. Kashshī, *Rijāl*, 6/11.

14. ʿĀmilī, *Wasāʾil*, 11/519, *Hadīth* #7.

15. Kashshī, *Rijāl*, 119/189; Ibn Saʿd, *Tabaqāt*, 5/121–43; Masʿūdī, *Murūj*, 3/193f.

16. Kashshī, *Rijāl*, 3/1.

17. Ibid., 3/2. See also E. W. Lane, *Arabic-English Lexicon*, bk. 1, part 2, p. 529, for a lexical explanation of the term *muhaddath*, which conforms with the signification of this term in Imamite *hadīth* in Kulaynī, *Kāfī*, *kitāb al-hujja*, *bāb*, 54, *Hadīth* #3.

18. Wansborough, *Sectarian Milieu*, pp. 88ff., alludes to this fundamental difference between the Sunnī and Shīʿī soteriology.

19. Kashshī, *Rijāl*, 76/131 and 84/139.

20. Ibid. In a note on Maytham al-Tammār (767/131), Kashshī reports that he claimed such esoteric knowledge and he was put to death. Similar treatment was suffered by Rushayd al-Hujrī, who, according to Kashshī, was taught ʿ*ilm al-balāya* by ʿAlī, and was known as Rushayd al-Balāya.

21. Ibid., 76/131; see also 84/139.

22. Ibid., 224–25/401–2.

23. Ibid., 324/588.

24. Abū al-Faraj, *Maqātil al-tālibiyyin*, 196; see also Tabarsī, *Ihtijāj*, 2/118; Majlisī, *Bihār*, 100/21.

25. Kashshī, *Rijāl*, 161/273. In another place Kashshī reports a similar request made by another disciple and the Imam referred him to Abū Basīr al-Murādī al-Asadī, 171/291. See also Qummī, *Muntahā*, 2/178.

26. Kashshī, *Rijāl*, 170/286.

27. Ibid., 224–25/401.

28. Ibid., 290f/511f. See also my article, "Abū al-Kattāb," in *Encyclopaedia Iranica*, 1/329–30.

29. Kashshī, *Rijāl*, 224–25/401.

30. Ibid., 290f./511f.

31. Ibid., 291–92/515; Nawbakhtī, *Firaq*, 38.

32. Nawbakhtī, *Firaq*, pp. 59ff.

33. Kashshī, *Rijāl*, 291f./551.

34. Ibid., 169/283, where al-Sādiq says: *halaka al-mutara'isūn fī adāynihim.* . . .

35. Ibid., 292–93/516.

36. Ibid., 132f.

37. Ibid., 133/210.

38. Ibid., 224/401.

39. See my preliminary study on Kashshī's *Rijāl*, "The Significance of Kashshī's *Rijāl* in Understanding the Early Role of the Shī'ite *Fuqahā'*," in *Logos Islamikos: Studia Islamica in honorem Georgii Michaelis Wickens*, ed. Roger M. Savory and Dionisius A. Agius (Toronto, 1984), pp. 183–206.

40. Kashshī, *Rijāl*, 331–32/406.

41. This point is well discussed by the *marja' al-taqlīd* of the majority of the Imamite Shī'ites, Āyat Allāh Abū al-Qāsim al-Mūsawī al-Khū'ī, in his lectures on *usūl al-fiqh* entitled *al-Tanqīh fī sharh al-'urwat al-wuthqā*, compiled by al-Mīrza 'Alī al-Gharawī al-Tabrīzī, 4 vols. (Najaf, 1386/1966–67), esp. vol. 1. In his main work on *Rijāl—Mu'jam al-rijāl al-hadīth*, 18 vols. (Najaf, 1970)—he summarizes his position on the *sunna* as a source in the *usūl al-fiqh* in the first volume, which constitutes an introduction to his work.

42. These are (1) *al-Kāfī fī 'ilm al-dīn*, by al-Kulaynī; (2) *Man lā yahduruh al-faqīh*, by Ibn Bābūya; (3) *Tahdhīb al-ahkām; and (4) al-Istibsār fī mā ikhtalafa min al-akhbār*, by Tūsī. For details on these books, see chap. 2, below.

43. Najāshī, *Rijāl*, p. 263. See also Scarcia Amoretti, "'Ilm al-ridjāl," *EI²*, 3/1150–52, and W. Madelung, "Al-Kashshī," *EI²*, 4/711–12.

44. Kashshī, *Rijāl*, 133/210.

45. Ibid., 238/431; 373/705; and 556/1050. In all, Kashshī counts eighteen among the prominent disciples of the Imams who formed the *ashāb al-ijmā'*.

46. Ibid., 138–39/221.

47. Ibid., 160/269.

48. Ibid., 210/370. According to Najāshī, *Rijāl*, p. 255, he was a Zaydī.

49. Kashshī, *Rijāl*, 158/262.

50. Ibid., 147/234.

51. Ibid., 145/231.

52. Ibid., 146/232.

53. Ibid., 156/257.

54. Ibid., 161/271.

55. Tūsī, *Fihrist*, p. 100.

56. Kashshī, *Rijāl*, 169/283.

57. Ibid., 162–63/275.
58. Ibid., 169/284.
59. Ibid., 163/276.
60. Ibid., 164/277.
61. Tūsī, *Fihrist*, pp. 72ff.; Qummī, *Kunā*, 2:96.
62. Mufīd, *Irshād*, p. 560.
63. Ibid., pp. 565–67.
64. See Appendix.
65. Kashshī, *Rijāl*, 459/871.
66. Sachedina, *Messianism*, pp. 78–108.
67. Al-Shashī al-Qaffāl, *Hilyat al-'ulamā'*, 8 vols. (Beirut, 1980), ed. Yāsin Ahmad Ibrāhim Darādka, 1/53.

Chapter 2

1. Sachedina, *Messianism*, pp. 86–99.
2. Suyūtī, *Tanwīr al-hawālik*, p. 6, mentions 'Umar among the early companions who had desired to put the *sunna* in writing. To that end he consulted other companions, who advised him to undertake the task. However, after a month, he decided against it, saying that the People of the Book had done the same thing and in the process they had neglected the Book of God and instead concentrated on nonscriptural writings.
3. Najāshī, *Rijāl*, p. 255. Al-Hakam b. 'Utayba was a Zaydī, belonging to the subdivision of Butriyya, or a Murji'ite, and a teacher of Zurāra b. A'yan. See Kashshī, *Rijāl*, 209–10/368–70.
4. Ibn Hajar, *Taqrīb*, 2/241; Najāshī, *Rijāl*, p. 3.
5. Tihrānī, *Dharī'a*, 1/17, citing al-Dhahabī's biographical note on Abān b. Taghlib.
6. These two works did not survive for posterity except in the form of disparate references to them by subsequent authors of biographical dictionaries. See Tūsī, *Fihrist*, p. 24; idem, *Rijāl*, p. 471. Tihrānī, *Dharī'a*, vol. 2, mentions 117 works on *asl*, with their authors.
7. Tūsī's Introduction to his *Rijāl*. However, Tūsī was not consistent in his classification. According to Khū'ī (present-day scholar of *rijāl*), a transmitter's name often occurs among those who reported directly from the Imams and at the same time the same name occurs among those who reported mediately. See Khū'ī, *Mu'jam*, 1/115–16.
8. Some of these technical phrases are explained in Shahīd Thānī, *Dirāya*, pp. 8ff, and others are explained by Abū Bakr Ahmad b. 'Ali al-Khatīb al-Baghdādī, *al-Kifāya fī 'ilm al-riwāya* (Hyderabad, Deccan, 1357/1938), p. 294.
9. See above, pp. 40ff. Kashshī, *Rijāl*, 224–25/401, and in other places, records many instances where the Imams who followed al-Sādiq had to correct misquoted or interpolated traditions. Thus, 'Alī al-Ridā had to defend the traditions of all the previous Imams, especially al-Sādiq, as being free of any interpolations. Furthermore, he declared that there was absolutely no disagreement between the communications of the Imams, which pointed to the controversies surrounding variants attributed to the Imams al-Bāqir and al-Sādiq.
10. Sachedina, *Messianism*, p. 98.
11. Tūsī regarded some of them as *thiqa* (reliable) because, he says, even when these transmitters became Fathite or Zaydī, and so on, their transmission was reliable. See his admission, in *Tahdhīb*, 7/101, of 'Ammār b. Mūsā al-Sābātī as an authentic transmitter, despite the fact of his having become a Fathite.
12. Kashshī, *Rijāl*, Index, pp. 4ff.
13. Tūsī, *Fihrist*, p. 161, mentions thirty. The present edition has thirty-five.
14. Kulaynī, *Kāfī*, 1/14.

15. Tūsī, *Fihrist*, p. 119.

16. Several individuals are mentioned as forming the *ahl al-ijmā'*, which was regarded as a prestigious position in the religious sphere and afforded these persons thorough reliability in the eyes of the community at large. See Qummī, *Muntahā*, 2/175.

17. *fāsiq*, "unrighteous." Mostly applied to a person who has acknowledged the authority of the Sharī'a, and has pledged to observe what it prescribes, and then fallen short of observance in respect of all, or some, of its ordinances (Lane, *Lexicon*, book I/6:2398).

18. Tūsī, *Talkhīs*, 2/245.

19. Najāshī, *Rijāl*, p. 185.

20. Ibid.; Tūsī, *Fihrist*, p. 119.

21. Tūsī, *Fihrist*, p. 184.

22. Tūsī, *Rijāl*, p. 482. Najāshī, *Rijāl*, p. 185, mentions that al-Samarrī informed the associates who had gathered in his home on that day that the death of Ibn Bābūya's father had occurred on that very day. This was taken to be a miraculous sign performed by the last agent of the twelfth Imam.

23. Najāshī, *Rijāl*, pp. 276–79.

24. Tūsī, *Fihrist*, p. 119, lists all the major works composed by Ibn Bābūya and his father, which had been collected by Mufīd, who then transmitted them to Tūsī.

25. Among his teachers and informants are persons with the surnames Sughdī, Farāghānī, Rūyānī, and so on—names derived from cities or regions in Khurāsān. There is also a person with the surname Sarandībī. Marw, Balkh, and Nishapur also appear in his *isnād*s.

26. Tūsī, *Ghayba*, p. 209.

27. Introduction to Ibn Bābūya, *Man lā yahduruh*, p. *adh*, citing al-Mīrza Husayn al-Nūrī, *Mustadrak wasā'il al-shī'a*.

28. Ibn Bābūya, *Man la yahduruh* 1/6. *Tayammum* is the ritual ablution performed with sand instead of water in certain cases.

29. Kulaynī, *Furū'*, 1/2.

30. Ibn al-Nadīm, *Fihrist*, p. 175.

31. Ibid., pp. 63–64.

32. Tūsī, *Tahdhīb*, 1/8–9.

33. Tūsī, *Fihrist*, p. 186.

34. For Tūsī's biography see Sachedina, *Messianism*, pp. 35–38.

35. Karājikī, *Kanz*, p. 186, mentions the fact of its being an abridgment when he says: "I excerpted it for our brothers from our Shaykh Mufīd's book."

36. Tūsī, *Fihrist*, p. 160; see also Khwānsārī, *Rawdāt*, pp. 168 and 561.

37. W. Madelung, "Imamism and Mu'tazilite Theology," in *Le Shī'īsme Imamite* (Strasbourg, 1970), p. 25.

38. Ibn Idrīs, *Sarā'ir*, p. 2, takes up Sharīf al-Murtadā's view on *khabar al-wāhid*.

39. Tūsī, *'Udda*, p. 51. Muhaqqiq al-Hillī, in his *Ma'ārij al-usūl*, p. 82, has elaborated on Tūsī's intent in this particular opinion.

40. A number of present-day Imamite jurists have regarded Ibn Idrīs's critique of Tūsī's works and his methodology as "liberating" Imamite jurisprudence, not allowing it to stagnate. See, for instance, Hakīm, *al-Usūl al-'āmma*, p. 600, where he discusses the reasons that led to the "closing" of the "gate of *ijtihād*" among the Sunnites, and contrasts it with Shī'ī scholars' attitude to *ijtihād* among whom Ibn Idrīs's contribution is highlighted as returning the confidence to the Shī'ī scholars to undertake investigation of the materials independent of the established works of jurisprudence.

41. Tūsī, *Tahdhīb*, 1/2ff.

42. Tūsī's preface to *Tahdhīb*, where he discusses his methodology.

43. Najāshī, *Rijāl*, pp. 2–3.

44. Bahrānī, in *Lu'lu' al-bahrayn*, p. 297, cites Ibn Idrīs's opinion and he vindicates

Tūsī's method in *Nihāya*. Bahrānī, who was an Akhbārī jurist, regarded *Nihāya* as the product of Akhbārī methodology.

45. *Sarā'ir*, p. 116.

46. Tūsī's Introduction to *Mabsūt*, p. 3.

47. Ibid.

48. Tihrānī, *Dharī'a*, 20/370. This book is also known as *al-Masā'il al-tabariyyāt*, because al-Nāsir wrote the original rulings from Āmul in Tabaristān where he was killed in 304/916–17.

49. Sharīf, *Intisār*, p. 91; Tūsī, *Khilāf*, p. 7.

50. In connection with the merits of the martyrdom commemoration, see Mahmoud Ayoub, *Redemptive Suffering in Islam* (The Hague, 1978), chap. 5, pp. 141–47.

51. Nūrī, *Mustadrak*, 3:506.

52. Ayoub, *Redemptive Suffering*, chap. 5, pp. 180–96.

53. The intellectual milieu from which Shī'ī Imāmism subsequently emerged has been studied by 'Abbas Iqbāl, *Khwāndān-i nawbakhtī* (Tehran, 1345/1966); W. Montgomery Watt, *The Formative Period of Islamic Thought* (Edinburgh, 1973), especially chap. 2 and 7; and M. G. S. Hodgson, *The Venture of Islam* (Chicago, 1974), vol. I, chap. 4.

54. Shahrastānī, *Milal*, 160–61.

55. This tendency continued long after the occultation, as shown by W. Madelung ("Imamism," pp. 17f.). He cites the traditionists of Rayy as opposed to the rational method in the explanation of religious doctrines.

56. Ibn Abī al-Hadīd, *Sharh NB*, 2/120, 128.

57. Shahrastānī, *Milal*, 124, 131.

58. Ibid., 116–17, 121.

59. Tūsī, *Talkhīs*, 2/94ff.; *Sharh Maqāsid*, 2/285.

60. Tūsī, *Talkhīs*, 98, 109f., 112.

61. Sachedina, *Messianism*, pp. 54–55.

62. Tūsī, *Fihrist*, pp. 19–21, 175; Nawbakhtī, *Firaq*, p. 9.

63. Tūsī, *Fihrist*, p. 175.

64. Ibid., p. 176.

65. Ibid., for his various titles on the subject of the Imamate.

66. 'Abd al-Jabbār al-Qazwīnī al-Rāzī, in the 6th/12th century, according to Madelung, used the *usūliyya* and *akhbāriyya* to designate the two groups among the Imamites who supported *al-'aql* and *al-sam'*, respectively, in their approach to the fundamentals of the faith. See "Imamism," pp. 20–21.

67. Ibn Hajar, *Lisān al-mīzān*, 4/224.

68. Tūsī's abridgment is entitled *Talkhīs al-shāfī*, which is now available in four volumes, published in Najaf, Iraq, 1394 A.H.

69. Hillī, *Mukhtasar*, p. 47.

70. See, for instance, his *Tanzīh*, p. 170, where he discusses the abdication of al-Hasan.

71. Ibn Abī al-Hadīd, *Sharh NB*, 1:41.

Chapter 3

1. The term used to indicate the illegitimacy of non-Imamite authority is *salātin al-jawr* (the "unjust" authorities), who, by definition, have claimed the *imāma* without proper designation by the Prophet. Thus, the Umayyads and the 'Abbāsids and nearly all the other well-known dynasties are all declared *salātin al-jawr*. See, for instance, Bahrānī, *Hadā'iq, kitāb al-tijāra*, 5/4, and other works on Imamite jurisprudence.

2. Mas'ūdī, *Murūj*, 5/422, mentions 'Awn b. 'Utba b. Mas'ūd who adhered to this principle.

3. *Sahīfa al-kāmila* (Tehran, 1375/1955), p. 364.

4. Regardless of the contradictory historical narratives touching on this period of general confusion and political turmoil in Islamic lands, it is plausible that the offer of the caliphate to Imam Ja'far al-Sādiq was not totally an invention to enhance al-Sādiq's prestige. By reason of his Fātimid lineage and his intellectual activities, he already enjoyed high regard. That there were pro-'Alid tendencies among the leaders of the 'Abbāsid movement cannot be doubted, because of the general sentiment among Muslims regarding the messianic role of the descendants of the Prophet at this time. Moreover, that there could have been attempts to seek an 'Alid candidate for the caliphate is not far-fetched; the sources do bear out the difficulty faced by 'Abbāsid missionaries in finding a single, universally recognized 'Alid leader for that purpose. Considering the prestige enjoyed by al-Sādiq at this time, it would therefore be probable that he was indeed approached to consider the caliphate. Fully aware of the political turmoil created by the overthrow of the Umayyads and disagreement among the 'Alids regarding the political direction of the Shī'ites, he understandably declined the offer and also warned his relatives and followers against an uprising.

5. See Wilferd Madelung, "A Treatise of the Sharīf al-Murtadā on the Legality of Working for the Government," *BSOAS,* 43 (1980), part 1, pp. 18–31.

6. In his *al-Masā'il al-hā'iriyāt*, pp. 312–13, Tūsī explains *amāna* as *taklīf* (religious-moral obligation imposed by God on humanity) and cites the Shī'ī opinion as the one in which *amāna* is equated with *wilāya*. However, Tūsī contends that such specification of *amāna* is unnecessary, because *taklīf* is general and it includes *wilāya*. In his Qur'ānic exegesis he explains *amāna* as the contract (*al-'aqd*) that humankind must fulfill because it has been entrusted to humankind by God (*Tibyān*, 8/368). He also cites the theological problems that Tabarsī points out, in addition to the differences of opinion on the interpretation of *amāna*. He cites several early authorities to show the difficulty of interpreting the verse with theological implications in the realm of human responsibility as the recipient of this "trust" (*Majma'*, 8/373–74; see also Zamakhsharī, *Kashshāf*, 3/276–77). However, as Tūsī explains in his *al-Masā'il al-hā'iriyāt*, it is in the early Imamite traditions dealing with the *wilāya* that the "trust" verse has been interpreted as pointing to the *wilāya* of 'Alī b. Abī Tālib. Tabātabā'ī's detailed exegesis is based on all these early materials, and his exegesis is based on these traditions regarding *wilāya*. Consequently, he explicitly equates the "trust" with *al-wilāyat al-ilāhiyya* and discusses the implications of its acceptance for humanity (*Mīzān*, 6/10ff.). The crux of the problem in the exegesis of the verse is that if humanity was the only creature of God who accepted the "trust," why should humanity be described as "tyrant" and "fool"? At this point, Tabātabā'ī's interpretation draws upon main tenets of Imamite theology, which regards the "trust" as a special favor to humanity entailing the enormous responsibility to stand by the obligation of guarding it.

7. See Wansborough, *Sectarian Milieu*, pp. 70–71, where he describes the Islamic concept of authority as "apostolic." The charismatic figure of the Prophet is depicted therein in an essentially public posture in the emergence of Islamic polity.

8. However, as Wansborough has pointed out in *Sectarian Milieu*, pp. 70ff., in the composition of Islamic salvation history, it was in Shī'ism that the *wilāya* of the Prophet, as I have elaborated it in this work, was repeatedly and consistently expressed by Shī'ī scholars; whereas in Sunnism the *wilāya* (authority in the form of *exemplum* [*imām*]) of the Prophet was located in the Sunna, which became the *imām* of the community in the absence of the charismatic authority of the Prophet. Norman Calder, in his dissertation "The Structure of Authority in Imāmī Shī'ī Jurisprudence" (London: School of Oriental and African Studies, 1980, completed under Dr. Wansborough's supervision), discusses the works of the Sunnī scholarly elite in chapter 1. He takes up the question of the authority of the Sunna in the development of the juristic authority of Sunnī scholars. There he makes the observation that

the Sunnī scholars also declared the Sunna as the Imam of the community. Calder attributes this development in Sunnī Islam to the breakdown of the caliphal authority and the polemics against the Shīʿī conception of the charismatic authority of the Imam. With the occultation of the twelfth Imam, a similar development occurred in the authority of the Shīʿī jurists in their juridical works, where the authority of the interpreters of the will of the Imam, the Shīʿī scholars, now located in the Sunna of the Imams, was attenuated by the *ijtihād* of the Shīʿī jurists.

9. Qurʾān 5:55. See Tūsī, *Tibyān*, 3/559.

10. Tūsī, *Tibyān*, 3/561; Kulaynī, *Kāfī*, 2/402, *Hadīth* #77; Tabātabāʾī, *Mīzān*, 6/1ff.; ʿAbd al-Husayn Sharaf al-Dīn al-Mūsawī, *al-Murājiʿāt* (Beirut, 1963), p. 180.

11. See, e.g., Tabarī, *Tafsīr*, 6/186ff.; Zamakhsharī, *Kashshāf*, 1/623–24; Baydāwī, *Anwār*, p. 154.

12. Tūsī, *Tibyān*, 3/559.

13. See, e.g., the Qurʾān, 47:7, 7:157, 59:8. See also Tabātabāʾī, *Mīzān*, 6/13. Tūsī, *Tibyān*, 3/565, alludes to this.

14. Tabarī, *Taʾrīkh*, 5/402. See also 5/357, where instead of *awlā*, *ahaqq* is used to signify the same conclusion of being "more entitled to *tasarruf*."

15. *Awlā* in this verse has been translated by Arberry and others as "nearer" and "closer." But, taking into consideration the Prophet's speech on the occasion of the Farewell Pilgrimage, where the same verse of the Qurʾān occurs in the form of a question by the Prophet to the Muslims, the implication is in the sense of being "more entitled." The Prophet asked the assembled pilgrims: "Am I not *awlā* to you than yourselves?" See Mufīd, *Irshād*, p. 161; Tabātabāʾī, *Mīzān*, 6/13ff.

16. Tabātabāʾī, *Mīzān*, 6/12–14. Among early works, besides Qurʾānic exegesis where *wilāya* occurs in the meaning of *wilāyat al-tasarruf*, one can cite the *Rijāl* of Kashshī, especially where he mentions the *wilāya* of Abū al-Khattāb, which was denounced by the Imam Jaʿfar al-Sādiq. See *Rijāl*, 296/523.

17. Kulaynī, *Kāfī*, 2/368, *Hadīth* #2.

18. Islamic soteriology retained the notion of messianic Imam, the Mahdī, at the popular level even when the tendency among Sunnī scholars was to emphasize membership in the *umma* as the correct expression of Islamic soteriology. The doctrine of the Imamate contributed significantly to seek the prophetic paradigm in the function of the Mahdī. However, the Sunnī equation of the Sunna with the Imam, and the declaration that the former was the prophetic paradigm in giving expression to the ideal public order, tended to oppose the notion of messianic soteriology. See Wansborough, *Sectarian Milieu*, pp. 88ff.

19. Tūsī, *Nihāya*, 1/358.

20. Mufīd, *Muqniʿa*, p. 129.

21. *Murūj*, 3/193ff. In a more explicit speech, al-Husayn, on his way to Iraq, reminds the soldiers of the Umayyad commander who had come to arrest him that the Prophet had required Muslims to challenge a *sultān jāʾir* who had ruled unjustly, breaking all the laws of God and opposing the Sunna of the Prophet. See Tabarī, *Taʾrīkh*, 5/403.

22. *Tanbīh al-umma*, p. 15.

23. Tūsī, *Nihāya*, 1/302–3; Mufīd, *Muqniʿa*, p. 129; Rāwandi, *Fiqh al-Qurʾān*, p. 358; Halabī, *Kāfī*, passim; ʿAllāma, *Muntahā*, 317.

24. Tūsī, *Mabsūt*, 4/70. In his *Khilāf*, he is very explicit when he says that if a person without any offspring leaves inheritance, it should not be transferred to the public treasury, because it belongs to the Imam, exclusively. But, according to all other *fuqahāʾ*, it should be transferred to the public treasury, because it belongs to the Muslims. According to Shāfiʿī, the Muslims inherit it on the grounds of their being members of the same group (*al-taʿsīb*); Abū Hanīfa agrees, on the basis of one of the two traditions reported by him (which includes Shāfiʿī's opinion), although the other tradition grounds it on friendship (*muwālāt*), not membership.

25. Tūsī, *Mabsūt*, 7/41.

26. See Fazlur Rahman, *Major Themes of the Qur'an* (Chicago, 1980), chap. 3, esp. p. 43.

27. *Muqnīʿa*, p. 129.

28. Ibid, p. 130.

29. Norman Calder in his above-cited dissertation has argued about the underlying motives of the Shīʿī jurists in these rulings dealing with the authority of the jurists and has discussed various exegetical and terminological stratagems employed by them to further their role as the *imām* of the Shīʿa. In the process, he has reduced the discussion of the development of authority in Imamite jurisprudence to motives that, if documented thoroughly, would appear contradictory as indicators of juristic authority. It will be difficult, for instance, to maintain that socially and politically active jurists like Sharīf al-Murtadā, and later on Muhaqqiq al-Karakī, who had probably vested interests in wielding more power, would include decisions that would eventually lead to the curbing of their authority in those spheres in which only the Imam could exercise political authority.

30. *Nihāya*, 1/302.

31. Muhaqqiq says that *al-sultān al-ʿādil* is "the one who is specified (*ʿayyina*) by the Imam *al-asl.*" See also his section on *wilāya* from *al-sultān al-ʿādil* in *Sharāʾiʿ*, 2/12.

32. ʿAllāma, *Tadhkira*, 2/595.

33. ʿAllāma, *Tahrīr*, p. 133.

34. In Islamic jurisprudence this is classified as *fard al-kifāya*, an obligation, the fulfillment of which by a sufficient member of individuals excuses other individuals from fulfilling it. For further discussion on *fard al-kifāya* and its translation as "representatively obligatory," see chap. 4, n. 66, below.

35. W. Montgomery Watt, "The Significance of the Theory of *Jihād*," in *Akten des VII. Kongresses für Arabistik und IslamWissenschaft* (Göttingen, 1974), p. 393.

36. Kulaynī, *Kāfī;* Tabarsī, *Ihtijāj*, 2/44; Majlisī, *Bihār*, 100/18. ʿAbbād al-Basrī on several occasions disagrees with Imam al-Sādiq on points of law. See Khūʾī, *Muʿjam*, 19/217.

37. Tūsī, *Mabsūt*, 2/9ff.

38. Ibid., 7/41. Tūsī makes this point in connection with *al-jirāh* (pl. of *jurh*, meaning "wounds" inflicted on bodies by iron instruments and the like), where he discusses the authority that can order another person to be killed. Tūsī clarifies the Imamite position and rules that *al-imām al-maʿsūm* does not order the killing of any person without a valid reason, because of his *ʿisma* (infallibility); whereas the jurists of other schools of law have permitted capital punishment, in accordance with their conception of Islamic leadership. The deputy (*khalīfa*) of the Imam, says Tūsī, is not protected by *ʿisma* and hence, he could commit an error of judgment. In this respect, the status of the deputy of the Imam is similar to that of the imām among the Sunnites.

39. *Maqātil al-tālibiyyin*, p. 196.

40. Tabarsī, *Ihitijāj*, 2/118; Majlisī, *Bihār*, 100/21.

41. ʿAllāma, *Tahrīr*, p. 133.

42. *Mabsūt*, 2/8. All other works on Imamite jurisprudence during the classical age maintain this exception. Ibn Idrīs, for instance, declares that *jihād* is not allowed without the order of the Imam or the one appointed by him for that purpose. When neither of these two is present, then it is not permissible to fight the enemy and conduct *jihād* under the leadership of unjust rulers. The exception to this prohibition, says Ibn Idrīs, is when Muslims are attacked by the enemy and when they need to defend Muslim territory and lives. See *Sarāʾir*, p. 156; Rāwandī, *Fiqh al-qurʾān*, pp. 330–32.

43. *Mabsūt*, 2/8.

44. *Sharāʾiʿ*, 1/307.

45. *Rawda*, 2/379.

46. ʿAmilī, *Wasāʾil*, 11/62, *Hadīth* #11.

47. Mufīd, *Irshād*, pp. 118–19.
48. ʿĀmilī, *Wasāʾil*, 11/33, *Hadīth* #4. *Jiwār* is living in the vicinity of the holy precinct in Mecca or Medina, for some considerable period of time and by deliberate choice, to perform acts of devotion and piety.
49. *Nihāya, kitāb al-jihād*, 1/292.
50. A. K. S. Lambton, "A Nineteenth-Century View of *Jihād*," *Studia Islamica*, 32 (1970) 181–92.
51. Etan Kohlberg, "The Development of the Imāmī Shīʿī Doctrine of *jihād*," *Zeitschrift der Deutschen Morgenländischen Gesellschaft*, 126 (1976) 64–86.
52. ʿĀmilī, *Wasāʾil*, 11/400, *Hadīth* #1.
53. Ibid., 11/464, *Hadīth* #19.
54. Ibid., 11/35, *Hadīth* #10. In *Khisāl*, 2/394, it is narrated that one should not kill a *kāfir* or a Sunnī in the *dār al-taqiyya* except when one is attacked by them or confronts a *kāfir* or a Sunnī who is engaged in spreading corruption, with the further condition that one is safe to do so without endangering one's life or the lives of one's associates. See also Majlisī, *Bihār*, 100/23.
55. ʿĀmilī, *Wasāʾil*, 11/483, *Hadīth*, #2.
56. *Rawda*, 2/379–81.
57. Sabzawārī, *Kifāya*, p. 74.
58. Halabī, *Ishārāt*, 130, mentions this third category of persons to engage in *jihād*, besides the Imam and his specially designated deputy. In *Ghunya*, p. 520, Abū al-Makārim mentions the third category, with the provision that, in addition, the members of the community ought to judge the situation as dangerous to their lives and property.
59. *Masālik, kitāb al-jihād*.

Chapter 4

1. *Fidya* signifies a ransom—a thing or a captive—given for a kidnaped person, who is therewith liberated. See Lane, *Lexicon*, book 1, part 6, p. 2354.
2. Ibn Manzūr, *Lisān al-ʿarab*, 13/458.
3. *Fitra* has been interpreted variously by Muslim theologians from the early days. It has been regarded as the natural basis of the true religion. According to Prophetic tradition, "Every child is born in the *fitra;* it is his parents who make him a Jew or a Christian or a Parsi. In the same way cattle give birth to calves without defects." The evolution of the term *fitra* to the meaning of "natural law" can be attributed to the theological dispute concerning the fate of children who die before reaching adulthood. The conflict between the tradition about the natural religion in which every child is born and the dogma of predestination was obvious and the theologians had to attempt exegetical resolution of the problem. In the end, the explanation that "every child is born with a disposition toward Islam" was favored over other explanations that support the predestinarian view of belief and unbelief. See A. J. Wensinck, *Muslim Creed* (London, 1965), pp. 42–44, 214–16. However, the concept of *fitra* also raised the fundamental question about the necessity of revelation when *fitra* could lead to salvation. This view found its exposition among the Muslim philosophers who relegated religion to "symbolic" status, being the form of truth intelligible to the average person (Wansborough, *Sectarian Milieu*, pp. 135–36).
4. Hourani, *Islamic Rationalism*, p. 3.
5. Ibid.
6. Even the Ashʿarī exegete Fakhr al-Dīn al-Rāzī, who maintains complete subordination of the human will to the divine will, recognizes two forms of "guidance": first, guidance by means of *dalīl* (demonstration) and *hujja* (proof or evidence), both activities of human rational faculty, which he considers limited; secondly, guidance through inner purification of

the soul and ascetic practices. He does not speak of revelation as a separate form of guidance; rather, as an Ash'arī, he considers that revelation (i.e., the will of God) superimposes all forms of guidance. See his *Tafsīr kabīr*, 1 (Cairo, 1938), pp. 9ff.

7. Some important observations in this connection have been made by Joseph Schacht, "Law and Justice," in *The Cambridge History of Islam*, vol. 2B, pp. 539–68.

8. See A. K. S. Lambton, *State and Government in Medieval Islam* (Oxford, 1981). The work is devoted to the study of Islamic political theory propounded by Muslim jurists who based their discussion on Qur'ānic exegetical materials, in addition to the traditional practice of the Prophet and the early community.

9. Najafī, *Jawāhir*, 40/8.

10. Shahīd Thānī, *Masālik, kitāb al-qadā'*, vol. 2.

11. Hillī, *Īdāh*, 4/293.

12. Fādil Hindī, *Kashf al-lithām*, vol. 2; Najafī, *Jawāhir*, 40/8–11.

13. Tūsī, *Mabsūt*, 2/17.

14. Abū al-Makārim, *Ghunya*, p. 560.

15. Karakī, *Maqāsid*, 2/304; Ansārī, *Makāsib*, 153.

16. Tūsī, *Nihāya*, 1/320.

17. Karakī, *Maqāsid*, 2/304.

18. Ibn Bābūya, *Man lā, yahduruh*, 3/4, *Hadīth* #1,82.

19. The theological implications are obvious in making it a religious office. Indeed, the Qur'ān treats the *qadā'* as one of the prophetic functions and requires anyone who undertakes it to be in possession of *taqwā*—i.e., spiritual-moral awareness.

20. Ibn Bābūya, *Muqni'*, p. 33.

21. Ibn Bābūya, *Man lā yahduruh*, 3/4 *Hadīth* #1 and 2.

22. Ibid., 3/4; also, *Muqni'*, p. 33, where the third category is not mentioned, probably because of the doubtful judgment regarding a person's ultimate destiny on the day of judgment.

23. Sallār, *Marāsim*, p. 592.

24. In *Islamic Messianism*, pp. 137ff., I have discussed the theological connection between the designation (*nass*) of the Imam and his knowledge.

25. 'Allāma, *Mukhtalaf*, 4/144; Sharīf Murtadā, *Intisār*, pp. 194–95.

26. 'Allāma, *Mukhtalaf*, 4/145.

27. Ibn Idrīs, *Sarā'ir*, p. 469.

28. Mūsawī, *Minhāj al-karāma*, 4/113.

29. I am not using the terms *usūlī-akhbārī* here as they came to be known in the Qājār period. As discussed in the Introduction, during the Qājār rule *usūlī* and *akhbārī* were the two schools of jurisprudence espousing one or the other method of deriving *shar'ī* laws. What I am implying here is that in inferring precepts from the authoritative sources of law—the Qur'ān and the Sunna—it was very soon found necessary to supplement *dalīl shar'ī* (explicit textual proof) with *dalīl 'aqlī* (rational proof), which laid the foundation of the elaborate *usūl al-fiqh* (principles employed to derive the laws of the Sharī'a), which entailed the methodology of deducing laws by utilizing both types of *dalīl: shar'ī* and *'aqlī*. However, from the early days of Imamite jurisprudence, jurists had shown their individual preference in using these two types of *dalīl*. Whereas some depended heavily on the *dalīl shar'ī*, especially on *riwāyāt* and *akhbār*, there were others who believed that in addition to the *dalīl shar'ī*, there were rational proofs deduced by the human faculty of reasoning, which could be used to supplement the essential preliminaries that underlie Qur'ānic legislation. The former group can be identified as *akhbārī*, whereas the latter group can be identified as *usūlī* in their jurisprudential methodology.

30. See chap. 1, above.

31. This is verse 46:6, which reads: "O believers, if an ungodly (*fāsiq*) man comes to you

with a tiding, verify it, lest you afflict a people unwittingly, and then repent of what you have done." The reverse of the command is deduced in this case, which would mean that a report by a single God-fearing (*'ādil*) person can be taken as reliable. See all the works on *usūl al-fiqh* in the section dealing with the accreditation of a "single-individual tradition" (*khabar al-wāhid*), e.g., Muhammad Bāqir Sadr, *Durūs fī 'ilm al-usūl*, 2/138.

32. Sadr, *Durūs*, 4/399ff.

33. Particularly, Shahīd Thānī, as cited by Khū'ī, *Mu'jam*, 13/31. See Khū'ī's own criticism *Tanqīh*, 1/143–44) of the *maqbūla* and the weakness of the *sanad* because of the lack of any authentification of 'Umar b. Hanzala in any communication from the Imams.

34. Most of the *usūl* works have a separate section discussing the methods of "accrediting" an important tradition that suffers the "wounds" of weakness in *sanad* or *matn*. See, e.g., Ansārī, *Rasā'il*, vol. 2.

35. *Tāghūt* signifies a devil or "idol" in general. Also, in this context, it would mean a person who has turned away from the good way, leader of error, one who transgresses the just limit, and so on. All these are used as descriptions of an oppressive ruler and as such *tāghūt* is a symbol of the caliphs in the *hadīth* under consideration. See also Lane, *Lexicon*, book 1, part 5, p. 1857.

36. Kulaynī, *Kāfī*, 1/113–5. Dāwūd b. Husayn in the *sanad* of this *maqbūla* tradition has been regarded "weak" by the *rijāl* scholars. See Khū'ī, *Mu'jam*, 7/101, where Khū'ī accepts the verdict of Tūsī and 'Allāma on Dāwūd's *wāqifī* ("stopping" with the Imamate of the seventh Imam, al-Kāzim) position on the question of the Imamate. As a requirement of *'adāla* in Imāmī traditions, a narrator must be an *imāmī*—possessing "sound belief" by accepting the Imamate of all the twelve Imams. In *Tanqīh*, 1/143–44, Khū'ī regards the tradition as being "weak" in transmission because of 'Umar b. Hanzala, who has not been accredited or praised in any communication.

37. *Islamic Messianism*, chap. 4, pp. 120ff.

38. 'Allāma, *Mukhtalaf*, 2/158; Halabī, *Ishārāt*, p. 131.

39. Tūsī, *Iqtisād*, p. 147.

40. 'Allāma, *Mukhtalaf*, 2/158.

41. Shahīd Thānī, *Rawda*, 2/410. See also Hillī, *Muhadhdhab, kitāb al-'amr;* Shahīd Awwal, *Durūs*, p. 165.

42. 'Allāma, *Mukhtalaf*, 2/159.

43. Shahīd Awwal, *Durūs*, p. 165.

44. Tūsī, *Iqtisād*, p. 150.

45. Tūsī, *Nihāya*, 1/302.

46. Tūsī, *Iqtisād*, 150–51. This is known as the principle of the correlation between the incumbency of an act and the incumbency of its prerequisites in *usūl al-fiqh*.

47. 'Allāma, *Mukhtalaf*, 2/159.

48. Ibid., 2/158, citing Halabī and Ibn Idrīs.

49. Sallār, *Marāsim*, p. 597.

50. For unlawful intercourse: "The adulterer and the adulteress, scourge ye each one of them with a hundred stripes" (24:2); false accusation: "And those who accuse honorable women but bring not four witnesses, scourge them with eighty stripes" (24:4); theft: "As for the thief, both male and female, cut off their hands" (5:38); and highway robbery: "The only reward for those who make war upon God and His Messenger and strive after corruption in the land will be that they will be killed or crucified" (5:33). The last citation is also used to justify capital punishment for rebellion.

51. Theoretically, *ta'zīr* crimes are those which bring injury to the social order as a result of the trouble they cause, and it was for this reason that their precise determination was left to the community and its representatives, the caliphs. See Ahmad 'Abd al-'Azīz al-Alfī, "Punishment in Islamic Criminal Law," in *The Islamic Criminal Justice System*, ed. M.

Cherif Bassiouni (New York, 1982), p. 227; see also S. M. Zwemer, *The Law of Apostasy in Islam* (London, 1924), esp. chap. 2. This latter work, although outdated and hostile in places, is still useful as a summary of laws of apostasy in Islam.

52. *Sunan Abī Dāwūd*, 4/126. See also a study on the sources of penal law by Taymour Kamil, "The Principle of Legality and its Application in Islamic Criminal Justice," in *Islamic Criminal Justice System*, pp. 149–69.

53. Shāfiʿī, *Kitāb al-umm*, 4/214–17. In his *War and Peace in the Law of Islam* (Baltimore, 1955), p. 150, Majid Khadduri cites 4:88–89 in support of the death penalty for those who apostatize or fall away from their religion, and maintains, without citing his sources, that all "commentators agree that a believer who turns back from his religion (*irtadda*), openly or secretly, must be killed if he persists in disbelief." He also maintains in a later passage that "traditions are more explicit in providing the death penalty for everyone who apostatizes from Islam." Actually, 4:88–89 deals with hypocrites, and all the major commentators state that the passage apparently refers to those Arabs who used to come to Medina and declare themselves to be Muslims and on their return to Mecca would revert to their pagan beliefs and engage in hostile acts against the Muslim community (Tabarī, *Tafsīr*, 5/124–25; Zamakhsharī, 1/550; Baydāwī, *Anwār*, p. 121). The Qurʾān says: "If they turn back [to enmity] then take them and kill them wherever ye find them, and choose no friend nor helper from among them" (4:89).

54. ʿAllāma, *Tadhkira*, 1/459.

55. Tūsī, *Nihāya*, 1/303.

56. Ibn Idrīs, *Sarāʾir*, pp. 160–61.

57. Ibid., p. 469.

58. Tūsī, *Nihāya*, 1/304, and all other subsequent works on Imamite *fiqh*. See, e.g., Shahīd Thānī, *Masālik*, *Kitāb al-ʾamr*. This work is regarded one of the important commentaries on Muhaqqiq al-Hillī's *Sharāʾiʿ*.

59. ʿAllāma, *Qawāʿid*, p. 45.

60. Muhaqqiq, *Sharāʾiʿ*, 1/344.

61. Shahīd Thānī, *Masālik*, 1/162.

62. See, for example, Najafī, *Jawāhir*, 21/386–90.

63. Ibn Idrīs, *Sarāʾir*, 2/469.

64. Fadīl Hindī, *Lithām*, 2/140.

65. Ibid.

66. *Fard al-kifāya* has been translated by Joseph Schacht, Ignac Goldziher, and others who follow them as "collective duty," in the sense that the fulfillment of it by a sufficient number of individuals excuses other individuals from fulfilling it. By translating as "collective duty" what they possibly meant was "duty to act on behalf [of the collectivity]," but that sense is not conveyed by this translation. *Kafā* itself means "to be capable of doing a task," but *kafāʾan* adds the idea of "doing it to save somebody else the trouble [or the frustration of proving to be incapable of it]." Moreover, *kifāya* has the sense of "sufficiency" and as such in *fiqh* texts implies an obligatory act that suffices and enables one to obtain what is sought, through the performance of a prescribed duty. In other words, it conveys the sense of "obligatory representational function" fulfilled by one or more in the community on behalf of everyone else. As such, its performance by that individual or individuals relieves others of that duty. In the legal context, then, it is more accurate to translate *fard al-kifāya* as "representatively obligatory," whereby an individual or some individuals take upon themselves to perform an "obligatory representational function," "representing" a body by a deputed duty. (I am grateful to Professor G. M. Wickens for pointing out this important aspect of the essential meaning of *kifāya*.)

67. Shahīd Thānī, *Bidāyat al-dirāya*, pp. 66–69, does not believe that *tawātur al-lafzī* is mentioned in any tradition except the one related by some sixty-two transmitters in which the

Prophet is believed to have declared: "The one who willfully attributes a falsehood to me will enter his resting place in the Fire."

68. Among non-Imamite jurists a *mursal* is a tradition that has been transmitted by a second generation Muslim by linking it to the Prophet without any mediator. Such are the traditions of Sa'īd b. Musayyib, whose *mursals*, according to Shāfi'ī, are like *musnads*—i.e., they have an adequate chain of transmission. However, Shāfi'ī has not accepted these traditions as documentation for making a legal decision. Both Abū Hanīfa and Mālik have regarded *mursal* traditions as authenticated for documentation purposes. See Yāsin Suwaylim Tāhā's commentary on Baydāwī's *Minhāj al wusūl ila 'ilm al-usūl* (Cairo, n.d.), 2/ 95–96.

69. Kashshī, *Rijāl*, p. 155.

70. Ibid., p. 344.

71. The Imamites have mentioned several ways of "discovering" the opinion of an Imam with certainty to arrive at consensus, which I have discussed in *Islamic Messianism*, chap. 4.

72. Much work is being done in this connection by Abū al-Qāsim Gurjī of Tehran University. See his introduction to the recent edition of Tūsī's *Khilāf* (Tehran, 1985).

73. Tūsī, *Mabsūt*, 8/99–100.

74. Mūsawī, *Miftāh al-karāma*, *Kitāb al-qadā'*, vol. 10.

75. 'Allāma, *Qawā'id*, 4/298.

76. Ibn Fahd al-Hillī, *Īdāh*, 4/298. On the question of whether the Prophet could write or not, Mufīd, *Awā'il*, pp. 111–12, argues that he could because God perfected everything in him, including the art of writing, when he was chosen to be the Prophet.

77. Tūsī, *Ghayba*, p. 231. For the discussion of this *hadīth*, see *Islamic Messianism*, pp. 99f.

78. *Al-tahkīm* is the authority vested in a person to adjudicate between two disputants. Such a person was required to possess the qualifications of *hākim al-shar'*. Shahīd Thānī maintains that *qādī al-tahkīm* is inconceivable during the occultation of the Imam, because the ruling of a well-qualified *mujtahid* is effective without being appointed in *tahkīm* (arbitration) by the people; otherwise the rulings of a *mujtahid* would not carry weight at all. Thus, *al-tahkīm* merges in the general *wilāya* of *qadā'*. See Sabzawārī, *Kifāyat al-ahkām*, pp. 261–62.

79. Najafī, *Jawāhir*, 40/15–20.

80. A logical conclusion of *'adāla* is *'isma*, although the latter is strictly said of the Imams and the prophets only.

81. Hakīm, *Usūl al-'āmma*, p. 571.

82. When there are problems that are not explicitly or implicitly deducible from the sources of the Sharī'a, jurists could refer to *al-usūl al-'amaliyya* (Practical Principles) in the *usūl al-fiqh*. These "Practical Principles" help the searcher to determine the correct solution to the given problem. Thus, for instance, the principle of *istislāh* refers to taking into consideration the "public interest" (*maslaha*), whereas *istishāb* is a principle that seeks to establish a "link" between a later set of circumstances with an earlier, and is based on the assumption that the rules applicable to certain conditions remain valid so long as it is not certain that these conditions have altered. For example, a thing that was ritually pure does not become impure on the basis of doubt generated by some lapse of time. In the case of such doubt, *istishāb* declares that a person confronted with such a situation should ignore the doubt and adhere to the previously held certainty. Similarly, the principle of *barā'a* among the "Practical Principles" is employed when there is explicit or implicit absence of a certain case. It signifies "absence of obligation" for Muslims in those cases where Shari'a has given no ruling. See the major works on *usūl al fiqh*, where *al-usūl al-'amaliyya* are treated in detail.

83. Khū'ī does not identify this *ijtihād* formally as *al-shar'ī*. But from his definition and

discussion it is obvious that it is *al-ijtihād al-shar'ī*, as explained by Hakīm in *Usūl al-'āmma*, p. 571.

84. Khū'ī, *Tanqīh*, 1/20–24.

85. Khwāja Tūsī, *Akhlāq-i nāsirī*, ed. Mujtabā Minuvī and 'Alīridā Haydarī (Tehran, 1360 A.H.), p. 75. I have used G. M. Wickens's translation, *Nasirean Ethics*, p. 52. The process is sometimes characterized as "identification of true knowledge with moral rectitude."

86. Sabzawārī, *Kifāyat al-ahkām*, pp. 261–62, cites the tradition of 'Umar b. Hanzala as supporting evidence.

87. Ibn Bābūya, *Man lā yahduruh*, 3/6, 9/2.

88. Ibid., 3/34–35, section 17, *Hadīth* #1.

89. In a long passage, the Akhbārī jurist Shaykh Yūsuf al-Bahrānī discusses the *'adāla* in connection with the *jum'a* (*Hadā'iq*, 4/112). It is significant that *muruwwa*, a pre-Islamic term, covers the natural human qualities that do not depend on revelation. The early Islamic writers used to contrast *muruwwa* and *dīn*, explaining how the latter built upon the former. This observation further strengthens the argument about the objective nature of good and evil as discussed earlier in this study.

90. Sabzawārī, *Dhakhīra*, *Kitāb al-salāt*, *al-maqsad al-thānī*.

91. Ibid.

92. Kāshif al-Ghitā', *Kashf*, p. 266; Shahīd Thānī, *Rawda*, 1/378.

93. Tūsī, *Mabsūt*, 8/99.

94. Ibn Hamza al-Tūsī, *Wasīla*, p. 697.

95. Shahīd Thānī, *Masālik*, *Kitāb al-qadā'*, vol. 2.

96. Sabzawārī, *Kifāyat al-ahkām*, pp. 261–62.

97. Muhaqqiq Hillī, *Sharā'i'*, 4/68. For the relevant *hadīth*, see Kulaynī, *Kāfī*, 7/412, *kitāb al-qadā' wa al-ahkām*.

98. 'Allāma, *Qawā'id*, p. 223.

99. See note 66, above.

100. Tūsī, *Mabsūt*, 8/84–86.

101. Tūsī, *Nihāya*, 1/304.

102. Ibn Idrīs, *Sarā'ir*, pp. 160–61.

103. Muhaqqiq Hillī, *Sharā'i'*, 2/12.

104. See, e.g., Karakī, *Maqāsid*, 2/304, where the author, Muhaqqiq al-Karakī (d. 937/1530–31), who was appointed the *Shaykh al-islām* during the reign of Shah Tahmāsp, the Safavid, explains *al-hākim* as being *al-imām al-'ādil* or the one appointed by him, including the well-qualified *faqīh*.

Chapter 5

1. Sīwurī, *Kanz*, 2/208–9.

2. Tūsī, *Nihāya*, *bāb al-awsiyā'*, 2/623.

3. 'Allāma, *Tadhkira*, 2/592.

4. Ibid., 2/595.

5. In the article on *al-jum'a*, in both the first and second editions of the EI, the Imamite viewpoint has not received proper attention. Although S.D. Goitein's article in the second edition (II/593) makes reference to the political connotation of this *salāt*, no reference has been made to the Shī'ī Imāmī opinion on the subject, despite the fact that the Imamate of the *jum'a* in the latter school is closely related to the question of the *wilāya* of the Imam in general, and to the Imamate of the twelfth Imam in particular.

6. See above, chap. 3, p. 89.

7. Tūsī, *Tahdhīb*, 3/20–21; *Mabsūt*, 1/143.

8. E.g., Sabzawārī, *Dhakhīra, kitāb al-salāt, al-maqsad al-thānī*, mentions seven conditions; Fāḍil Hindī, *Lithām*, p. 261, mentions six.

9. Ibn Babuya, *Man lā yahduruh*, 1/245. See also Tūsī, *Tahdhīb*, 324, regarding congregational worship and its excellencies.

10. Tūsī, *Khilāf*, 1/250.

11. Tūsī, *Mabsūt*, 1/143.

12. Lane, *Lexicon*, book 1, part 5, p. 1975. Tūsī, *Mabsūt*, 1/157, discusses the qualifications of the *imām al-jamāʿa* of the five daily worship services.

13. Ḥakīm, *Mustamsak*, 7/367.

14. Tūsī, *Khilāf*, 1/235. The jurists usually speak of two *khutba*s (*khutbatān*), which are pronounced by the *khatīb*, who sits between them. In practice, it is one *khutba* in two sections with a pause between them, when the *khatīb* sits. See the article on *khutba* in the *Shorter Encyclopaedia of Islam*, pp. 258–59.

15. Ibn Bābūya, *ʿUyūn akhbār*, 2/109ff.

16. Tūsī, *Khilāf*, 1/235.

17. Tūsī, *Mabsūt*, 1/147.

18. Najafī, *Jawāhir*, 11/161–62.

19. Tūsī, *Nihāya*, 1/106. In *Tahdhīb*, 3/16, Tūsī argues for the incumbency of the *jumʿa*, if the conditions were favorable during the period of *taqiyya*.

20. Tūsī, *Tahdhīb*, 3/239.

21. ʿĀmilī, *Wasāʾil*, 5/14ff.

22. Najafī, *Jawāhir*, 11/161. Shaykh Jaʿfar in *Kashf al-ghitāʾ*, p. 252, makes a strong case for the imamate of the *jumʿa* to be reckoned as an official position and argues that the reports that urge the Shīʿa to perform the *jumʿa* were related under *taqiyya*.

23. In this connection an article by Joseph Eliash, "The Ithnā ʿAsharī Juristic Theory of Political and Legal Authority," in *SI* (1969), is an important contribution, although he seems to be too schematic in dividing the juridical and constitutional authority of the Imam so clearly. Such a clear division of the Imam's *wilāya* is not possible in light of the overlapping of certain functions with theologico-political connotation. Such is the case with *al-hukm* and also *al-jumʿa*.

24. Tūsī, *Khilāf*, 1/248.

25. See above, p. 180.

26. Tūsī, *Khilāf*, 1/248.

27. Tūsī, *Nihāya*, 1/106.

28. Ibid., 1/302.

29. Mufīd, *Muqniʿa*, p. 27.

30. Muhaqqiq Ḥillī, *Sharāʾiʿ*, 1/98.

31. *Ghunya*, p. 496.

32. ʿAllāma, *Tadhkira*, 1/145; see also his *Taḥrīr*, pp. 43ff. In his *Qawāʿid, kitāb al-salāt, maqsad thālith*, he maintains the same opinion, but the way in which he speaks of the deputy of the Imam corroborates the point that any person who can meet the conditions for such a position can regard his position as appointed by or confirmed by the Imam.

33. Sabzawārī, *Dhakhīra, kitāb al-salāt*.

34. Shahīd Thānī, *Rawda*, 1/299.

35. Ibid.

36. Ibid., 1/299–300.

37. Shahīd Thānī, *Masālik, kitāb al-salāt*, volume 1.

38. Shahīd Awwal, *Dhikrā*, p. 231. This is mentioned as the ninth requirement in the deputy of *al-sultān al-ʿādil* who must undertake to lead the Friday service.

39. Mufīd, *Irshād*, p. 674.

40. Tūsī, *Nihāya*, 1/103; *Tibyān*, 10/8; *Khilāf*, 1/235.

41. Nūrī, *Mustadrak, Bāb salāt al-jum'a, Hadīth* #5.
42. Khwansārī, *Rawdāt*, 1/299.
43. Sallār al-Daylamī, *Marāsim*, manuscript at Tehran University Library, fl. 19ff.
44. Ibn Idrīs, *Sarā'ir*, p. 161.
45. 'Allāma, *Mukhtalaf*, 1108–9.
46. Fādil Hindī, *Lithām*, 1/247.
47. Halabī, *Ishārāt*, p. 122, states explicitly that the *jum'a* is incumbent with the presence of *imām al-asl* or the one appointed by him and deputized by him by virtue of his worthiness and the authentic qualities in him.
48. *Lithām*, 1/247–48.
49. Karakī, *Maqāsid*, 1/131.
50. Ibid.
51. Sabzawārī, *Dhakhīra, Kitāb al-salāt, Al-maqsad al-thānī*.
52. Sabzawārī, *Kifāyat al-ahkām*, p. 20.
53. Majlisī, *Bihār*, 89/146–47.
54. Ibid., 89/143.
55. Najafī, *Jawāhir*, 11/178.
56. See Sabzawārī, *Dhakhīra, Kitāb al-salāt*, where he cites Ibn Bābūya's opinion.
57. Kāshānī, *Mafātīh, Miftāh* #6, 1/22–23.
58. Shaykah Ja'far, *Kashf al-ghitā'*, p. 251.
59. Ibid., pp. 251–52.
60. Bihbanānī, *Sharh Mafātīh*, fl. 82, 83–84.
61. Bahrānī, *Hadā'iq*, 4/92–97.
62. Najafī, *Jawāhir*, 11/190–91.
63. 'Allāma, *Tadhkira*, 1/241.
64. Shahīd Thānī, *Masālik, Kitāb al-zakāt*.
65. *Da'ā'im*, 1/312.
66. *Jawāhir*, 15/417–22.
67. Ibid.
68. Majlisī, *Bihār*, 6/60.
69. See above, chap. 3, p. 101.
70. Yazdī, *'Urwa*, p. 361.
71. Najafī, *Jawāhir*, 22/155–56.
72. Ansārī, *Makāsib*, p. 153. See also the article "Ansārī, Shaykh Murtadā," in *EI*[2], vol. 5 (Suppl.), pp. 75–77, by Abdul-Hadi Hairi, who makes references to the *'urf*, but not in the sense that Ansārī intends from its usage. Certainly, *'urf* in the meaning of "social conventions and common practices" has been known to Muslim jurists from the early days of Islamic jurisprudence. At times it was equated with *al-ijmā'* ("consensus") (Hakīm, *al-Usūl al-'āmma*, pp. 419–26). *'Urf*, although not recognized as the principal source (*al-asl*), has been considered an importance source for legal decisions on matters not covered by the Islamic Sharī'a. It serves the function of a minor term of a syllogism to deduce a legal decision. Ansārī uses *'urf* in the meaning of the language of the people, which ought to be understood and analyzed before giving any legal decision. It does not in any way legitimize *'urf* in the sense understood by Hairi.
73. Ghazālī, *Mustasfā*, 2/123.
74. Kulaynī, *Kāfī*, 1/40ff.
75. Ibid., 1/53.
76. Mufīd, *Awā'il*, pp. 10–11; *Fusūl*, p. 79.
77. Mufid, *Fusūl*, pp. 78–79. Martin J. DeDermott has published a detailed study of Mufīd's theology, *The Theology of al-Shaikh al-Mufid* (Beirut, 1978). His translations of the theological texts of Imamite theologians in this work suffer from inaccuracies and omissions of phrases important in the explication of theological statements. This shortcoming renders

the work unreliable in important places. Moreover, there seems to be misunderstanding of the *kalām* terminology as used by the theologians of the classical age. For instance, *taklīf* is always translated "moral obligation." This is a gross misinterpretation of the concept.

78. This is the definition adopted by Khū'ī in *Tanqīh*, 1/79. He is fully aware of the Akhbārī criticism of *taqlīd* and in his effort to reconcile the Usūlī and Akhbārī positions on both *ijtihād* and *taqlīd* he has not only rejected some previously held definitions of these terms, but has also replaced them with more correct and rationally acceptable definitions based on *al-'urf* and other juridical principles. See, e.g., his correction of Ākhund's definition of *taqlīd* in *Kifāya* (pp. 79–80).

79. See the article *taklīd* in the *Shorter Encylcopaedia of Islam*, pp. 562–64.

80. Khū'ī *Tanqīh*, 1.62ff. It is interesting that the Imamite jurists of Aleppo (a subbranch of the Baghdad school) used to regard *ijtihād* as a personally and individually incumbent obligation (*wājib 'aynī*), thereby making all believers individually responsible to investigate jurisprudence thoroughly in order to be able to put confidence in the correctness of rulings. This opinion flies in the face of reason, as Hilla and Isfahān jurists later demonstrated in their works.

81. Khū'ī, *Tanqīh*, 1/143.

82. The problem of *taqlīd* has been treated by Narāqī, *'Awā'id*, p. 191, and also in his *Minhāj al-ahkām*, where he refutes the Akhbārī opinion that *taqlīd* is based on *al-zann*. Among modern jurists I have found that almost everyone has taken up the issue of *taqlīd* at the beginning of the *fiqh* text. See, e.g., detailed discussion on *taqlīd* by Shaykh 'Alī, *al-Nūr al-sāti'*; vol. 2 (585 pages) deals with *taqlīd*.

83. Ansārī, *Makāsib*, p. 153.

84. For the Practical Principles, see Muzaffar, *Usūl*, vols. 3 and 4.

85. Ansārī, *Makāsib*, p. 153.

86. Majlisī, *Bihār*, 6/60, cites the long tradition on the authority of Kulaynī and Ibn Bābūya in his section on the *imāma*.

87. Ansārī, *Makāsib*, p. 153.

88. Kulaynī, *Kāfī*, 1/342.

89. Ibid., 1/46.

90. 'Āmilī, *Wasā'il*, 18/66.

91. Kulaynī, *Kāfī*, 1/67. See above, chap. 4, for the complete text of the *hadīth*.

92. Tūsī, *Ghayba*, pp. 176ff. I have discussed this rescript in my *Islamic Messianism*, pp. 100–101, where I have explained the problem, together with the preserved texts and variations in them, reflecting some intentional tampering with the rescript.

93. Ansārī, *Makāsib*, p. 154.

94. Ibid.

95. Ibid., p. 155.

96. *Nahj al-balāgha*, speech #172.

97. See Muhammad Jawād Maghniyya, *al-Khumaynī wa al-dawlat al-islāmiyya* (Beirut, 1979), who takes up this question of *wilāya* of a jurist and challenges interpretations that limit it to the jurists as a class.

98. Khumaynī, *al-Hukūmat al-islāmiyya* (Najaf, 1971).

99. Iranian Constitution, Article 56.

100. Ibid., Article 5.

101. See an important work calling for reform in the institution of *marja' al-taqlīd* in the 1960s entitled, *Marja'iyyat wa rūhāniyyat*, especially the essay by Murtadā Mutahharī advocating centralization of Imamite religious authority, pp. 167–91.

102. *Al-Badr al-zāhir*, pp. 52–57.

103. Najafī, *Jawāhir*, 22/155–56.

104. See Appendix.

105. The treatment of the *fuqahā'* as a special class has been a point of controversy

among Imamite scholars in modern times. There are those who believe that *al-wilāyat al-'āmma* must be assumed by a *faqīh* to issue legal decisions. But there are others who believe that it is not necessary that such a person should be a *faqīh;* rather, he should be acknowledged and accepted by the Shī'a as a worthy person to administer the affairs of the community. The people should have confidence in him as the chosen leader. The *fuqahā'*, then, should take upon themselves to assist the government in providing necessary guidance to enact laws in conformity with the principles of Islam. For a government to be Islamic, there is no necessity for a jurist to be the head of state. This was the main difference between the views of the late Āyatullāh Sharī'atmadārī and Āyatullāh Khumaynī. The former upheld views similar to those held by Maghniyya, *al-Khumaynī wa al-dawlat al-islāmiyya*, pp. 59–71; whereas the latter confined the *wilāya* to the jurist only.

106. Najafi, *Jawāhir*, 11/178.

107. *'Awā'id*, pp. 178–79.

Appendix

1. Published in *JNES*, vol. 39 (1980), no. 4, pp. 275–89.

2. Tūsī, *Mabsūt*, 2/64.

3. Tūsī, *Tahdhīb*, 4/128.

4. Tūsī, *Khilāf*, 2/123–26.

5. Najafi, *Jawāhir*, 16/174–77.

6. Sachedina, "*Al-khums*," p. 278, and the Sunnī sources cited there in n. 23.

7. *'Allāma*, *Mukhtalaf*, 2/36–40, is the longest section devoted to the question of *al-khums* during the *ghayba*.

8. Ibn Idrīs, *Sarā'ir*, p. 117.

9. *Mukhtalaf*, 2/37.

10. Ibid., 2/39. All these views are discussed in greater detail in Tūsī, *Mabsūt*, 1/264. See also Sachedina, "*Al-khums*," pp. 287–88, for modern critique in the demonstrative jurisprudence of Hakīm, *Mustamsak*, vol. 9/579f.

11. *Mukhtalaf*, 2/39.

12. Ibid.

13. Sallār, *Marāsim*, p. 580.

14. Ibn Hamza, *Wasīla*, p. 716; also, *Mukhtalaf*, 2/39.

15. Ibn Idrīs, *Sarā'ir*, p. 116.

16. Muhaqqiq Hillī, *Sharā'i'*, p. 97; Najafi, *Jawāhir*, 16/177–80.

17. *Tadhkira*, 1/255.

18. *Qawā'id, kitāb al-zakāt, bāb al-thālith*.

19. *Mukhtalaf*, 2/40. See also his *Muntahā*, p. 555.

20. Shahīd Awwal, *Lum'a;* Shahīd Thānī, *Rawda*, 2/79–80.

21. *Jawāhir*, 16/177–80.

22. *Kifāyat al-ahkām*, p. 45.

23. *Dhakhīra, kitāb al-zakāt*.

24. *Kashf al-ghitā'*, pp. 362–63, 404.

25. *Jawāhir*, 16/177–80.

26. Ibid., 16/156–65.

27. *Hadā'iq*, 5/91–94.

28. *'Urwa*, pp. 385–86.

29. *Mustamsak*, 9/582–84.

SELECTED BIBLIOGRAPHY

Ākhund, Muhammad Kāzim b. al-Husayn al-Khurāsānī al-. *Kifāyat al-usūl*. Najaf, n.d.

'Alī, al-Shaykh. *al-Nūr al-sātiʿ fī al-fiqh al-nāfiʿ*. Najaf, 1383/1963.

'Allāma al-Hillī, al-Hasan b. Yūsuf b. al-Mutahhar al-. *Mukhtalaf al-shīʿa fī ahkām al-sharīʿa*. Litho., 1323/1905.

———. *Muntahā al-matlab fī tahqīq al-madhhab*. Litho., 1316/1898.

———. *Qawāʿid al-ahkām fī maʿrifat al-halāl wa al-harām*. Litho., n.d.

———. *Tabsirat al-mutaʿallimīn*. Najaf, 1382/1962.

———. *Tadhkirat al-fuqahāʾ*. 2 vols. Tabrīz, litho., 1276/1859.

———. *Tahrīr al-ahkām al-sharʿiyya ʿalā madhhab al-imāmiyya*. Litho., 1314/1896.

'Āmilī, Muhammad al-Jawād b. Muhammad al-Husaynī al-. *Miftāh al-karāma fī sharh qawāʿid al-ʿallāma*. Tehran, 1378/1958.

'Āmilī, Muhammad b. al-Hasan al-Hurr al-. *Wasāʾil al-shīʿa ilā tahsīl masāʾil al-sharʿiyya*. 20 vols. Beirut, 1381/1971.

'Āmilī, Muhsin b. ʿAbd al-Karīm al-Husyanī al-. *Aʿyān al-shīʿa*. Damascus, 1354/1935– .

Ansārī, al-Shaykh Murtadā al-. *Farāʾid al-usūl* (also known as *al-Rasāʾil al-arbaʿa*). Qumm, 1374/1954.

———. *al-Makāsib* (also known as *al-Matājir*). Tabrīz, 1375/1955.

———. *al-Makāsib*. 9 vols. With commentary by al-Sayyid Muhammad Kalāntar. Najaf, 1392/1972.

Āqā Buzurg al-Tihrānī, Muhammad Muhsin. *al-Dharīʿa ilā tasānīf al-shīʿa*. 20 vols. Tehran, 1388/1968.

———. *Tabaqāt aʿlām al-shīʿa*. 4 vols. Najaf, 1374/1954–.

Ardabīlī, Ahmad b. Muhammad al-Muqaddas al-. *Sharh al-irshād al-ʿallāma*. Litho., n.d.

Bahr al-ʿulūm, al-Sayyid Muhammad al-Mahdī. *Bulghat al-faqīh*. 4 vols. Najaf, 1388/1968.

———. *al-Fawāʾid al-rijāliyya* (also known as *Rijāl al-sayyid bahr al-ʿulūm*). 3 vols. Najaf, 1385/1965.

Bahrānī, Yūsuf b. Ahmad b. Ibrāhim al-. *al-Hadāʾiq al-nādira fī ahkām al-ʿitrat al-tāhira*. Litho., 1315/1897.

———. *Luʾluʾa al-bahrayn fī al-ijāzāt wa tarājīm rijāl al-hadīth*. Najaf, 1353/1934.

Bihbahānī, Muhammad Bāqir b. Muhammad Akmal al-Wahīd al-. "Sharh mafātīh al-sharāʾiʿ fī fiqh al-imāmiyya." Manuscript in 2 vols. Written in 1239/1823 at Āyatullāh Mīlānī's library in Mashhad, Iran.

Fādil al-Hindī, al-Mullā Bahāʾ al-Dīn Muhammad b. al-Hasan al-Isfahānī al-. *Kashf al-lithām ʿan qawāʿid al-ahkām sharʿ*. 2 vols. Litho., 1271/1854.

Hakīm, Muhsin b. Mahdī al-Tabātabā'ī al-. *Mustamsak al-'urwat al-wuthqā.* 14 vols. Najaf, 1391/1971.

Hakīm, Muhammad Taqī, al-. *al-Usūl al-'āmma li al-fiqh al-muqārin.* Beirut, 1983.

Halabī, Abū al-Majd al-. "Ishārāt at-sabq fī al-taklīf al-shar'ī." Manuscript in Astaneh Quds-i Radawī Library, Mashhad, Iran.

Halabī, Taqī b. al-Najm Abū Salah al-. "al-Kāfī fī al-fiqh." Manuscript in Astāneh Quds-i Radawī Library, Mashhad, Iran.

Hillī, Ibn Fahd al-. "al-Muhadhdhab al-bāri' fī sharh mukhtasar al-nāfi'." Manuscript in Astāneh Quds-i Radawī Library, Mashhad, Iran.

Hillī, Muhammad b. al-Hasan b. Yūsuf al-. *Īdāh al-fawā'id fī sharh ishkālāt al-qawā'id.* 4 vols. Qumm, 1389/1969.

Ibn Abī al-Hadīd. *Sharh nahj al-balāgha.* 20 vols. Cairo, 1959–64.

Ibn al-Athīr, 'Izz al-dīn. *al-Kāmil fī al-ta'rīkh.* 12 vols. Beirut, 1965.

Ibn al-Barrāj. *Kitāb jawāhir al-fiqh* (part of *al-Jawāmi' al-fiqhiyya*). Litho., 1276/1859.

————. *Sharh jumal al-'ilm wa al-'amal* (of al-Sharīf al-Murtadā). Mashhad, 1352/1974.

Ibn Bābūya, Muhammad b. 'Alī. *'Ilal al-sharā'i'.* Najaf, 1963.

————. *Ma'āni al-akhbār.* Tehran, 1379/1959.

————. *Man lā yahduruh al-faqīh.* 4 vols. Najaf, 1378/1958.

Ibn Hajar al-'Asqalānī, Shihāb al-Din Abū al-Fadl Ahmad b. 'Alī al-Kinānī. *al-Durar al-kāmina fī a'yān al-mā' al-thāmina.* Cairo, 1966.

————. *Lisān al-mīzān.* Beirut, 1971.

————. *Taqrīb al-tahdhīb.* Beirut, 1975.

Ibn Hamza al-Tūsī, 'Imād al-Dīn Muhammad b. 'Alī. *al-Wasīla fī al-fiqh* (part of *al-Jawāmi' al-fiqhiyya*). Litho., 1276/1859.

Ibn Idrīs al-Hillī, Muhammad b. Mansūr b. Ahmad. *Kitāb al-sarā'ir.* Litho., 1270/1853.

Ibn Sa'd. *Tabaqāt al-kabīr.* Leiden, 1905–6.

Ibn Shahrāshūb, Muhammad b. 'Alī. *Ma'ālim al-'ulamā'.* Najaf, 1948.

Ibn Zuhra, Abū al-Makārim Hamza b. 'Alī al-Husaynī al-Halabī. *Kitāb al-Ghunya fī al-usūl* (part of *al-Jawāmi' al-fiqhiyya*). Litho., 1276/1859.

————. *Kitāb al-Ghunya fī al-furū'* (part of *al-Jawāmi' al-fiqhiyya*). Litho., 1276/1859.

Isfahānī, Abū al-Faraj al-. *Maqātil al-talībiyīn.* Najaf, 1965.

Jawāmi' al-fiqhiyya al-. Litho., 1276/1859.

Karakī, 'Alī b. 'Abd al-'Ālī al-Muhaqqiq al-. *Jāmi' al-maqāsid fī sharh al-qawā'id.* 2 vols. Tehran, reprinted, litho., 1363/1943.

Kāshānī, al-Mullā Muhsin al-Fayd al-. *Mafātīh al-sharā'i' fī fiqh al-imāmiyya.* Beirut, 1969.

Kāshif al-Ghīta, 'Alī b. Muhammad Ridā b. Hādī al-. *Adwār 'ilm al-fiqh wa atwāruh.* Beirut, 1399/1979.

Kāshif al-Ghitā', al-Shaykh Ja'far al-Kabīr. *Kashf al-ghitā' 'an mubhamāt sharī'at al-ghurra.* Litho., n.d.

Kashshī, Abū 'Amr 'Umar b. 'Abd al-'Azīz al-. *Ikhtiyār ma'rifat al-rijāl.* Mashhad, 1348/1964.

Kāzimī, Sadr al-Dīn al-. *Hāmish al-lu'lu'.* n.p., n.d.

Khwānsārī, Aqā Husayn al-. *Mashāriq al-shumūs fī sharh al-durūs.* Litho., 1112/1700.

Khwānsārī, al-Sayyid Muhammad Bāqir al-Mūsawī al-. *Rawdāt al-jannāt fī ahwāl al-'ulamā' wa al-sādāt.* 8 vols. Tehran, 1389/1969.

Khū'ī, al-Sayyid Abū al-Qāsim al-Mūsawī al-. *Minhāj al-sālihīn.* 4 vols. Beirut, 1393/1983.

————. *Mu'jam al-rijāl al-hadīth.* 18 vols. Najaf, 1970.

————. *al-Tanqīh fī sharh al-'urwat al-wuthqā.* Lectures compiled by al-Mīrza 'Alī al-Gharawī al-Tabrīzī. 4 vols. Najaf, 1386/1966.

Khumaynī, al-Sayyid Rūh Allāh al-Mūsawī al-. *al-Hukūmat al-islāmiyya.* Najaf, 1970.

————. *Kitāb al-bay'.* 5 vols. Qumm, 1387/1962.

————. *al-Makāsib al-muharrama.* 2 vols. Qumm, 1386/1961.

———. *Tahrīr al-wasīla*. 2 vols. Najaf, 1392/1972.

Kulaynī, Muhammad b. Ya'qūb b. Ishaq al-. *al-Usūl min al-kāfī*. 4 vols. Tehran, 1392/1972.

———. *Furū' al-kāfī*. Tehran, 1375/1955.

Majlisī, Muhammad Bāqir al-. *'Ayn al-hayāt*. Tehran, 1373/1953.

———. *Bihār al-anwār*. 102 vols. Tehran, 1384/1964.

Mas'ūdī, 'Alī b. al-Husayn al-. *Murūj al-dhahab wa ma'ādin al-jawhar*. 4 vols. Beirut, 1966.

Mīlānī, al-Sayyid Muhammad Hādī al-. *Muhādarāt fī fiqh al-imāmiyya*. 4 vols. Mashhad, 1974.

Mufīd, Muhammad b. Muhammad al-Nu'mān al-. *al-Irshād*. Tehran, 1351/1972.

———. *al-Muqni'a fī masā'il al-halāl wa al-harām*. Litho., n.d.

Muhaqqiq al-Hillī, Abū al-Qāsim Ja'far b. al-Hasan al-. *Sharā'i' al-islām*. Beirut, 1978.

———. *al-Mukhtasar al-nāfi'*. Litho., n.d.

Mūsawī, al-Sayyid Muhammad b. al-Sayyid 'Atiyya al-. *Minhāj al-karāma fī sharh tahdhīb al-'allāma*. n.p., n.d.

Nā'inī, al-Mīrza Muhammad Husayn al-. *Tanbīh al-umma wa tanzīh al-milla*. Tehran, 1955.

Najafī, al-Shaykh Muhammad Hasan al-. *Jawāhir al-kalām fī sharh sharā'i' al-islām*. 42 vols. Tehran, 1392/1972.

Najāshī, Ahmad b. 'Alī b. al-'Abbās al-. *Fihrist asmā' musannifī al-shī'a* (also known as *Rijal*). n.p., n.d.

Narāqī, Ahmad b. Muhammad Mahdī al-. *'Awā'id al-ayyām fī bayān qawā'id al-ahkām*. Litho., 1245/1828.

———. *Mustanad al-shī'a fī ahkām al-sharī'a*. Litho., n.d.

Nawbakhtī, al-Hasan b. Musa al-. *Firaq al-shī'a*. Istanbul, 1931.

Nūrī, al-Mīrza Husayn al-. *Mustadrak wasā'il al-shī'a*. Tehran, 1321/1903.

Qadī al-Nu'mān, al-. *Da'ā'im al-islām*. 2 vols. Cairo, 1963.

Qummī, al-Shaykh 'Abbas al-. *al-Kunā wa al-alqāb*. 3 vols. Najaf, 1389/1969.

Rāwandī, Qutb al-Dīn Abū al-Husayn Sa'd b. Hibat Allāh al-. *Fiqh al-qur'ān*. Qumm, 1397/1976.

Sabzawārī, Muhammad Bāqir al-Muhaqqiq al-. *Dhakhīrat al-ma'ād fī sharh al-irshād*. Tehran, litho., 1274/1857.

———. *Kifāyat al-ahkām*. Tehran, litho., 1261/1845.

Sadr, Muhammad Bāqir al-. *Durūs fī 'ilm al-usūl*. Cairo, 1978.

Sallār al-Daylamī, Hamza b. 'Abd al-'Azīz al-Tabaristānī al-. *al-Marāsim fī al-fīqh* (also known as *al-Risāla*) (part of *al-Jawāmi' al-fiqhiyya*). Litho., 1276/1859.

Shahīd al-Awwal, Muhammad b. Makkī al-'Amilī al-. *Dhikrā al-shī'a fī ahkām al-sharī'a*. Litho., 1272/1855.

———. *al-Lum'a al-dimashqiyya*. Najaf, 1386/1966.

Shahīd al-Thānī, Zayn al-Dīn b. Ahmad al-Shāmī al-'Āmilī al-. *Masālik al-afhām fī sharh sharā'i' al-islām fī masā'il al-halāl wa al-harām*. 2 vols. Litho., 1283/1866.

———. *al-Rawdat al-bahiyya fī sharh al-lum'at al-dimashqiyya*. 8 vols. Najaf, 1386/1966.

———. *Sharh al-bidāya fī 'ilm al-dirāya*. Tehran, 1402/1982.

Sharīf al-Murtadā, Abū al-Qāsim 'Alī b. al-Husayn al-Mūsawī al-. *Kitāb al-intisār* (part of *al-Jawāmi' al-fiqhiyya*). Litho., 1276/1895.

———. *al-Masā'il al-nāsiriyyāt* (part of *al-Jawāmi' al-fiqhiyya*). Litho., 1276/1896.

Sīwurī, Jamāl al-Dīn al-Miqdād b. 'Abd Allāh al-. *Kanz al-'irfān fī fiqh al-qur'ān*. Tehran, 1384/1964.

Suyūtī, Jalal al-Dīn 'Abd al-Rahmān b. Abī Bakr al-Khadīrī al-. *Bughayt al-wi'ā fī tabaqāt al-lughawiyyīn wa al-nuhā*. Cairo, 1964.

———. *Tanwīr al-hawālik: sharh 'alā muwatta' mālik*. Beirut, 1970.

Tabarsī, Abū Mansūr Ahmad b. 'Alī b. Abī Tālib al-. *al-Ihtijāj*. Beirut, 1966.

Tabātabā'ī, al-Sayyid 'Alī b. Muhammad 'Alī. *al-Riyād al-masā'il fī fiqh al-ahkām bi al-dalā'il*. 2 vols. Litho., n.p., n.d.

Tūsī, Muhammad b. al-Hasan al-. *al-Fihrist.* Najaf, 1961.
――――. *al-Ghayba.* Najaf, 1965.
――――. *al-Iqtisād al-hādī ilā tarīq al-rashād.* Qumm, 1400/1980.
――――. *al-Istibsār fīmā ikhtalafa min al-akhbār.* Najaf, 1966.
――――. *al-Khilāf fī al-fiqh.* Tehran, 1382/1962.
――――. *al-Mabsūt fī fiqh al-imāmiyya.* 8 vols. Tehran, 1378/1958.
――――. *al-Nihāya fī mujarrad al-fiqh wa al-fatāwā.* Beirut, 1970.
――――. *Rijāl al-tūsī.* Najaf, 1381/1961.
――――. *Tahdhīb al-ahkām fī sharh al-muqniʿa.* 10 vols. Najaf, 1959.
――――. *Tafsīr al-tibyān.* 10 vols. Najaf, 1957.
――――. *Talkhīs al-shāfī.* 4 vols. Qumm, 1394/1974.
Tūsī, Nasīr al-Dīn Muhammad. *Nasirean Ethics.* Trans. and annotated by G. M. Wickens. London, 1964.
――――. *Akhlāq-i Nāsirī.* Ed. Mujtabā Mīnuvī and ʿAliridā Haydarī. Tehran, 1360 Sh. 1981.
Rāzī, Fakhr al-Din al-. *Tafsīr al-kabīr.* Cairo, 1938.
Shahrastānī, Muhammad b. ʿAbd al-Karīm al-. *al-Milal wa al-nihal.* Cairo, 1948–49.
Shāshī al-Qaffāl, Sayf al-Dīn Abī Bakr Muhammad b. Ahmad al-. *Hilyat al-ʿulamāʾ fī maʿrifat madhāhib al-fuqahāʾ.* Beirut, 1980.
Yazdī, al-Sayyid Muhammad Kāzim al-Tabātabāʾī al-. *al-ʿUrwat al-wuthqā.* Najaf, 1348/1929.

INDEX